Transmission lines

This rigorous treatment of transmission lines presents all the essential concepts in a clear and straightforward manner. Key principles are demonstrated by numerous practical worked examples and illustrations, and complex mathematics is avoided throughout.

Early chapters cover pulse propagation, sinusoidal waves and coupled lines, all set within the context of a simple loss-less equivalent circuit. Later chapters then develop this basic model by demonstrating the derivation of circuit parameters, and the use of Maxwell's equations to extend this theory to major transmission lines. Finally, a discussion of photonic concepts and properties provides valuable insights into the fundamental physics underpinning transmission lines.

Covering DC to optical frequencies, this accessible text is an invaluable resource for students, researchers, and professionals in electrical, RF, and microwave engineering.

Richard Collier is a former Director of the Electronic Engineering Laboratory of the University of Kent, and a former Senior Research Associate and Affiliated Lecturer at the Cavendish Laboratory, University of Cambridge. He is a Chartered Engineer, and a Fellow of the IET.

The Cambridge RF and Microwave Engineering Series

Series Editor
Steve C. Cripps, Distinguished Research Professor, Cardiff University

Peter Aaen, Jaime Plá and John Wood, *Modeling and Characterization of RF and Microwave Power FETs*

Dominique Schreurs, Máirtín O'Droma, Anthony A. Goacher and Michael Gadringer, *RF Amplifier Behavioral Modeling*

Fan Yang and Yahya Rahmat-Samii, *Electromagnetic Band Gap Structures in Antenna Engineering*

Enrico Rubiola, *Phase Noise and Frequency Stability in Oscillators*

Earl McCune, *Practical Digital Wireless Signals*

Stepan Lucyszyn, *Advanced RF MEMS*

Patrick Roblin, *Nonlinear FR Circuits and the Large-Signal Network Analyzer*

Matthias Rudolph, Christian Fager and David E. Root, *Nonlinear Transistor Model Parameter Extraction Techniques*

John L. B. Walker, *Handbook of RF and Microwave Solid-State Power Amplifiers*

Anh-Vu H. Pham, Morgan J. Chen and Kunia Aihara, *LCP for Microwave Packages and Modules*

Sorin Voinigescu, *High-Frequency Integrated Circuits*

Richard Collier, *Transmission Lines*

Forthcoming

David E. Root, Jason Horn, Jan Verspecht and Mihai Marcu, *X-Parameters*

Richard Carter, *Theory and Design of Microwave Tubes*

Nuno Borges Carvalho and Dominique Schreurs, *Microwave and Wireless Measurement Techniques*

Valeria Teppati, Andrea Ferrero and Mohamed Sayed, *Modern RF and Microwave Measurement Techniques*

Transmission Lines

Equivalent Circuits, Electromagnetic Theory, and Photons

RICHARD COLLIER
University of Cambridge

CAMBRIDGE UNIVERSITY PRESS
Cambridge, New York, Melbourne, Madrid, Cape Town,
Singapore, São Paulo, Delhi, Mexico City

Cambridge University Press
The Edinburgh Building, Cambridge CB2 8RU, UK

Published in the United States of America by
Cambridge University Press, New York

www.cambridge.org
Information on this title: www.cambridge.org/9781107026001

First published 2013

Printed and bound in the United Kingdom by the MPG Books Group

A catalogue record for this publication is available from the British Library

Library of Congress Cataloguing in Publication data

Collier, R. J. (Richard J.)
Transmission lines : equivalent circuits, electromagnetic theory, and photons / Richard Collier, University of Cambridge.
pages cm. – (Cambridge RF and microwave engineering series)
Includes bibliographical references.
ISBN 978-1-107-02600-1 (Hardback)
1. Telecommunication lines – Textbooks. 2. Photons – Textbooks. I. Title.
TK5103.15.C65 2013
621.382–dc23

2012032044

ISBN 978-1-107-02600-1 Hardback

A man that looks on glass
   On it may stay his eye;
Or if he pleaseth, through it pass,
   And then the heaven espy.

George Herbert 1593–1633

"This book presents a new and refreshing look at the subject of electromagnetic transmission lines. The clarity of the explanations given in the book indicate Dr. Collier's many years of teaching this subject to both undergraduate and graduate level university students. It is an ideal reference book for this subject, and should be read by both scientists and electronic/electrical engineers needing to use and understand transmission lines."

Nick Ridler, IET Fellow

Contents

Preface

The use of transmission lines has increased considerably since the author began his lectures on them at the University of Kent at Canterbury in October 1968. Now the mighty internet involves huge lengths of optical fibres, estimated at over 750 million miles, and similar lengths of copper cables. The ubiquitous mobile phones and personal computers contain circuits using microstrip, coplanar waveguide and stripline. However, despite all these widespread modern applications of transmission lines, the basic principles have remained the same. So much so, that the many classic textbooks on this subject have been essential reading for nearly a hundred years. It is not the purpose of this book to repeat the content of these standard works but to present the material in a form which students may find more digestible. Also this is an age where mathematical calculations are relatively simple to perform on modern personal computers and so there is less need for much of the advanced mathematics of earlier years. The aim of this book is to introduce the reader to a wide range of transmission line topics using a straightforward mathematical treatment which is linked to a large number of graphs illustrating the text. Although the professional worker in this field would use a computer program to solve most transmission line problems, the value of this book is that it provides exact solutions to many simple problems which can be used to verify the more sophisticated computer solutions. The treatment of the material will also encourage 'back-of-envelope' calculations which may save hours of computer usage. The author is aware of the hundreds of books published on every aspect of transmission lines and the myriads of scientific publications which appear in an ever increasing number of journals. To help the reader get started on exploring any topic in greater depth, this book contains comments on many of these specialist books at the end of each chapter. Following this will be the reader's daunting task to search through the scientific literature for even more information. It is the author's hope that this book will establish some of the basic principles of this extensive subject which make the use of some of these scientific papers more profitable.

Initially, transmission lines are described in this book in terms of an equivalent circuit containing two distributed elements. The first three chapters use this circuit to illustrate many of the features of transmission lines. Chapter 1 consists mainly of the author's lectures to first year undergraduate computer science students at the University of Kent. For this reason it is all about step waves and pulses on transmission lines and avoids the use of Laplace transforms. This book introduces digital signals at several stages as they are by far the majority of the traffic on modern lines. The second chapter, on mainly sinusoidal waves, was given to electronic engineering students at the same university. This chapter covers the Smith chart and scattering matrices and their use in circuit analysis. Finally, the third chapter introduces the reader to coupling between transmission lines, including some unique circuits which use coupled waves.

Although these first three chapters are sufficient for many transmission line problems, there are some basic principles which this treatment omits. The most obvious ones are the values of both the velocity of propagation and the characteristic impedance which are just stated in the early chapters. Less obvious are the higher order modes of propagation which can exist on all transmission lines. So Chapter 4 covers the derivation of the capacitances and inductances needed to calculate the velocities of propagation and the characteristic impedances of many transmission lines. The method mainly uses just two line integrals from electromagnetism to derive the static fields required for the analysis. Chapter 5 uses Maxwell's equations to derive the electromagnetic wave picture of the lines. In particular it shows that the lines have multiple modes of propagation and it introduces metallic and dielectric waveguides which cannot be adequately described using a simple equivalent circuit. The treatment of Maxwell's equations is somewhat brief, as fuller descriptions are readily available elsewhere. However, the analysis of the various problems will illustrate how these important equations are used. The topic of attenuation was intentionally omitted up to this point, as it complicates the material in the earlier chapters. Chapter 6 is entirely devoted to this topic and includes both the skin effect and dispersion and the way they modify pulse shapes.

This is the point where most textbooks end, but with the rise of electrodynamics and quantum electrodynamics, the interest in the photon has greatly increased in recent years. This elusive fundamental particle or packet of energy is the basic component of all electromagnetic waves. So this book has included some thoughts on photons which bring out a few of the basic processes going on when a wave propagates. Chapter 7 concentrates on the two properties of photons: that they travel in straight lines and at the velocity of light. Many of the transmission lines are revisited to show that complex solutions of Maxwell's equations can be broken down into the propagation of plane waves. This is further developed by considering plane waves passing through dielectric and resistive films. This topic was studied by the author whilst he was working in the Microelectronics Research Laboratory in the Cavendish Laboratory at the University of Cambridge. Finally, the book ends in Chapter 8 with a close look at how photons interact with the guiding structures of transmission lines. Some of the comments in this part will prove interesting to anyone involved in photon propagation. There are various small sections at the end of this chapter on current hot-topics which could prove useful as a starting point for those interested in such areas.

The author wishes to thank the Cambridge University Press for publishing this book and in particular Julie Lancashire for commissioning the work and Elizabeth

Horne for sorting out the text. He would also like to thank the many colleagues and students at the University of Kent who made helpful comments about some of the content of the first six chapters. In particular, from the University of Cambridge, he would like to thank Professor Richard Philips and Dr David Hasko from the Cavendish Laboratory, as well as Chris Nickerby and Nilpesh Patel from Corpus Christi College for their help with the last two chapters. He would also like to acknowledge many helpful comments from Dr Nick Ridler of the National Physical Laboratory and Dr David Williams of Hitachi Cambridge.

Finally, I should like to acknowledge the loving support of my wife Ruth, who has helped to keep me going during the years needed to produce this book.

Part 1

Transmission lines using a distributed equivalent circuit

1 Pulses on transmission lines

The term 'transmission line' is not uniquely defined and is usually taken to mean a length of line joining a source to a termination. A simple example of such a transmission line is shown in Figure 1.1.

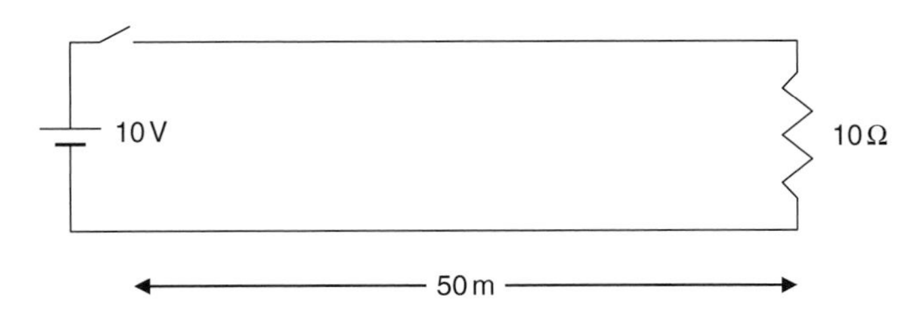

Figure 1.1 A simple transmission line 50 m long joining a 10 V source to a 10 Ω termination.

This transmission line might consist of just two parallel wires or a coaxial cable or something more unusual. Using ordinary equivalent circuit theory, which has widespread use in both Electrical and Electronic Engineering, a current of 1 A flows in the whole circuit, the moment the switch is closed. This is only true if every element in the circuit is considered to be a lumped element. Now a lumped element is a circuit component in which a current is instantaneously produced as soon as a voltage is applied. The battery and the resistor may well have a small delay before any current appears, but in this introductory chapter it will be assumed these effects can be neglected. However, the 50 m length of line cannot be described as a lumped element as it takes a finite time for a voltage introduced at one end to propagate to the other. Only if this time is much smaller than any other transient being considered can the transmission line effects described in this chapter be neglected. In order to analyse this propagation, a distributed circuit is needed which gives some considerable insight into the performance of transmission lines. This circuit approach is limited to the use of voltages and currents, which are inappropriate for transmission lines like waveguides and optical fibres. In the later chapters, two alternative descriptions of the lines will be given; one in terms of electromagnetic fields and the other in terms of photons. All three descriptions reveal unique aspects of the propagation along transmission lines and together they give a more complete picture.

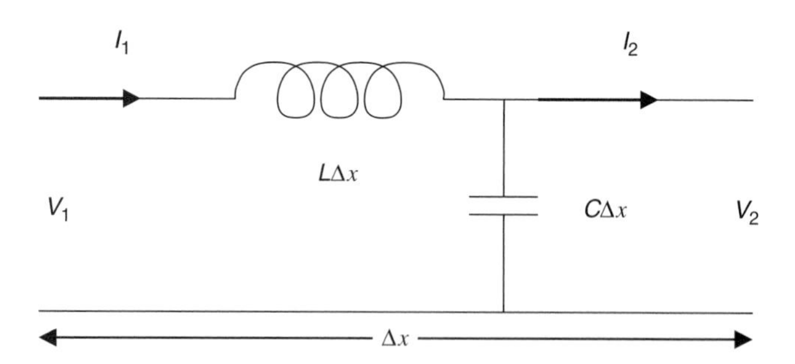

Figure 1.2 An equivalent circuit of a small length, Δx, of loss-less transmission line.

The distributed circuit method applies to those transmission lines which have two or more conductors. However, this chapter will discuss the simplest case of just two conductors, which is the most common configuration. There are two assumptions that are needed before a distributed circuit can be developed. The first is that the conductors must be good conductors so that the voltage does not vary around the cross-section of an individual line, or in other words, it is an equipotential surface. Secondly, the lines will be assumed to be free from losses so that signals can travel down them without attenuation. In Chapter 6, the effects of various types of losses on the operation of the lines will be introduced. Electrically, there are only two things left to consider; they are the capacitance and the inductance of the line. Since a transmission line normally is assumed to be uniform, both of these quantities increase with length, so it is common practice to define them for a one-metre length of line. They are then called the distributed capacitance, C, and distributed inductance, L, and have the units of Fm^{-1} and Hm^{-1} respectively.

In practice, these parameters can easily be measured using a one-metre length of the line. If the far end is open-circuited, then the capacitance, C, can be measured at the near end, with some corrections for the capacitance of the open circuit. Similarly, if the far end is short-circuited, the inductance, L, can also be measured, again with an appropriate correction for the inductance of the short circuit. These measurements can be made over a wide range of frequencies and, to get started on this basic description, these distributed parameters will be assumed to be constant with frequency. In Chapter 6 the variation of these parameters with frequency, called dispersion, will be discussed. So an equivalent circuit for a short length of line, Δx, is fairly simple and is shown in Figure 1.2.

The short length of line is important because, as $\Delta x \to 0$, the two distributed elements become effectively lumped elements again and so ordinary circuit equations can legitimately be applied. This sounds like a circular argument, but the principle being applied is to do with the transit time of signals. If a length of line is small enough, this transit time can be neglected and so ordinary circuit theory can be used. In the electromagnetic field description, given in later chapters, this somewhat circuitous argument is not required!

1.1 Velocity and characteristic impedance

Imagine then a voltage source, V_1, is connected to the left-hand side of a short length of transmission line, as shown in Figure 1.2, which also causes a current, I_1, to flow as shown. As the signal travels down the short length, two changes occur. Firstly, there is a voltage drop across the inductance and secondly, a current loss through the capacitor. So at the right-hand side the voltage and current become:

$$V_2 = V_1 - L\Delta x\frac{\partial I}{\partial t} \text{ and } I_2 = I_1 - C\Delta x\frac{\partial V}{\partial t}, \tag{1.1}$$

where these are obtained from the usual circuit laws for capacitances and inductances. These equations would be the same for an equivalent circuit with the inductance on the right-hand side of the capacitance.

The equations can be rearranged as follows:

$$V_2 - V_1 = \Delta V = -L\Delta x\frac{\partial I}{\partial t} \text{ and } I_2 - I_1 = \Delta I = -C\Delta x\frac{\partial V}{\partial t},$$

$$\frac{\Delta V}{\Delta x} = -L\frac{\partial I}{\partial t} \text{ and } \frac{\Delta I}{\Delta x} = -C\frac{\partial V}{\partial t}, \tag{1.2}$$

and in the limit as $\Delta x \to 0$:

$$\frac{\partial V}{\partial x} = -L\frac{\partial I}{\partial t} \text{ and } \frac{\partial I}{\partial x} = -C\frac{\partial V}{\partial t}. \tag{1.3}$$

These are called the Telegraphists' equations and are useful in linking the voltage and current on a transmission line. However, since they are cross-linked in these two variables, it is not possible directly to eliminate one or other of them to find a solution. The normal route to a solution is to differentiate each of them with respect to both time and distance:

$$\frac{\partial^2 V}{\partial x^2} = -L\frac{\partial^2 I}{\partial x \partial t} \text{ and } \frac{\partial^2 V}{\partial t \partial x} = -L\frac{\partial^2 I}{\partial t^2},$$

$$\frac{\partial^2 I}{\partial x^2} = -C\frac{\partial^2 V}{\partial x \partial t} \text{ and } \frac{\partial^2 I}{\partial t \partial x} = -C\frac{\partial^2 V}{\partial t^2}. \tag{1.4}$$

Since the parameters of space and time are independent, the order of the differentiation is not important. So eliminating the mixed differentials gives:

$$\frac{\partial^2 V}{\partial x^2} = LC\frac{\partial^2 V}{\partial t^2} \text{ and } \frac{\partial^2 I}{\partial x^2} = CL\frac{\partial^2 I}{\partial t^2}. \tag{1.5}$$

These equations are called wave equations because their solutions are waves. Both the voltage and the current obey the same equation in this simple case. The solutions of these wave equations are any functions of the variable:

$$t \pm \frac{x}{v}, \text{i.e. } V = f\left(t \pm \frac{x}{v}\right) \text{ and } I = g\left(t \pm \frac{x}{v}\right), \tag{1.6}$$

where v is a constant.

Substituting the voltage function into the wave equations gives

$$\frac{1}{v^2}f''\left(t \pm \frac{x}{v}\right) = LCf''\left(t \pm \frac{x}{v}\right).$$

So the constant, v, is given by:

$$v = \frac{1}{\sqrt{LC}}. \tag{1.7}$$

By examining Equations (1.6) and taking the minus sign it can be seen that any function of t is delayed in the positive x direction. This function is called a forward wave and its velocity is given by v in Equation (1.7). The positive sign is for waves moving in the negative x direction and these are called backward or reflected waves. Now the link between the voltage and the current waves is found by using the Telegraphists' equations in (1.3). Substituting the functions given in Equation (1.6) gives

$$\pm\frac{1}{v}f'\left(t \pm \frac{x}{v}\right) = -Lg'\left(t \pm \frac{x}{v}\right) \text{ and } \pm\frac{1}{v}g'\left(t \pm \frac{x}{v}\right) = -Cf'\left(t \pm \frac{x}{v}\right).$$

Integrating both sides of the equations with respect to time gives

$$\pm\frac{1}{v}f\left(t \pm \frac{x}{v}\right) = -Lg\left(t \pm \frac{x}{v}\right) \text{ and } \pm\frac{1}{v}g\left(t \pm \frac{x}{v}\right) = -Cf\left(t \pm \frac{x}{v}\right). \tag{1.8}$$

Then using Equations (1.6) and (1.7) gives

$$\frac{V}{I} = \pm\sqrt{\frac{L}{C}}$$

from both equations in (1.8).

The positive sign relates to the forward waves and the negative sign to the reverse waves. This ratio of voltage to current has the units of ohms in this case and is normally given the symbol Z_0 and called the characteristic impedance of the transmission line. By denoting a subscript *plus* to forward waves and a subscript *minus* to reverse waves gives

$$Z_0 = \sqrt{\frac{L}{C}} = \frac{V_+}{I_+} = -\frac{V_-}{I_-}. \tag{1.9}$$

The negative sign in Equation (1.9) is because the wave is travelling in the reverse direction.

1.2 Reflection coefficient

The next concept to consider is the reflection of waves. This is often caused by a sudden change of impedance along a transmission line. The simplest case is a line terminated with an impedance, Z_L, which will cause reflections because the total voltage across the impedance, V_L, and the current through it, I_L, is given by Ohm's Law as

$$\frac{V_L}{I_L} = Z_L = \frac{V_+ + V_-}{I_+ + I_-}. \tag{1.10}$$

This assumes that Z_L has small dimensions so that there is zero transit time between the arrival of the wave and the appearance of a current in the impedance. In other words it is subject to the normal circuit laws. The presence of the reflected wave enables Ohm's Law to be obeyed both at the termination and in the two waves given in Equation (1.9). In the time domain, a termination may not always be described as a simple impedance, Z_L, whereas in the frequency domain it is always possible. In many of the examples that follow, Z_L is a pure resistance, and so Equation (1.10) is valid for both domains. However, for the later examples, 1.9 onwards, more complex time domain expressions are developed.

A special case for Equation (1.10) is when:

$$Z_L = Z_0. \tag{1.11}$$

Then Z_L is called a matched termination and, since it is equal to the characteristic impedance, no reflections occur. A useful measure of the amount of reflection is the ratio, ρ, of the reflected to the incident voltage wave:

$$\rho = \frac{V_-}{V_+} = -\frac{I_-}{I_+}. \tag{1.12}$$

In the third part of Equation (1.12) the negative sign is because of the negative sign in Equation (1.9). So the current reflects in an equal and opposite way to the voltage. Substituting Equation (1.12) into Equation (1.10) gives

$$Z_L = \frac{V_+(1+\rho)}{\frac{V_+}{Z_0}(1-\rho)} \quad \text{or} \quad \rho = \frac{Z_L - Z_0}{Z_L + Z_0}. \tag{1.13}$$

It is useful to realise the significance of ρ or, as it is commonly called, the reflection coefficient. In the frequency domain, both Z_L and Z_0 can be complex, making the reflection coefficient complex as well.

If we limit the discussion to real values of Z_0 and values of Z_L where the real part is positive, then

$$|\rho| \leq 1 \quad \text{and} \quad -\pi \leq \angle\rho \leq \pi. \tag{1.14}$$

For example, some typical values are:

Z_L/Z_0	$\lvert\rho\rvert$	$\angle\rho$
0 (short circuit)	1	$\pm\pi$
∞ (open circuit)	1	0
1 (matched load)	0	indeterminate
ja (inductance)	1	π to 0
$-jb$ (capacitance)	1	$-\pi$ to 0
2 (resistance $>Z_0$)	$\frac{1}{3}$	0
0.5 (resistance $<Z_0$)	$\frac{1}{3}$	$\pm\pi$

where a and b are constants.

Only when there is a resistive part to Z_L does the amplitude of ρ go below unity. When Z_L is reactive the phase or argument of ρ can be as large as π, due to the bilinear nature of Equation (1.13). So far in the discussion, the nature of the voltage waveform has been left totally general. For instance, it could be a voltage step, which would occur when a battery is connected, or it could be a sine wave, or a pulse or some rarer wave like a bi-pulse. To illustrate these waveforms in the time domain, a series of problems will now be briefly described, starting with only resistive terminations and ending with more complex terminations. The next chapter will discuss problems involving sinusoidal waves.

1.3 Step waves incident on resistive terminations

Example 1.1 A powerful ten-volt battery is suddenly connected to a 100 m long transmission line. At the far end of the line is a short circuit. If the velocity of propagation is 2.10^8 ms^{-1} and the characteristic impedance is 50 Ω, find the current in the short circuit after 5 μs, assuming the battery has zero internal resistance. See Figure 1.3.

Solution to Example 1.1

This problem would have a simple solution in circuit theory, as the current would instantaneously be infinite! However, in transmission line theory, this is not the correct solution. Initially the battery sends a voltage step of amplitude 10 V and

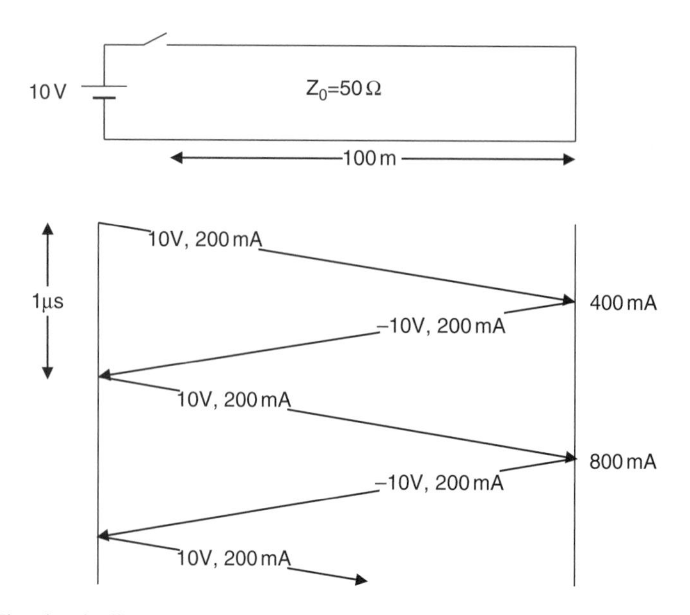

Figure 1.3 The circuit diagram and wave diagram for Example 1.1.

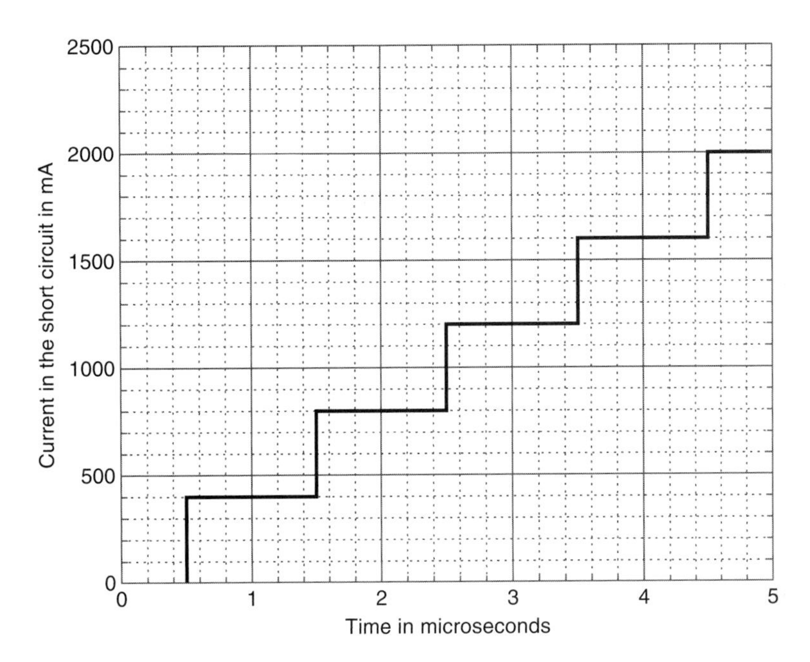

Figure 1.4 The current in the short circuit against time for Example 1.1.

current 200 mA (see Equation (1.9)) which takes 0.5 μs to reach the short circuit. Using Equations (1.12) and (1.13), this step is reflected so that reverse wave has a voltage of –10 V and a current of +200 mA. Thus the current in the short circuit jumps up to a total of 400 mA but with no overall voltage as the two waves superimpose so as to cancel their voltages and add their currents. The reflected wave returns after a similar reflection at the battery in 1.5 μs to add a further 400 mA to the current in the short circuit. This is shown in the wave diagram in Figure 1.3 where time is the coordinate vertically downwards. So just after 4.5 μs the current will be 2 A, as shown in Figure 1.4. Obviously the current will be limited by several factors, for instance, if the maximum current that the battery could supply was 40 A, then this current would be reached in 100 μs.

The essential thing to notice from this problem is that the transmission line limits the initial supply of current from the source and also that the step wave goes back and forth, endlessly delivering increases in current until a limit is reached. The battery supplies the original 200 mA continuously and the current builds up because of the wave motion, which keeps increasing the battery current in steps. So a 'long' short circuit could prevent dangerous currents for a short period.

It is useful to note that the battery is effectively 'seeing' a changing resistive load which reduces in value with time, as shown in Figure 1.5. If the line had been very long the battery would have just supplied 200 mA to the line. However, the multiple reflections in this example result in an increasing current being drawn from the battery.

Finally, the wave induces a positive current in one of the lines and a negative current in the other line. Thus the battery is sending out a current from one of its

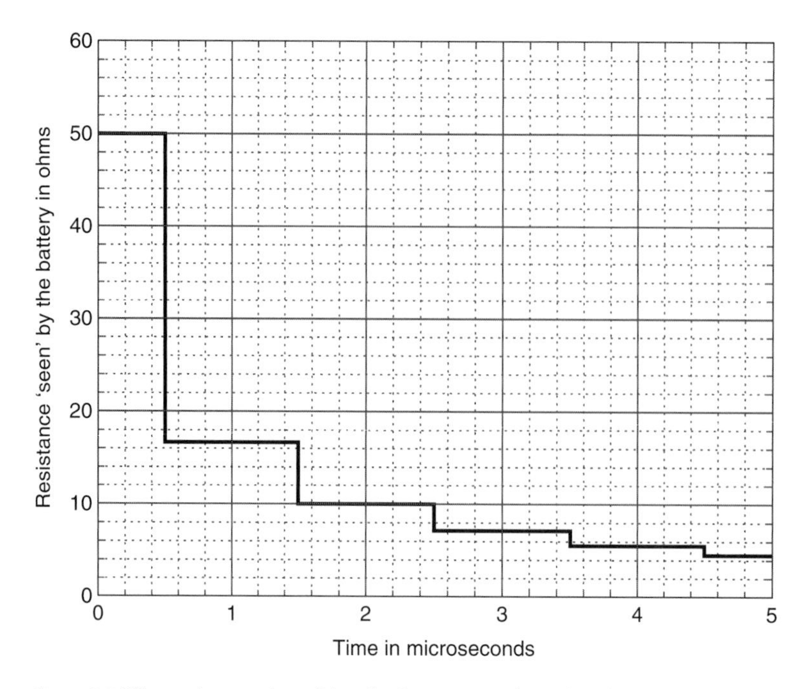

Figure 1.5 The resistance 'seen' by the battery against time in Example 1.1.

terminals and receiving a current at the other, as the wave progresses. It is usually easier to think of the upper wire in Figure 1.3 carrying the positive current of 200 mA and the lower or return wire a negative current of 200 mA.

Example 1.2 Using the same circuit as in Example 1.1, the switch is closed at $t = 0$ as before, but then opened at $t = 5\,\mu\text{s}$. Find the current waveform in the short circuit for the next $5\,\mu\text{s}$.

Solution to Example 1.2

This problem again would have a simple solution in circuit theory: the current would be zero. However, the wave theory does not give that answer. When the switch is opened, the wave is trapped in the circuit and cannot escape. When it reflects from the switch, which is now effectively an open circuit, the total wave goes on reflecting back and forth forever. In practice there will be some loss mechanisms, which will reduce the wave amplitude eventually to zero, but in this special case with no losses the circuit becomes a square wave oscillator. The solution is easier to see if all the five waves are added together at the moment the switch is opened. The total wave approaching the short circuit has an amplitude of 50 V and carries a current of 1 A. The total wave departing from the short circuit has an amplitude of –50 V and also carries a current of 1 A. These two

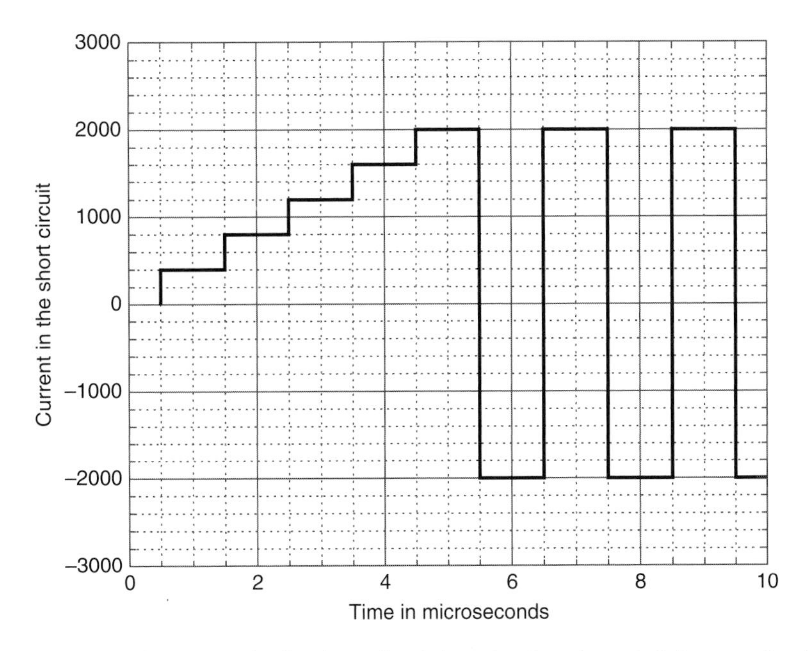

Figure 1.6 The current in the short circuit in mA, after the switch is opened at 5 μs.

sections form a wave of duration 1 μs. The front of this wave now reflects from the open switch with a reflection coefficient of +1 rather than –1. At the moment when the front of this wave begins to arrive back at the short circuit, the back of the wave leaves the short circuit. So the net effect is that the current in the short circuit goes to –2 A for 1 μs and then to +2 A and so on. The voltage at the switch is also a square wave, starting at 5 μs, and it has an amplitude of 100 V. The total energy trapped in the circuit is 5.10^{-5} J and for 1 μs will have a power of 50 W. The current variation with time is shown in Figure 1.6. Again, the result is surprising and not predicted by circuit theory. It also illustrates a typical problem when circuits are switched off, in that the trapped energy can generate a considerable voltage; in this case the 10 V battery when disconnected leaves behind a 100 V square wave.

Example 1.3 A 49 V battery is connected to a 300 Ω resistor via 50 m of transmission line. The velocity of propagation along the line is 1.10^8 ms^{-1} and its characteristic impedance is 50 Ω. Find the time that elapses before the voltage across the resistor is within 1% of its final steady state value, assuming the internal impedance of the battery is negligible. The circuit diagram is shown in Figure 1.7.

Solution to Example 1.3

The final voltage is the easy part: it must be 49 V! When the battery is connected a wave of amplitude 49 V sets off towards the 300 Ω resistor.

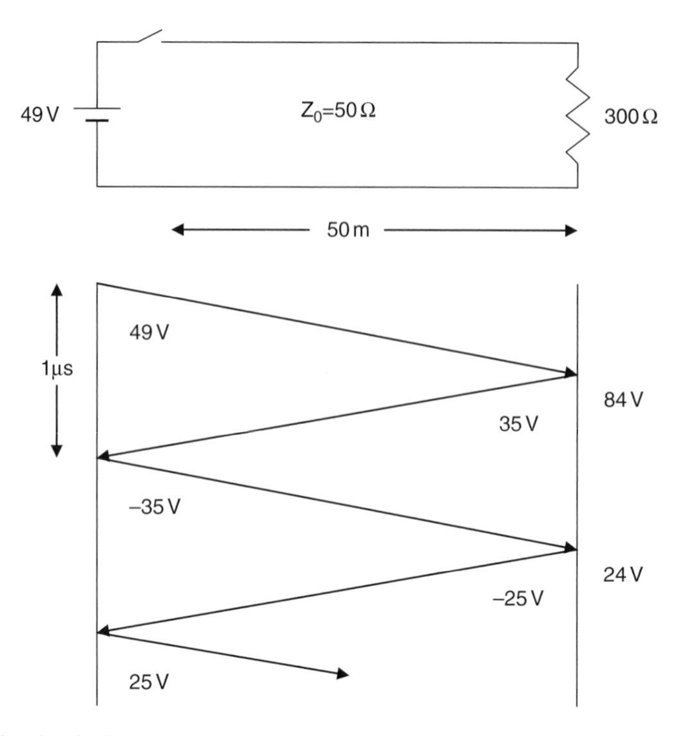

Figure 1.7 The circuit diagram and wave diagram for Example 1.3.

It takes just 0.5 μs to arrive. Using Equation (1.13), the reflection coefficient is $\frac{5}{7}$ so the incident wave is partly reflected as a 35 V wave. The sum of these waves, 84 V, will be the voltage across the resistor until the next wave arrives. This will be a wave of amplitude −35 V due to the negligible impedance of the battery. After reflection, this will become –25 V, making the new voltage across the resistor 24 V, as shown in the wave diagram in Figure 1.7. In Figure 1.8 is shown the voltage across the resistor for 10 μs. However, given 10 μs or more, the voltage is very near the result which would have been obtained from simple circuit theory.

This example shows that when a circuit is switched on, there is a wave, which reflects many times until the circuit reaches its steady state. However, in the steady state, these waves are still there. If we list the forward waves, that is those coming from the battery towards the 300 Ω resistor, they are:

$$49\ \text{V},\quad -35\ \text{V},\ 25\ \text{V},\quad -17.857\ \text{V},\ 12.755\ \text{V},\quad -9.111\ \text{V},\ \text{etc.}$$

Since the overall reflection coefficient is $-\frac{5}{7}$ they are reduced by this factor each time. So the voltages can also be written as

$$49\left(1-\frac{5}{7}+\left(\frac{5}{7}\right)^2-\left(\frac{5}{7}\right)^3+\left(\frac{5}{7}\right)^4-\cdots\right).$$

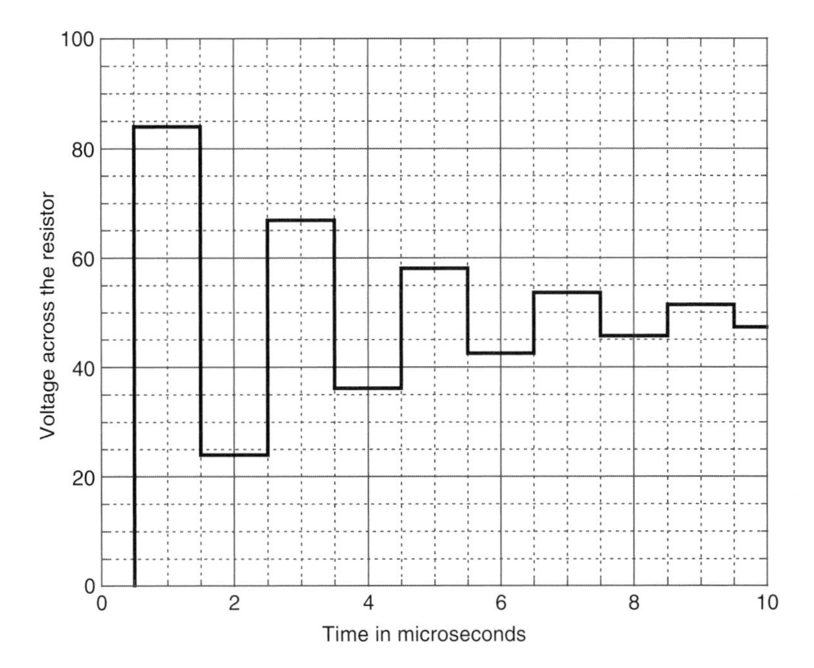

Figure 1.8 The voltage across the 300 Ω resistor against time in Example 1.3.

This can be summed, since it is a geometric series, to give 28.583 V. Similarly, the reflected waves are

$$35 \text{ V}, \quad -25 \text{ V}, \quad 17.857 \text{ V}, \quad -12.755 \text{ V}, \quad 9.111 \text{ V}, \quad -6.508 \text{ V}, \text{ etc.}$$

Again, these can be summed to give 20.416 V. It can be seen from the two lists of voltages above that their sum is 49 V. To be within 1% of the final voltage, the factor $\left(\frac{5}{7}\right)^n \leq 0.01$ and this occurs for $n = 14$ or after 14.5 μs.

The significance of these results is that these waves keep on travelling long after the wave nature of the results appears to have faded away leaving the usual circuit result. The final current in the 300 Ω resistor is 163.333 mA. The forward current is 571.666 mA and the current associated with the reflection is – 408.333 mA using Equation (1.9). The final current is the sum of these two currents. Although an ammeter would only read 163.333 mA in the steady state, in reality it is measuring the total induced current caused by all the waves. So in any circuit, what may appear to be just voltages and currents will be in reality the sum of many waves. If this circuit is subsequently disturbed, as in Example 1.2, then these waves will reappear.

Example 1.4 The circuit in Example 1.3 can be modified to reduce multiple reflections by inserting a 50 Ω resistor in series with the battery. Find the voltage waveform across both the resistors for the first 2.5 μs after the switch is closed. The circuit is shown in Figure 1.9.

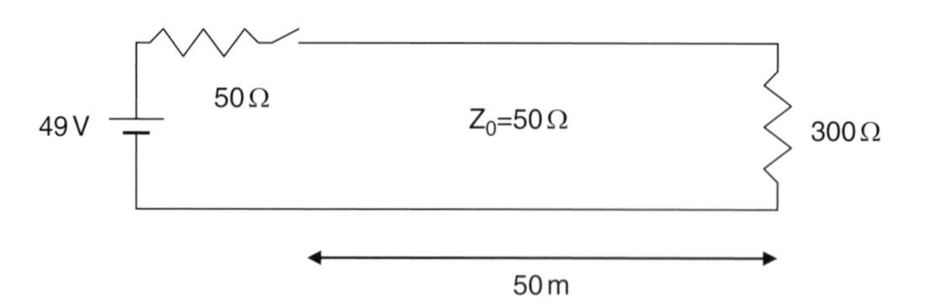

Figure 1.9 The circuit diagram for Example 1.4.

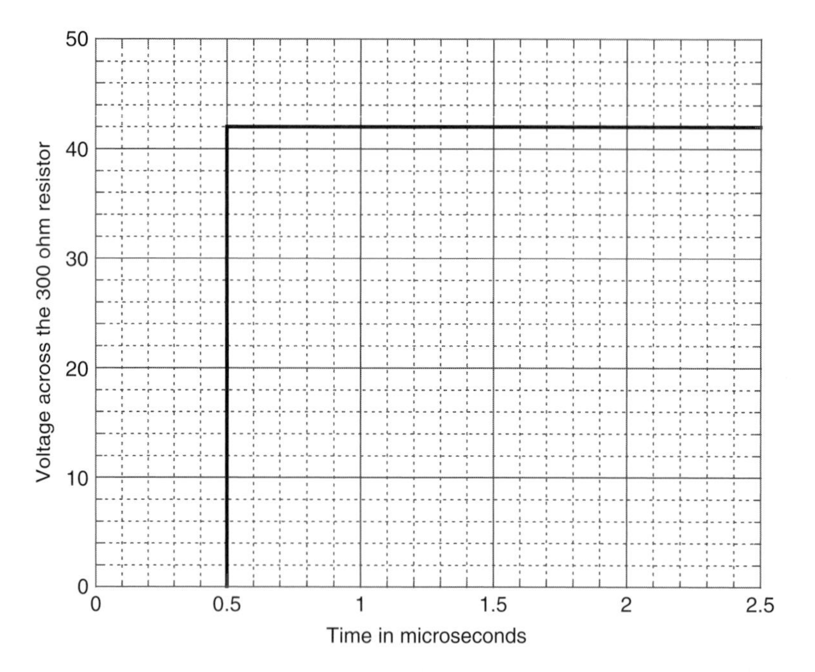

Figure 1.10 The voltage across the 300 Ω resistor against time in Example 1.4.

Solution to Example 1.4

The battery initially 'sees' the new 50 Ω in series with the 50 Ω transmission line. So the forward wave is $\frac{49}{2}V$ in amplitude. After 0.5 μs the wave reflects with the same reflection coefficient and becomes $\frac{35}{2}V$. This gives the final voltage across the 300 Ω resistor as 42 V, see Figure 1.10. This voltage is expected from circuit theory, as it is the fraction of 49 V which would appear across a 300 Ω resistor in series with 50 Ω.

So the mystery is at the battery end. Initially the voltage across the 50 Ω is $\frac{49}{2}V$ and when the reflected wave arrives after 1 μs the wave will not be reflected back as this is the matched case described in Equation (1.11). By inspecting the circuit in Figure 1.9, it can be seen that the reflected wave reduces the voltage across this resistor down to 7 V, that is $\frac{49}{2} - \frac{35}{2}V$. One way of looking at this is that the reflected wave does not 'see' the battery, as its impedance is zero. So if the battery

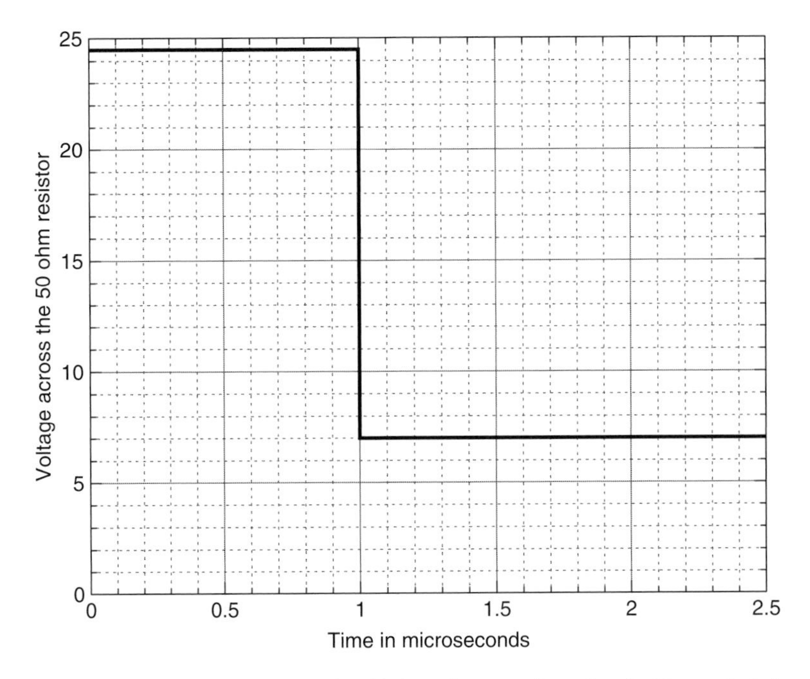

Figure 1.11 The voltage across the 50 Ω resistor against time in Example 1.4.

is removed from the circuit for the reflected wave, the positive end of the source resistor becomes connected to the lower or negative wire. In other words the two voltages have opposite polarity, see Figure 1.11.

Clearly the addition of the 50 Ω resistor has reduced reflections to the minimum. However, the waves are still present in the steady state as described at the end of the last example. Not surprisingly, most circuit designers try to match the internal impedance of the source to the characteristic impedance of the transmission line. In addition, if the terminating impedance is also matched, there is a further reduction of reflections. In many circuits the value of the characteristic impedance of the transmission line is chosen to be 50 Ω and this is often the case for coaxial cables. Other transmission lines may have a different characteristic impedance to meet a specific design criterion.

1.4 Pulses incident on resistive terminations

Example 1.5 A pulse generator is connected to a 10 m long transmission line. The pulse generator produces a single pulse of duration 1 μs and amplitude 9 V. The velocity of propagation along the transmission line is 1.10^8 ms^{-1} and the characteristic impedance is 50 Ω. At the end of the transmission line is a terminating resistor of 100 Ω. The internal impedance of the pulse generator is 25 Ω. Find the waveform of the voltage across the 100 Ω resistor. The circuit diagram is shown in Figure 1.12.

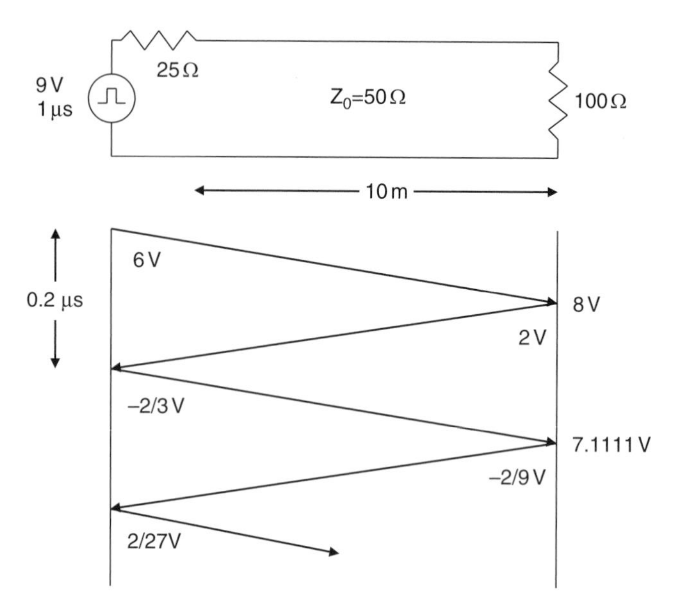

Figure 1.12 The circuit diagram and wave diagram for Example 1.5.

Solution to Example 1.5

The resistances have been chosen to be mismatches so as to illustrate their effect on pulses. The solution is obtained by first solving for a step wave of 9 V at $t = 0$ followed by a step wave of −9 V at $t = 1$ μs and then superimposing the two results. Using Equation (1.13), the reflection coefficients for the 100 Ω and 25 Ω resistors are $\frac{1}{3}$ and $-\frac{1}{3}$, so the round trip reduces a wave by a factor of $-\frac{1}{9}$. See Figure 1.12. The amplitude of the first step is 6 V due to the voltage drop across the internal impedance of 25 Ω. So the waves arriving at the 100 Ω resistor are

$$6\text{V},\ \frac{-6}{9}\text{V},\ \frac{6}{81}\text{V},\ \frac{-6}{729}\text{V},\quad \text{etc.}$$

Adding up these voltages using the geometric series gives 5.4 V. The reflected waves will be similar but only one third i.e. 1.8 V, so the final voltage across the 100 Ω resistor will be 7.2 V. However, when the negative step arrives the process reverses and the result can be seen in Figure 1.13. The transit time along the transmission line is 0.1 μs.

Example 1.6 Repeat Example 1.5, but with a pulse width of 0.1 μs.

Solution to Example 1.6

The pulses are now resolved as they are equal to the transit time. This is shown in Figure 1.14.

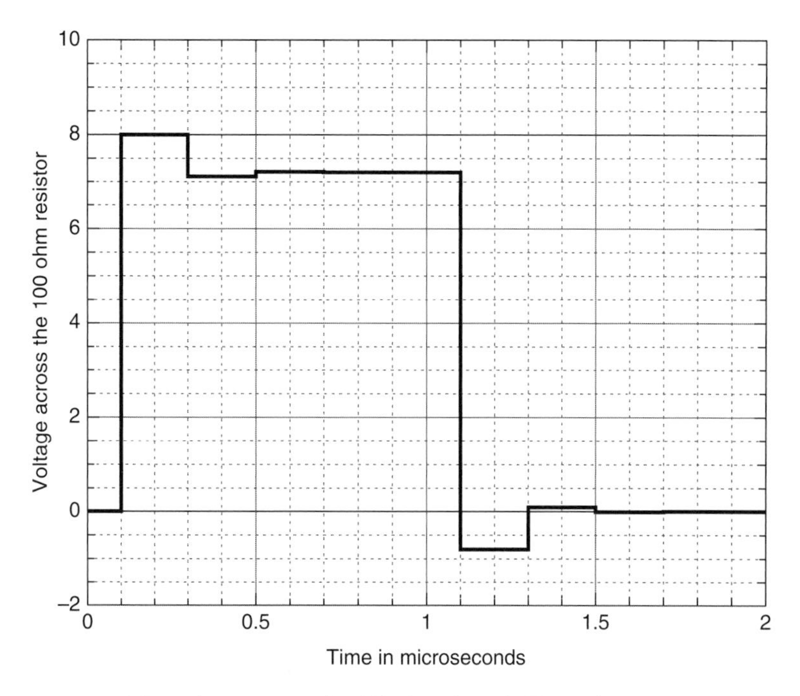

Figure 1.13 The voltage across the 100 Ω resistor in Example 1.5.

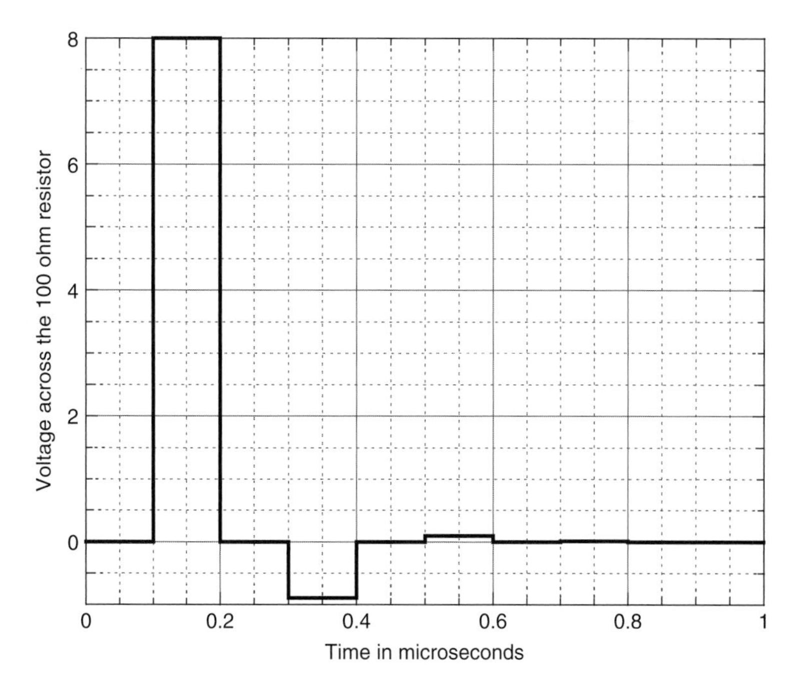

Figure 1.14 The voltage across the 100 Ω resistor against time for Example 1.6.

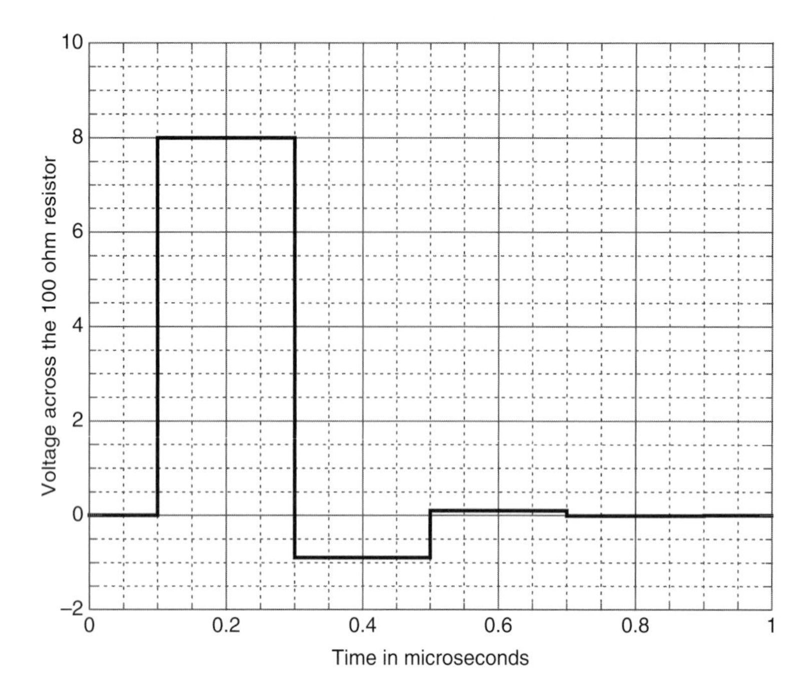

Figure 1.15 The voltage across the 100 Ω resistor against time for Example 1.7.

Example 1.7 Repeat Example 1.5 but with a pulse width equal to 0.2 μs.

Solution to Example 1.7

This is a special case where the pulse width is equal to twice the transit time. In Example 1.5, the pulse width was much greater than the transit time, i.e. ten times it, and in Example 1.6, the pulse width was equal to the transit time. These examples illustrate all the possible waveforms often seen in practical circuits where impedances are not properly matched. This special case is similar to Example 1.6 except that the separate pulses now join up to form a 'decaying pulse'. The solution is shown in Figure 1.15.

Example 1.8 Repeat the same problem as Example 1.7 but observe the voltage waveform 5 m along the line.

Solution to Example 1.8

This problem completes the set of problems involving resistors and steps and pulses. The voltage waveform halfway along the line in Figure 1.16, derived from

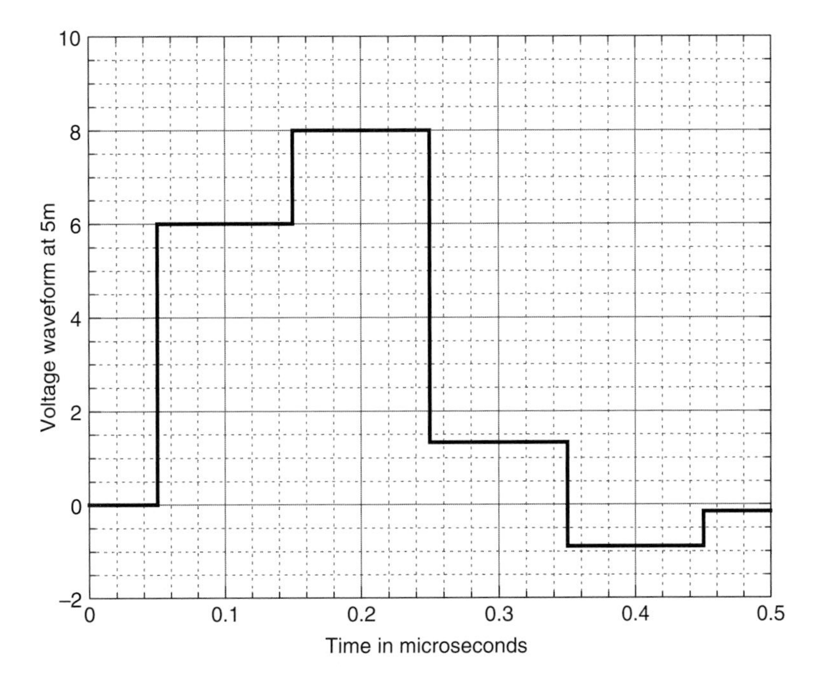

Figure 1.16 The voltage waveform at 5 m along the line for Example 1.8.

the wave diagram in Figure 1.12, shows the incident and reflected pulses. In a practical situation, it might be easy to sample the voltages at such a point and the complex nature of the waveform shows how carefully the measurements need analysing. Many a new person involved with measurements would say that the pulse generator was not working if the waveform in Figure 1.16 appeared. After twisting and bending all the connections and switching the pulse generator on and off, there may even be the odd thump on the side of the generator to improve things! In fact the waveform is not uncommon and shows how important matching is to avoid these multiple reflections. For instance, if the sample at 5 m was taken to get a single pulse of 0.2 μs long; clearly this is not the case.

1.5 Step waves incident on a capacitor

Before some examples of the reflections from a capacitor can be described, some theoretical considerations have to be made. Starting from Equation (1.10), but avoiding impedance: $V_C = V_+ + V_-$ and $I_C = I_+ + I_-$, where V_C and I_C are now the voltage across and the current through a capacitor at the end of a transmission line. If Q is the charge on this capacitor then, using the usual circuit equations:

$$Q = CV_C \text{ and } \frac{dQ}{dt} = I_C = C\frac{dV_C}{dt}.$$

So combining these equations gives:

$$I_C = I_+ + I_- = \frac{V_+}{Z_0} - \frac{V_-}{Z_0} = C\frac{dV_+}{dt} + C\frac{dV_-}{dt}.$$

Now, rewriting these equations just in terms of voltages:

$$\frac{dV_-}{dt} + \frac{V_-}{CZ_0} + \frac{dV_+}{dt} - \frac{V_+}{CZ_0} = 0. \tag{1.15}$$

This first order differential equation for V_- in terms of the incident voltage wave V_+ has the solution

$$V_- = \exp\left(-\frac{t}{CZ_0}\right)\int\left(\frac{V_+}{CZ_0} - \frac{dV_+}{dt}\right)\exp\left(\frac{t}{CZ_0}\right)dt + D\exp\left(-\frac{t}{CZ_0}\right), \tag{1.16}$$

where D is a constant to be determined by the boundary conditions.

In order to simplify the solutions, the first example will involve only one reflection. After that, more complex examples will show the unusual results that occur when multiple reflections are involved.

Example 1.9 A 10 volt battery with an internal impedance of 50 Ω is connected at $t = 0$ to a transmission line. The characteristic impedance of the line is 50 Ω and the length is 3 m. At the end of the line is a 200 pF capacitor and the velocity of propagation along the line is 3.10^8 ms^{-1}. Calculate the voltage waveform at the capacitor from $t = 0$ to $t = 70$ ns. The circuit diagram is shown in Figure 1.17.

Solution to Example 1.9

In this particular case $V_+ = 5$ V as the internal impedance of the battery will reduce the incident voltage wave to half the battery voltage. Since V_+ is constant, and $CZ_0 = 10$ ns, taking the unit of time as 1 ns, using Equation (1.16) and ignoring the delay between the battery and the capacitor:

$$V_- = \exp\left(-\frac{t}{10}\right)\int\frac{5}{10}\exp\left(\frac{t}{10}\right)dt + D\exp\left(-\frac{t}{10}\right),$$

which gives

$$V_- = 5 + D\exp\left(-\frac{t}{10}\right). \tag{1.17}$$

Now the initial charge on the capacitor will be zero and so the initial voltage will be the same. So using Equation (1.17) gives

$$V_C = V_+ + V_- = 5 + 5 + D. \tag{1.18}$$

An inspection of Equation (1.18) shows that the initial value of V_C is zero if $D = -10$ V and the final value will be 10 V. Finally, the transit time from the battery

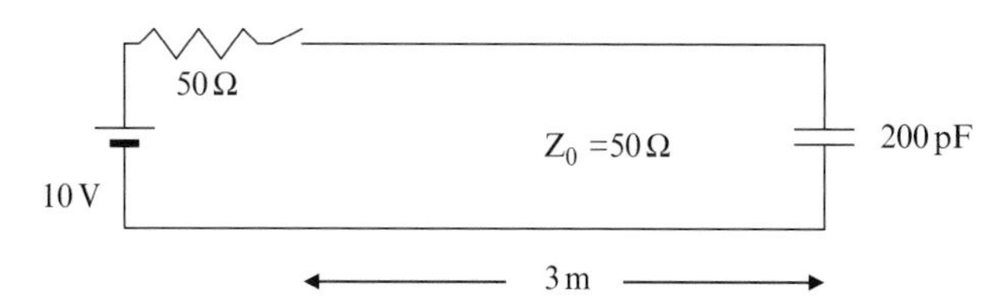

Figure 1.17 The circuit diagram for Example 1.9.

needs to be added to Equations (1.17) and (1.18) to account for the delay as the 5 V step travels down the transmission line. The delay is 10 ns, so the complete solution to this problem is

$$V_- = 5 - 10 \exp - \left(\frac{t-10}{10}\right), \quad V_C = 10\left(1 - \exp - \left(\frac{t-10}{10}\right)\right), \tag{1.19}$$

where $V_- = 0$ and $V_C = 0$ for $t < 10$ ns.

Now the reflected wave, V_-, is then absorbed in the 50 Ω internal impedance of the battery some 20 ns later. This result is very similar to the one obtained from ordinary circuit theory except for the delay. The final voltage across the capacitor is 10 V and the time constant is CZ_0, which in this case is the same as the time constant normally taken from circuit theory as RC where R is the internal impedance of the battery. The graph of the second equation in (1.19) is shown in Figure 1.18.

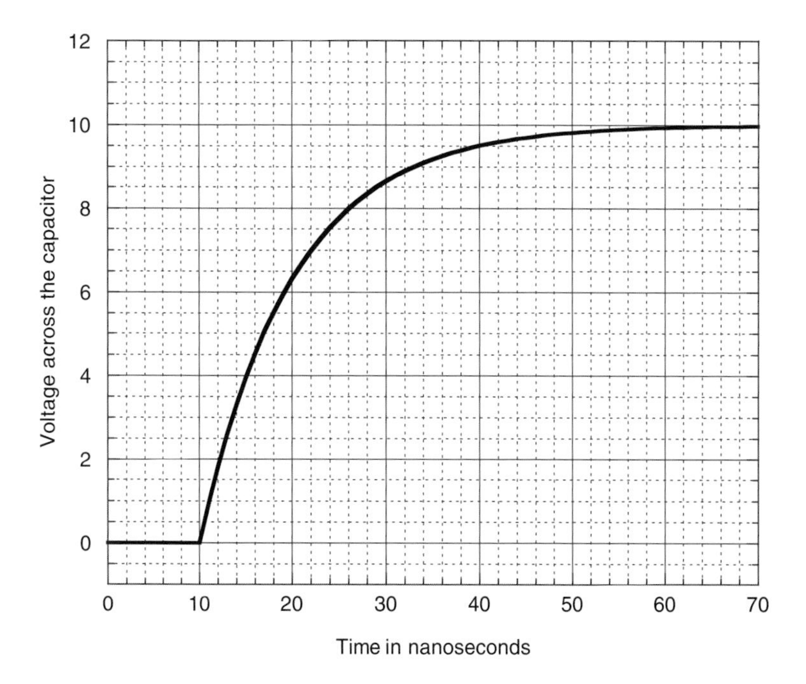

Figure 1.18 The voltage across the capacitor for Example 1.9.

1.6 A pulse incident on a resistor and a capacitor in parallel

Combining some of the previous examples, a more complicated circuit can now be analysed. This one involves a pulse incident on a capacitor and a resistor in parallel. The theory developed in Section 1.5 can now be modified to include the extra current induced in the resistor. A typical circuit is shown in Figure 1.19, which follows in Example 1.10. Beginning with Equation (1.15), the total current in the termination, I_L, now becomes

$$I_L = I_+ + I_- = \frac{V_+}{Z_0} - \frac{V_-}{Z_0} = C\frac{dV_+}{dt} + C\frac{dV_-}{dt} + \frac{V_+}{R} + \frac{V_-}{R}, \tag{1.20}$$

where the last two terms arise from the current induced in the resistor due to both the incident and reflected waves. Rearranging the last two terms of this equation into a first order differential equation gives

$$\frac{dV_-}{dt} + \frac{V_-}{CRZ_0}(R + Z_0) = \frac{V_+}{CRZ_0}(R - Z_0) - \frac{dV_+}{dt}. \tag{1.21}$$

By putting

$$A = \frac{(R + Z_0)}{CRZ_0} \text{ and } B = \frac{(R - Z_0)}{CRZ_0} \tag{1.22}$$

and using Equation (1.16), the solution of Equation (1.21) becomes

$$V_- = \exp(-At)\int\left(BV_+ - \frac{dV_+}{dt}\right)\exp(At)dt + D\exp(-At), \tag{1.23}$$

where D is a constant to be determined by the boundary conditions.

Example 1.10 A pulse incident on a resistor and a capacitor in parallel: A pulse generator is matched to a 50 Ω transmission line and this line is terminated with a 100 Ω resistor in parallel with a 200 pF capacitor. The length of the line is 3 m and the velocity of propagation is 3.10^8 ms^{-1}. The circuit is shown in Figure 1.19. Calculate the voltage wave across the termination and the reflected wave, which arrives back at the generator, for the first 100 ns after a 10 V, 40 ns pulse begins to leave the generator.

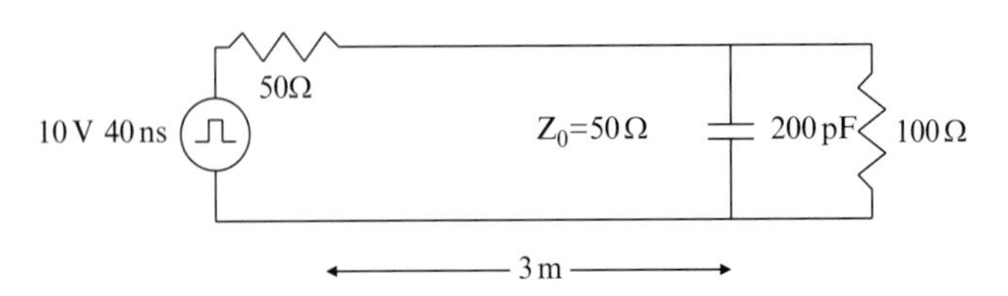

Figure 1.19 The circuit for Example 1.10. The components are assumed to be physically small compared to the length of the line.

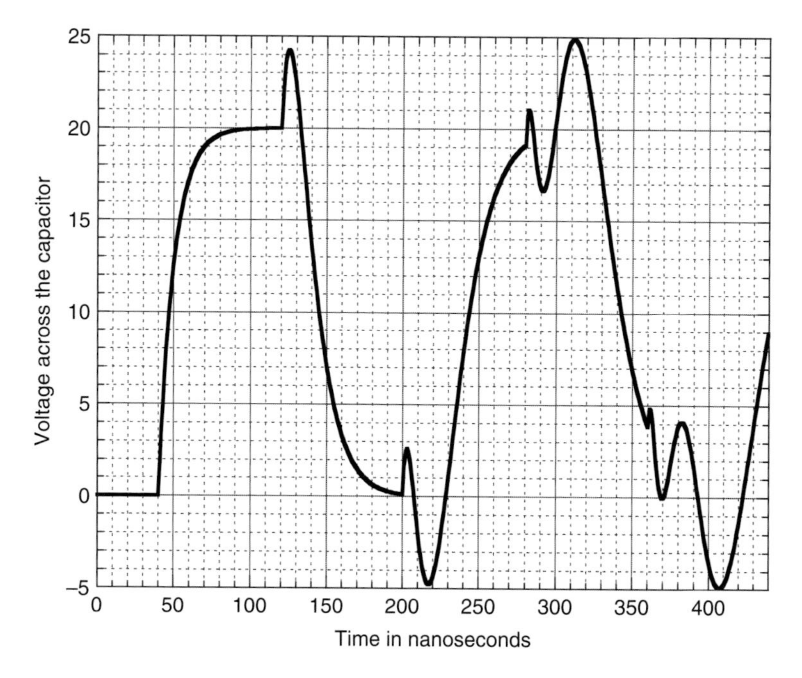

Figure 1.23 The voltage across the capacitor against time for zero source resistance.

The voltages at the end of each transient are given by the non-transient terms in the equations. They are respectively 16.7 V at $t = 120$ ns; 5.55 V at $t = 200$ ns; 13.0 V at $t = 280$ ns; 9.02 V at $t = 360$ ns; and 11.3 V at $t = 440$ ns, which are generated by taking the terms in

$$\frac{50}{3}\left(1 - \frac{2}{3} + \frac{4}{9} - \frac{8}{27} + \frac{16}{81} - \cdots\right) \text{V}.$$

Summing the series to infinity gives 10 V, which is the expected answer from simple circuit theory, however it will need over ten reflections to be within 1% of the final voltage and will take over 800 ns. It is interesting to note that if the length of line is reduced to zero, the time constant reverts back to 2 ns. So the capacitor would charge in only say 20 ns rather than 800 ns.

Finally, just to complete the view of this problem, when the source resistance is reduced to zero, the energy is no longer lost in the circuit and as a result builds up in an increasingly more complex way. The results are shown in Figure 1.23 and the waveform is an 'oscillation', but rich in harmonics and almost chaotic. This is clearly undesirable in most circumstances and yet could occur in some circuits.

Example 1.12 Two capacitors in parallel – a circuit conundrum: A capacitor, C_1, is charged to a voltage, V_1, and then connected in parallel to a second capacitor, C_2. Find the final voltage across the two capacitors. The circuit diagram is shown in Figure 1.24, where a transmission line has been used to connect the two capacitors.

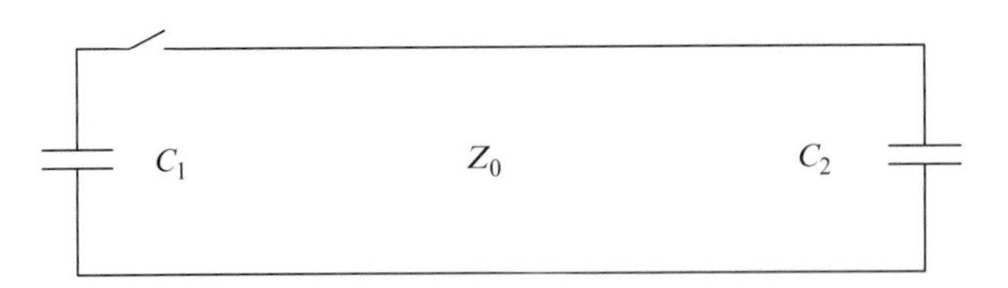

Figure 1.24 The circuit diagram for Example 1.12.

Some considerations of Example 1.12

The solution to this problem using circuit theory gives two answers dependent on the assumptions made.

Assumption 1: energy is conserved

$$\frac{1}{2}C_1V_1^2 = \frac{1}{2}(C_1 + C_2)V_F^2. \tag{1.36}$$

So the final voltage is

$$V_F = \sqrt{\frac{C_1}{C_1 + C_2}}V_1. \tag{1.37}$$

Assumption 2: charge is conserved

$$Q_1 = C_1V_1 = (C_1 + C_2)V_F. \tag{1.38}$$

So the final voltage is

$$V_F = \frac{C_1}{C_1 + C_2}V_1. \tag{1.39}$$

Equations (1.37) and (1.39) cannot both be correct and one of the assumptions must be wrong; this is the conundrum. It is usually resolved in favour of the second assumption, by arguing that some energy must be lost when the switch is closed via perhaps a spark or electromagnetic radiation. However, if Equation (1.39) is correct, then the loss of energy is very precisely given by

$$\frac{1}{2}C_1V_1^2 - \frac{1}{2}(C_1 + C_2)V_F^2 = \frac{1}{2}\frac{C_1C_2V_1^2}{C_1 + C_2}. \tag{1.40}$$

According to circuit theory this must be dissipated as soon as the switch is closed. Something is clearly wrong and transmission line theory does give some clues as to the answer. If there are no loss mechanisms in the circuit, then the energy cannot be lost as suggested in the above discussion. However, transmission line theory

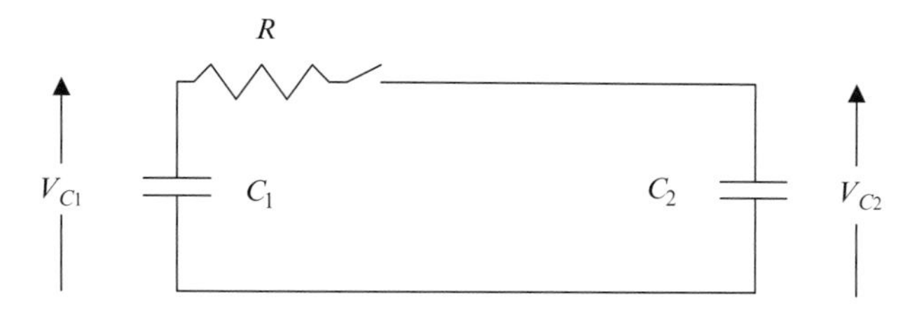

Figure 1.25 Modified circuit diagram for Example 1.12 showing the addition of a resistor, R.

suggests that the energy will go on reflecting between the two capacitors with ever increasing complexity. On the other hand, if there are some losses, the system will reach the steady state predicted by Assumption 2.

If the distance between the capacitors is small, these losses can be represented by a resistor and analysed using circuit theory. Using the diagram in Figure 1.25, it can be seen that a current, I, that flows in the circuit after the switch has closed is given by the following:

$$I = I_0 \exp\left(-\frac{t}{AR}\right), \quad \text{where } A = \frac{C_1 C_2}{C_1 + C_2} \text{ and } I_0 = \frac{V_1}{R}, \tag{1.41}$$

where the capacitors have been combined in series.

Integrating this current to find the voltage V_{C1} gives

$$V_{C1} = -\frac{1}{C_1}\int I \mathrm{d}t = \frac{ARI_0}{C_1} \exp\left(-\frac{t}{AR}\right) + D = \frac{AV_1}{C_1} \exp\left(-\frac{t}{AR}\right) + D,$$

where D is a constant. Using the initial conditions at $t = 0$:

$$V_{C1} = V_1 = \frac{AV_1}{C_1} + D, \text{ hence } D = \frac{V_1 C_1}{C_1 + C_2} \text{ and } V_{C1} = \frac{V_1 C_1}{C_1 + C_2} + \frac{V_1 C_2}{C_1 + C_2} \exp\left(-\frac{t}{AR}\right). \tag{1.42}$$

This equation gives the same final voltage as Equation (1.39), which confirms the second assumption. Solving for V_{C2} gives

$$V_{C2} = \frac{V_1 C_1}{C_1 + C_2}\left(1 - \exp\left(-\frac{t}{AR}\right)\right)$$

and hence the voltage across the resistor is

$$V_R = V_{C1} - V_{C2} = V_1 \exp\left(-\frac{t}{AR}\right).$$

Using this result, the energy lost in the resistor is given by

$$\int_0^\infty V_R I \mathrm{d}t = V_1^2 \int_0^\infty \frac{1}{R} \exp\left(-\frac{2t}{AR}\right) \mathrm{d}t = \frac{V_1^2 A}{2} = \frac{1}{2}\frac{C_1 C_2 V_1^2}{C_1 + C_2}. \tag{1.43}$$

This is the same as Equation (1.40). So the energy lost in the resistor is not a function of the resistance. The same final voltages occur, whatever the value of the resistance. All the resistance determines is the time taken to arrive at this final voltage. So, in summary, if no resistance is present, the solution may look like Figure 1.23. If the line is very long the individual reflections will be totally separated out. However, it is not possible to have a circuit completely free from losses and so it will settle down to the solution which assumed the preservation of charge. The only extra item will be that the total capacitance will need to include the capacitance of the length of the transmission line.

So a feature of many of these problems is that when a circuit contains no losses, the energy in it can be reflected back and forth for ever. As soon as some losses are introduced, these oscillations decay away and the final result is the one predicted by ordinary circuit theory. The value of the resistance determines how long it will take before this happens. Comparing Examples 1.1 and 1.3 illustrates this feature for step waves.

1.8 Step waves incident on inductors

As in the previous section on capacitors, some theoretical consideration must be made before any examples can be considered. Starting from Equation (1.10), $V_\mathrm{L} = V_+ + V_-$ and $I_\mathrm{L} = I_+ + I_-$, where V_L and I_L are now the voltage across and the current through an inductor at the end of a transmission line. Using the usual circuit law for an inductor:

$$V_\mathrm{L} = L\frac{\mathrm{d}I_\mathrm{L}}{\mathrm{d}t}, \text{ then}$$

$$V_+ + V_- = L\frac{\mathrm{d}I_+}{\mathrm{d}t} + L\frac{\mathrm{d}I_-}{\mathrm{d}t}.$$

Now, rewriting these equations in terms of just voltages,

$$\frac{\mathrm{d}V_-}{\mathrm{d}t} + \frac{Z_0 V_-}{L} - \frac{\mathrm{d}V_+}{\mathrm{d}t} + \frac{Z_0 V_+}{L} = 0. \tag{1.44}$$

This is a first order differential equation for V_- similar to Equation (1.15) for reflections from a capacitor. The solution is given below:

$$V_- = \exp\left(-\frac{Z_0 t}{L}\right)\int\left(\frac{\mathrm{d}V_+}{\mathrm{d}t} - \frac{Z_0 V_+}{L}\right)\exp\left(\frac{Z_0 t}{L}\right)\mathrm{d}t + E\exp\left(-\frac{Z_0 t}{L}\right), \tag{1.45}$$

where E is a constant to be determined by the boundary conditions. In the example which follows, only one reflection will be considered. The more complex case of multiple reflections discussed for capacitors is similar for inductors as the equations that govern their responses are also similar.

Example 1.13 This example has been included for completeness and involves one reflection from an inductor. A 10 volt battery with an internal impedance of 50 Ω is connected at $t = 0$ to a transmission line. The characteristic impedance of the line is 50 Ω and the length is 3 m. At the end of the line is a 0.5 μH inductor and the velocity of propagation along the line is 3.10^8 ms^{-1}. Calculate the voltage waveform at the inductor from $t = 0$ to $t = 70$ ns. The circuit diagram is shown in Figure 1.26.

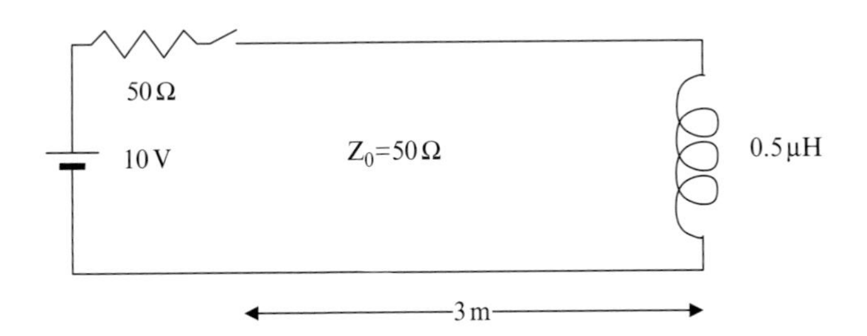

Figure 1.26 The circuit diagram for Example 1.13.

Solution to Example 1.13

As in Example 1.9, the voltage step arriving at the inductor will be only 5 V as the internal impedance of the battery will also have 5 V across it. So $V_+ = 5$ V and the time constant L/Z_0 is 10 ns. Taking the unit of time as 1 ns, Equation (1.45) becomes

$$V_- = \exp\left(\frac{-t}{10}\right)\int -\frac{5}{10}\exp\frac{t}{10}\mathrm{d}t + E\exp\left(-\frac{t}{10}\right), \tag{1.46}$$

which gives

$$V_- = -5 + E\exp\left(-\frac{t}{10}\right) \text{ and } V_\mathrm{L} = E\exp\left(-\frac{t}{10}\right). \tag{1.47}$$

In contrast to the capacitor example, the initial current in the inductor is zero and so it appears as an open circuit. So the initial voltage across the inductor will be twice the incident step, that is 10 V, and hence $E = 10$ V. Adding in the delay of 10 ns into the solution gives

$$V_- = 10\exp - \left(\frac{t-10}{10}\right) - 5 \text{ and } V_\mathrm{L} = 10\exp - \left(\frac{t-10}{10}\right), \tag{1.48}$$

where $V_- = 0$ and $V_\mathrm{L} = 0$ for $t < 10$.

The reflected wave, V_-, is then absorbed by the 50 Ω internal impedance of the battery, beginning 20 ns later. Again this result is similar to that expected from circuit theory, except for the delay. The final voltage across the inductor is zero, and the graph of the second equation in (1.48) is shown in Figure 1.27.

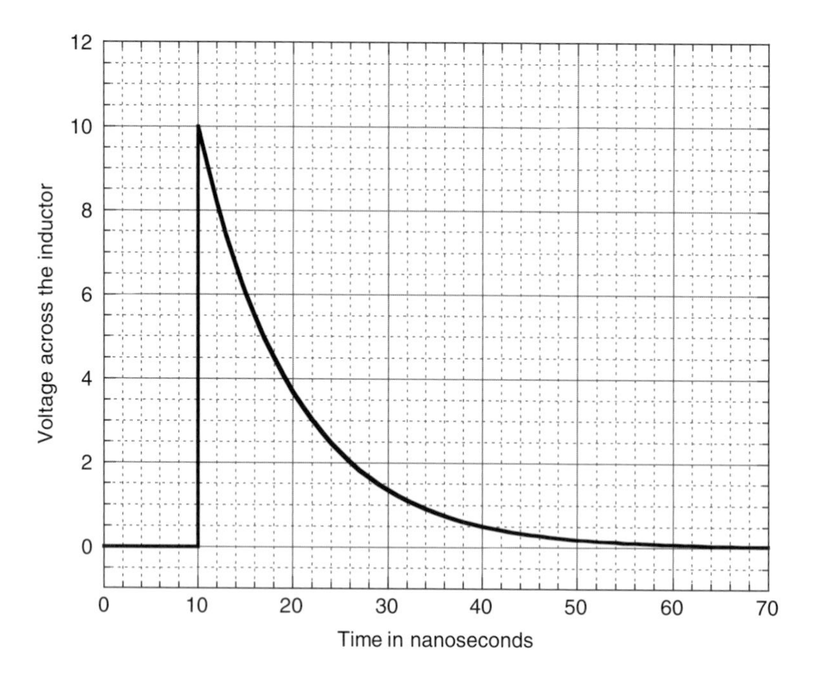

Figure 1.27 The voltage across the inductor against time for Example 1.13.

The waveform in Figure 1.27 is often interpreted as an unwanted 'spike' by many observers unaware of transmission line theory.

1.9 Conclusions on the use of circuit theory and transmission line theory

Finally, some general remarks about the boundary between circuit theory and transmission line theory can now be formulated. Clearly, when the time period of a pulse or other waveform is much longer than any transit time in a circuit, then the transmission line effects described in this chapter do not appear and simple circuit theory is adequate to predict the electrical performance of a circuit. As a guide to design, perhaps a safety margin might usefully be put at

$$\text{time period of a pulse} = 100 \times \text{transit time of circuit.} \tag{1.49}$$

To illustrate this a graph showing the boundary between circuit theory and transmission line theory is shown in Figure 1.28. The velocity of propagation has been taken as 3.10^8 ms^{-1} and the frequency has been taken as the inverse of the time period. As expected, very low frequencies like mains frequencies are adequately described by circuit theory up to pylon lines 100 km long. However, circuits involving frequencies in the low MHz range are not covered by circuit theory if their length is above one metre. In the microwave range, it is common to have circuits in an integrated form that need only circuit theory as well as much longer circuits requiring the full transmission line theory. Since the physical size of

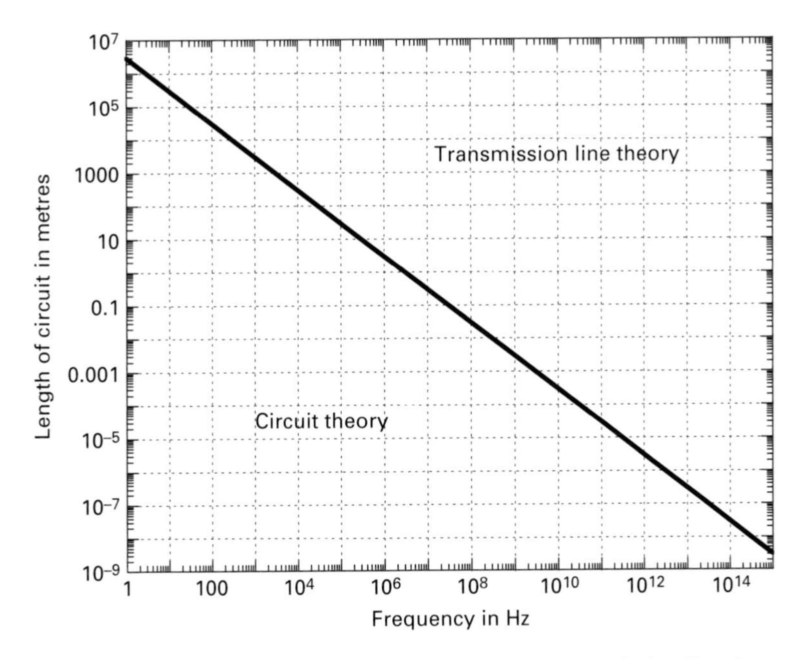

Figure 1.28 The boundary between circuit theory and transmission line theory.

circuits is limited by the atomic structure much below 1 nm, Figure 1.28 has a lower limit for circuit length.

Having discussed this demarcation between the two theories, it must be added that even in circuits, where ordinary circuit theory is sufficient, there is still energy stored in these lines. When these circuits are switched, this energy will also be changed and so any simple equivalent circuit will need to be modified to account for this phenomenon.

This chapter has been concerned with signals in the time domain on loss-less transmission lines. In Chapter 3 the topic of pulses on loss-less coupled lines is discussed and, finally, Chapter 6 contains examples of the effects of attenuation on pulses. The reflection from capacitors is also discussed in Chapter 8 using photons.

1.10 Further reading

S. Ramo, J. R. Whinnery and T. Van Duzer *Fields and Waves in Communication Electronics*, Third edition, New York, Wiley, 1993. Chapter 5, sections 5.4 to 5.7.

J. D. Krauss *Electromagnetics*, Fourth edition, New York, McGraw-Hill, 1992. Chapter 12, sections 10.1 and 10.2.

D. K. Cheng *Field and Wave Electromagnetics*, Second edition, New York, Addison-Wesley, 1989. Chapter 9, section 9–5.

H. A. Haus and J. R. Melcher *Electromagnetic Fields and Energy*, London, Prentice Hall, 1989. Chapter 14, sections 14.3 and 14.4.

P. A. Rizzi *Microwave Engineering*, London, Prentice Hall, 1988. Chapter 3, section 3.2.

R. E. Matick *Transmission Lines for Digital and Communication Networks*, New York, McGraw-Hill, 1969. Chapter 5.

P. C. Magnusson, G. C. Alexander and V. K. Tripathi *Transmission Lines and Wave Propagation*, New York, CRC Press, 1992. Chapter 3.

R. E. Collin *Foundations for Microwave Engineering*, New York, McGraw-Hill, 1992. Chapter 3, sections 3.1 to 3.3.

T. H. Lee *Planar Microwave Engineering, a Practical Guide to Theory, Measurements and Circuits*, Cambridge, Cambridge University Press, 2004. Chapter 8, section 8.2.2.

2 Sine waves and networks

The last chapter introduced the equivalent circuit for a transmission line and, by using various problems as illustrations, showed how step waves and pulses can be reflected from some circuit elements. This discussion was in the time domain, or in other words all the graphs had time along the x-axis. At the beginning of this chapter, the discussion will move into the frequency domain, where many of the results will have frequency along the x-axis. There is a practical problem with the frequency domain. It is assumed in this domain that the circuit has reached its steady state, or that the single frequency source has been switched on for much longer than any transients. However, many of the interesting effects in the frequency domain occur because of changes from this steady state. For example, some of these circuits are used in communications where amplitude, phase and frequency may all be modulated in order to transmit information. Later in this chapter, some of the problems considered will cover this transient aspect and the x-axis will again become time. However, before that, some basic theory and definitions need to be described, which are exclusive to sine waves.

2.1 Sine waves

From Equation (1.6) in the last chapter, any function of $t \pm x/v$ is a solution of the wave equation. For sine waves it is convenient to choose the function to be

$$\sin\omega\left(t \pm \frac{x}{v}\right) = \sin\left(\omega t \pm \frac{\omega x}{v}\right). \tag{2.1}$$

This function represents a travelling sine wave and the parameter in the brackets is the phase or argument of the sine function. The angular frequency, ω, represents the rate of change of this argument with time and ω/v represents the rate of change of the argument with distance. The latter quantity, in transmission line theory, is called the phase constant, β. Later on in this book, in electromagnetic wave theory, it will appear again as the wavenumber, k. Both are the same quantity and

$$\beta = k = \frac{\omega}{v} = \frac{2\pi f}{f\lambda} = \frac{2\pi}{\lambda}; \text{ see Figure 2.1.} \tag{2.2}$$

So the source in Figure 2.1 has an output of $\sin(\omega t)$ and at a distance down the line, D, the signal will be delayed in phase and be $\sin(\omega t - \beta D)$. The negative sign is

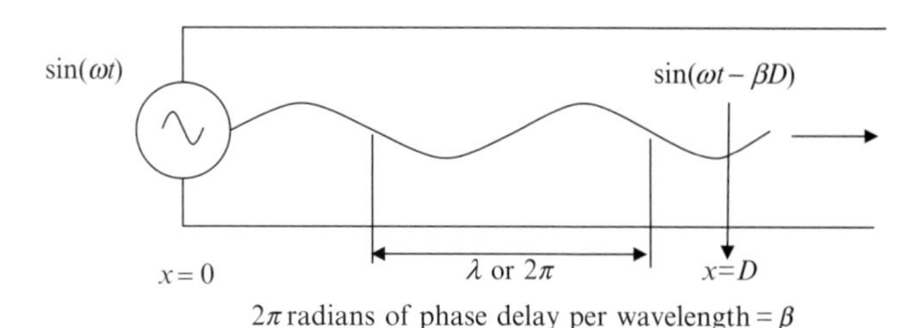

Figure 2.1 Showing the delay in phase of 2π radians per wavelength for a sine wave travelling along a transmission line.

taken for the positive x direction as the wave moves along the line and it will clearly take a finite time for the wave to arrive at any point.

In many texts a complex form is used to represent travelling sine waves:

$$\exp \mathrm{j}(\omega t \pm \beta x) = \cos(\omega t \pm \beta x) + \mathrm{j}\sin(\omega t \pm \beta x). \tag{2.3}$$

So Equation (2.1) is the imaginary part of Equation (2.3). Since in the frequency domain the frequency is common throughout, it is often separated from the delay:

$$\exp(\mathrm{j}\omega t) \times \exp(\pm \mathrm{j}\beta x). \tag{2.4}$$

Since the first part of (2.4) is sometimes called a phasor and is a common term in most equations, it is often omitted in many texts, including this one from now on. So, using the same subscripts as in Chapter 1, a wave travelling in the positive x direction or forward wave is

$$V_+ = V_1 \exp(-\mathrm{j}\beta x) \tag{2.5}$$

and a reverse or backward wave is

$$V_- = V_2 \exp(+\mathrm{j}\beta x), \tag{2.6}$$

where the voltages V_1 and V_2 are the amplitudes of the waves. To obtain a sine wave from these equations, the imaginary part must be taken and the time variation reintroduced.

2.2 Reflections from impedances

All of the theory described in Chapter 1 still applies and, in particular, unique values of the reflection coefficient exist for all values of the terminating impedance, Z_L. For example, if the terminating impedance is an inductance, L, then the reflection coefficient is given by Equation (1.13) as

$$\rho = \frac{\mathrm{j}\omega L - Z_0}{\mathrm{j}\omega L + Z_0} \text{ and } |\rho| = 1 \text{ and the phase of } \rho = \angle\rho = 2\tan^{-1}\left(\frac{Z_0}{\omega L}\right). \tag{2.7}$$

So a sine wave is totally reflected from an inductance since there is no absorption of energy, but with an advance in phase. Using the same technique, a capacitor has the following reflection coefficient:

$$\rho = \frac{\frac{1}{j\omega C} - Z_0}{\frac{1}{j\omega C} + Z_0} \text{ and } |\rho| = 1 \text{ and the phase of } \rho = \angle\rho = -2\tan^{-1}(\omega C Z_0). \quad (2.8)$$

Again there is no absorption of energy so the wave is totally reflected with a delay in phase. In a similar manner, Equation (2.8) can be derived from Equation (1.15) by replacing the time differentials with $j\omega$.

2.3 Power in waves

If there is a resistive part to the termination, then the wave is no longer totally reflected but partly absorbed. In particular, when a matched load is at the end of a transmission line, then all the incident power is absorbed. This gives a method for finding the power in a wave. The voltage across the terminating impedance, Z_0, and the incident voltage will both be V_1 and so the incident power and the power in the termination will be

$$\frac{V_1^2}{2Z_0} \text{ or } \frac{I_1^2 Z_0}{2}. \quad (2.9)$$

By similar arguments, in general, the reflected power will be

$$\frac{V_2^2}{2Z_0} \text{ or } \frac{I_2^2 Z_0}{2}, \text{ where } V_2 \text{ is the reflected voltage.} \quad (2.10)$$

Using Equation (1.12), these expressions can be linked to the reflection coefficient in the following way:

$$|\rho|^2 = \frac{|V_-|^2}{|V_+|^2} = \frac{V_2^2}{V_1^2} = \frac{\text{reflected power}}{\text{incident power}}. \quad (2.11)$$

Thus the modulus of the reflection coefficient squared is a measure of what fraction of the incident power is reflected by the termination. Also the fraction of the incident power that is absorbed by the termination is

$$1 - |\rho|^2. \quad (2.12)$$

2.4 Voltage standing wave ratio

When a wave is reflected, the voltage at a distance D from the termination is given by

$$V = V_1 \exp(j\beta D) + \rho V_1 \exp(-j\beta D).$$

The signs of the phase have been chosen as the point D is away from the termination and going back towards the source. If the reflection coefficient is expressed as

$$|\rho| \exp(j\angle\rho)$$

then this voltage becomes

$$V = V_1 \exp(j\beta D) + |\rho|V_1 \exp(j(\angle\rho - \beta D)), \tag{2.13}$$

where the phase is measured relative to the termination.

This voltage has a maximum, V_{MAX}, when the two exponential arguments are equal, i.e. when

$$\beta D = \angle\rho - \beta D + 2n\pi, \text{ where } n = 0, 1, 2, \ldots$$

$$\text{or } 2\beta D - \angle\rho = 2n\pi. \tag{2.14}$$

The value of the maximum voltage is given by

$$V_{MAX} = V_1 + |\rho|V_1 = V_1(1 + |\rho|). \tag{2.15}$$

Similarly, the voltage will be a minimum, V_{MIN}, when

$$2\beta D - \angle\rho = \pi + 2n\pi. \tag{2.16}$$

The value of the minimum voltage is given by

$$V_{MIN} = V_1 - |\rho|V_1 = V_1(1 - |\rho|). \tag{2.17}$$

Now the positions of these maxima and minima are fixed along the transmission line relative to the position of the termination. So when two waves travel in opposite directions, their total amplitude also varies in a wave-like manner going from V_{MAX} to V_{MIN}. The shape of this wave is not usually a true sinusoid. Since the position of these points is fixed with reference to the termination, this pattern is called a standing wave. However, it should be noted that this standing wave is still a combination of two travelling waves. In many applications it is useful to be able to measure the impedance at the end of a transmission line. This is usually done by analysing the reflection from this impedance. A simple technique is to measure the ratio of the maximum voltage to the minimum voltage, by sliding a detector along the transmission line. This ratio is called the voltage standing wave ratio, and is designated as *VSWR*:

$$VSWR = \frac{V_{MAX}}{V_{MIN}} = \frac{1 + |\rho|}{1 - |\rho|}. \tag{2.18}$$

By rearranging Equation (2.18):

$$|\rho| = \frac{VSWR - 1}{VSWR + 1}. \tag{2.19}$$

From Equation (2.19) and a measurement of the *VSWR*, the modulus of the reflection coefficient can be found. Then from the position of either a maximum or minimum, using Equations (2.14) and (2.16), the argument of the reflection coefficient can also be found. Finally, using Equation (1.13), a value of Z_L can be evaluated. Modern measurement techniques avoid movement along the line by using sampling at two points and comparing the results with a reference signal.

All this is now computerised and available as a piece of measurement equipment called an automatic network analyser.

2.5 The input impedance of a length of line

One of the unusual features of transmission lines is that impedances can be changed by adding a length of line to them. For example, if an impedance, Z_L, is connected to a length, D, of transmission line with a characteristic impedance, Z_0, then Equation (2.13) gives the voltage at the input of such a line as

$$V = V_1 \exp(\mathrm{j}\beta D) + \rho V_1 \exp(-\mathrm{j}\beta D)$$

and the current will be similar:

$$I = I_1 \exp(\mathrm{j}\beta D) - \rho I_1 \exp(-\mathrm{j}\beta D).$$

So the ratio of these two gives the input impedance of a length of transmission line:

$$Z_{IN} = \frac{V_1 \exp(\mathrm{j}\beta D) + \rho V_1 \exp(-\mathrm{j}\beta D)}{I_1 \exp(\mathrm{j}\beta D) - \rho I_1 \exp(-\mathrm{j}\beta D)}. \tag{2.20}$$

This can be re-expressed using Equations (1.9) and (1.13) as

$$Z_{IN} = Z_0\left(\frac{(Z_L + Z_0)\exp(\mathrm{j}\beta D) + (Z_L - Z_0)\exp(-\mathrm{j}\beta D)}{(Z_L + Z_0)\exp(\mathrm{j}\beta D) - (Z_L - Z_0)\exp(-\mathrm{j}\beta D)}\right)$$

and then expanding the complex exponentials:

$$Z_{IN} = Z_0\left(\frac{Z_L\cos(\beta D) + \mathrm{j}Z_0\sin(\beta D)}{Z_0\cos(\beta D) + \mathrm{j}Z_L\sin(\beta D)}\right)$$

which gives

$$Z_{IN} = Z_0\left(\frac{Z_L + \mathrm{j}Z_0\tan(\beta D)}{Z_0 + \mathrm{j}Z_L\tan(\beta D)}\right). \tag{2.21}$$

This equation shows the great range of impedances that are possible, using transmission lines. For instance, if $Z_L = 0$, i.e. a short circuit termination, then

$$Z_{IN} = \mathrm{j}Z_0\tan(\beta D), \tag{2.22}$$

which gives every value of reactance. A similar result is given if $Z_L \Rightarrow \infty$, i.e. an open circuit termination, when

$$Z_{IN} = -\mathrm{j}Z_0\cot(\beta D). \tag{2.23}$$

Special cases occur when $D = \frac{\lambda}{4} + \frac{n\lambda}{2}$ or $\beta D = \frac{\pi}{2} + n\pi$, where $n = 0, 1, 2, \ldots$, then

$$Z_{IN} = \frac{Z_0^2}{Z_L}. \tag{2.24}$$

Finally, when $D = \frac{n\lambda}{2}$ or $\beta D = n\pi$, where $n = 0, 1, 2, \ldots$, then

$$Z_{IN} = Z_L. \tag{2.25}$$

Representing these results is often done using the Smith chart described in the next section. After that, there will be some examples of their common use.

2.6 The Smith chart

In the field of microwave measurements, a common way of representing transmission line parameters in the frequency domain is to use a Smith chart. To introduce the Smith chart, it is convenient to begin with a typical measurement problem. Often the input impedance of a device cannot be measured directly as a transmission line is needed to connect the device to some impedance measuring equipment. This means that the impedance measured is not the device impedance, Z_L, but Z_{IN}, which is given in Equation (2.21) in the previous section. Figure 2.2 shows a typical arrangement for this measurement.

Now a wave coming from the measurement port will be delayed in phase an amount βD before it arrives at the device to be measured. After reflection it will be delayed a further βD in phase before arriving back at the measurement port. If there are no losses in the transmission line, the only error in the measurement of the reflection coefficient, ρ, will be a total phase of $2\beta D$. Or, in mathematical terms,

$$\rho = \rho_L \exp(-j2\beta D). \tag{2.26}$$

So assuming the value of βD is known, a correction can be made to compensate for the length of line involved. In practice the measurement equipment may well be a network analyser, which will have an inbuilt computerised error correction for this phase, that is, assuming the analyser is set up for the correct transmission line. Now the reflection coefficients in Equation (2.26) can be displayed on an Argand diagram as shown in Figure 2.3.

Using Equation (1.13) and the discussion in Section 1.2, it can be seen that as $\rho_L = \frac{Z_L - Z_0}{Z_L + Z_0}$, then for $Z_L = 0$ that is a short circuit $\rho_L = -1$ and for $Z_L = \infty$ that is an

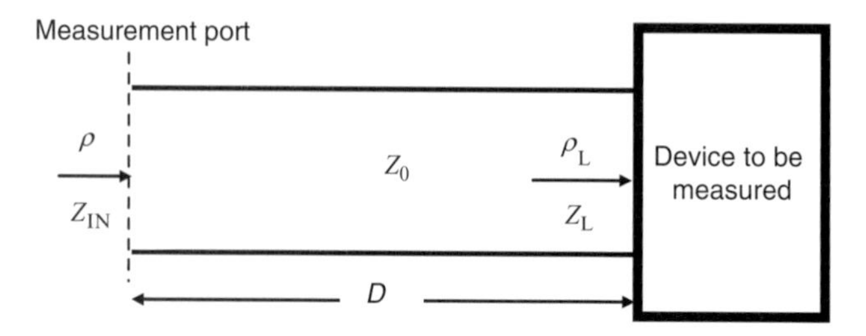

Figure 2.2 A typical arrangement for an impedance measurement. The device to be measured is connected to a suitable measurement equipment via a length of transmission line.

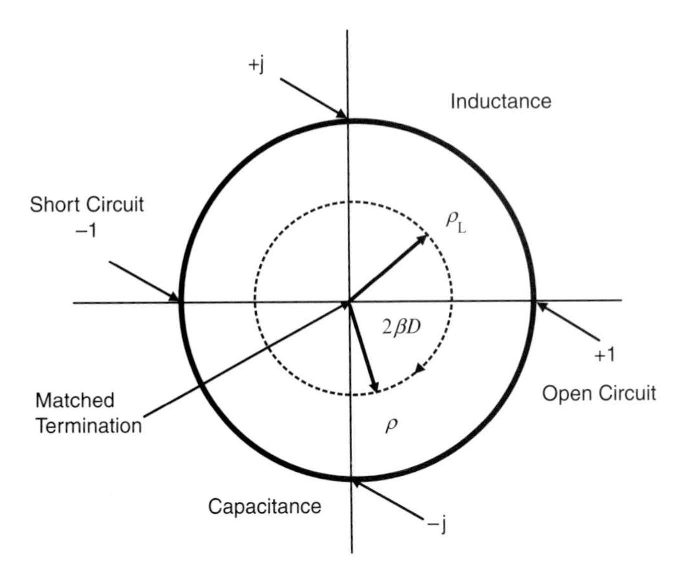

Figure 2.3 An Argand diagram for the two reflection coefficients in Equation (2.26). The phase shift of $2\beta D$ is between ρ_L and ρ.

open circuit $\rho_L = +1$. These are both shown in Figure 2.3. The Smith chart uses this Argand diagram for the reflection coefficient and superimposes on it the values of the normalised impedance, i.e. Z_L/Z_0. So the equations for the reflection coefficient and Z_{IN} become

$$\rho_L = \frac{\frac{Z_L}{Z_0} - 1}{\frac{Z_L}{Z_0} + 1} \quad \text{and} \quad \frac{Z_{IN}}{Z_0} = \left(\frac{\frac{Z_L}{Z_0} + \mathrm{j}\tan(\beta D)}{1 + \mathrm{j}\frac{Z_L}{Z_0}\tan(\beta D)} \right). \tag{2.27}$$

Now the normalised impedance can be divided into its resistive and reactive parts as

$$\frac{Z_L}{Z_0} = r + \mathrm{j}x. \tag{2.28}$$

These two parts are shown in Figure 2.4 and this is the basic Smith chart. The next figure, 2.5, shows the complete Smith chart. The advantage of the chart is that for every value of the reflection coefficient, the two components of the normalised impedance can be read off from the circles and arcs superimposed on the Argand diagram. In the top half of the chart, the reactive part of Z_L is positive, that is it has an inductive component. The lower part, being negative, represents impedances with a capacitative component.

A complete rotation on the chart is equivalent to moving half a wavelength along the line, as Equation (2.25) shows that $Z_{IN} = Z_L$. In other words, after half a wavelength the value of the impedance returns to the same value. A rotation of 180° on the chart is equivalent to moving a quarter of a wavelength down the line

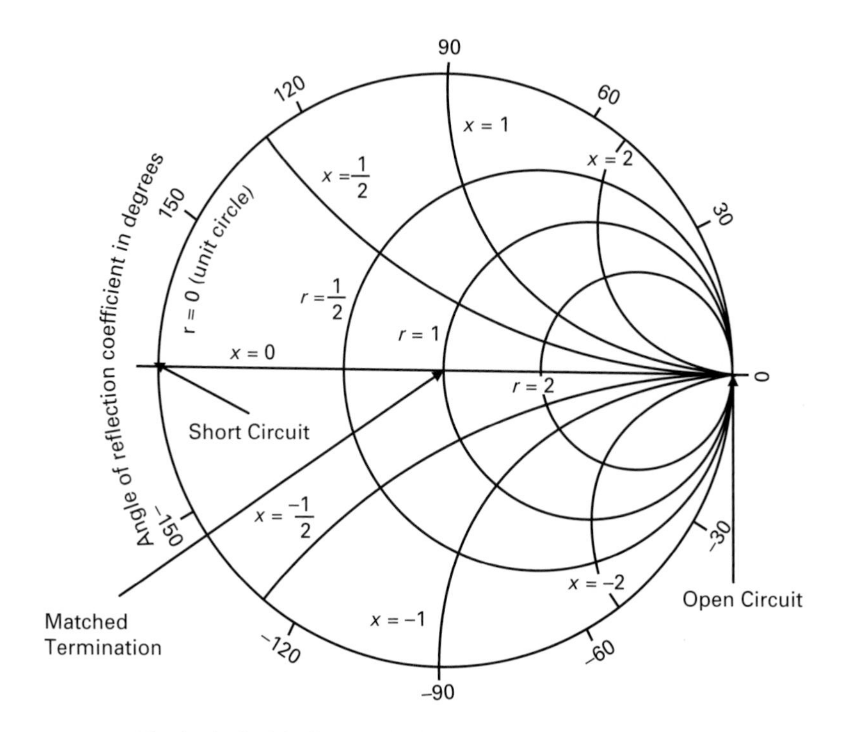

Figure 2.4 The basic Smith chart. This is the same Argand diagram as in Figure 2.3, except now just the real and imaginary parts of the impedance are shown. This figure is taken from S. Ramo, J. R. Whinnery and T. Van Duzer, *Fields and Waves in Communication Electronics*, Third edition, 1994, John Wiley and Sons. This material is reproduced with permission of John Wiley and Sons.

and this is described by Equation (2.24). Clearly an open circuit becomes a short circuit and vice versa over this distance.

Example 2.1 Single stub matching using the Smith chart: An impedance of 20 + j10 Ω is connected to a transmission line. The characteristic impedance of the line is 50 Ω and the velocity of propagation is $3.10^8\ \mathrm{ms}^{-1}$. The impedance needs to be matched to the line at 10 GHz. Design a single stub to achieve this.

Solution to Example 2.1

The technique involves adding a length of line, which is usually terminated in an open or short circuit, at the correct position between the impedance and the source or generator. This extra section of line is called a stub. The circuit is shown in Figure 2.6. The impedance chosen for this example has a normalised value of 0.4+j0.2, see Equation (2.28).

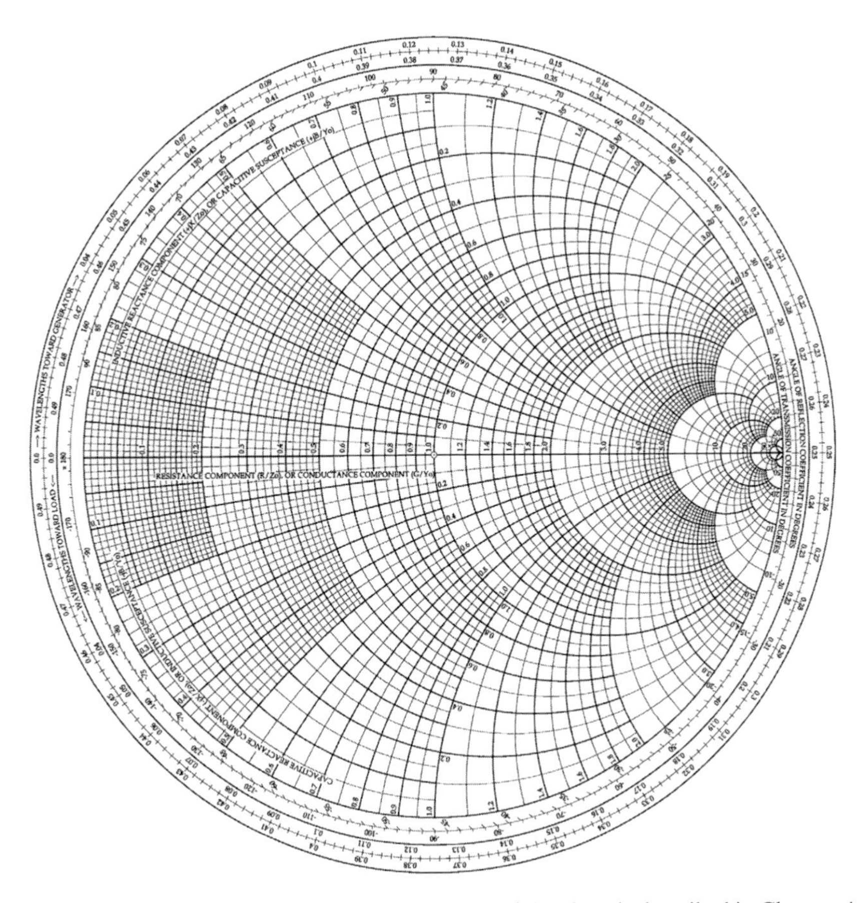

Figure 2.5 The complete Smith chart. The theory of the chart is described in Chapter 4 and the next example gives one of its common uses. Other uses are described in the references at the end of this chapter.

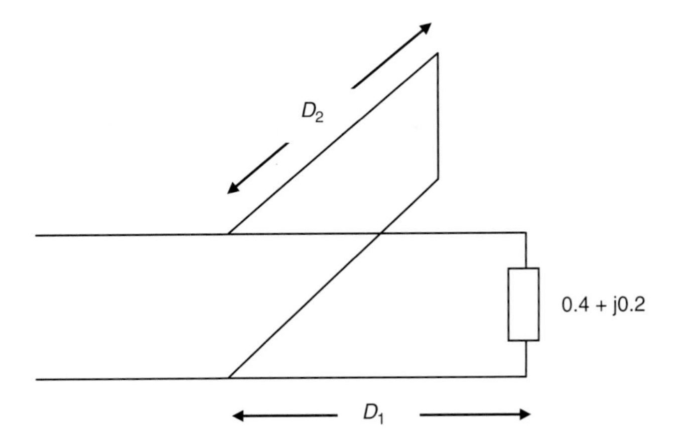

Figure 2.6 The circuit for single stub matching of a normalised impedance 0.4+j0.2 to the transmission line.

Now the stub and the impedance are combined in parallel and so admittances rather than impedances are more useful at this point. The solution will be done in two sections. The first will be using the equations already derived in this chapter and the second will be using the Smith chart. It is hoped that this will show how the chart relates to the theory. The numbers have been chosen to make the calculations simple.

The solution via equations

The solution via equations begins by finding the normalised admittance of this impedance:

$$\frac{1}{0.4+\mathrm{j}0.2}=\frac{0.4-\mathrm{j}0.2}{(0.4+\mathrm{j}0.2)(0.4-\mathrm{j}0.2)}=2-\mathrm{j}.$$

The next step is to move down the transmission line a distance D_1 so that this impedance has the form $1\,\pm\,\mathrm{j}x$. Using Equation (2.21) for the input impedance in a normalised form:

$$\frac{Z_{\mathrm{IN}}}{Z_0}=\left(\frac{\dfrac{Z_{\mathrm{L}}}{Z_0}+\mathrm{j}\,\tan(\beta D_1)}{1+\mathrm{j}\dfrac{Z_{\mathrm{L}}}{Z_0}\,\tan(\beta D_1)}\right),$$

the normalised admittance becomes

$$\frac{Z_0}{Z_{\mathrm{IN}}}=\left(\frac{\dfrac{Z_0}{Z_{\mathrm{L}}}+\mathrm{j}\,\tan(\beta D_1)}{1+\mathrm{j}\dfrac{Z_0}{Z_{\mathrm{L}}}\,\tan(\beta D_1)}\right).\tag{2.29}$$

Substituting in the values gives

$$1\pm\mathrm{j}x=\left(\frac{2-\mathrm{j}+\mathrm{j}T}{1+\mathrm{j}(2-\mathrm{j})T}\right),\quad\text{where } T=\ \tan(\beta D_1).$$

Rearranging the right-hand side gives

$$1\pm\mathrm{j}x=\frac{2+2T^2}{(1+T)^2+4T^2}+\mathrm{j}\frac{T^2-4T-1}{(1+T)^2+4T^2}.$$

Equating the real and imaginary parts gives

$$T=\frac{1}{3}\ \text{or}\ -1\ \text{and hence}\ x=-1\ \text{or}\ 1.$$

Choosing the positive value of T as that which represents the direction towards the source or generator:

$$T=\frac{1}{3}=\ \tan(\beta D_1)\ \text{so}\ \beta D_1=0.3218\ \text{or}\ 18.435^\circ.$$

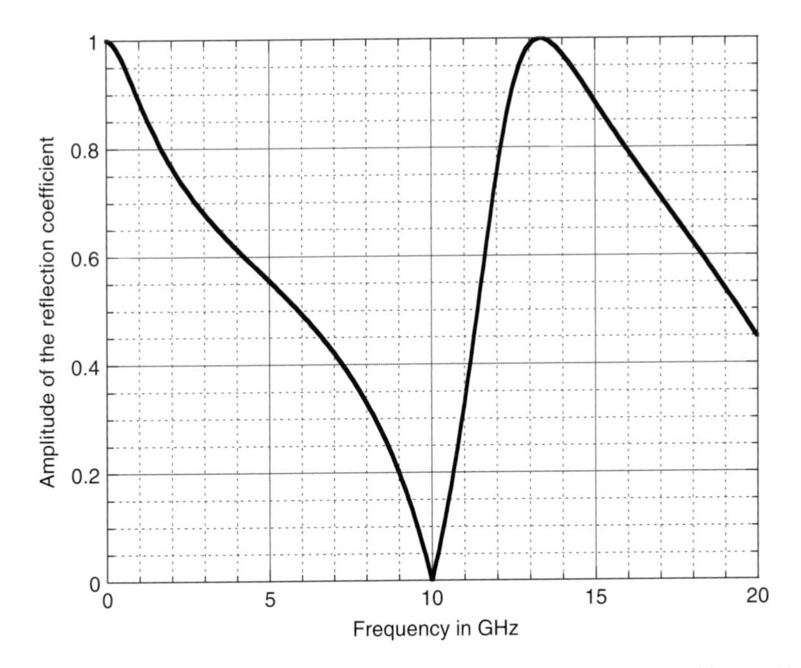

Figure 2.7 The frequency response of a single short-circuit stub matching at 10 GHz.

Now the frequency is 10 GHz and the velocity of propagation is $3.10^8\,\mathrm{ms}^{-1}$, so the wavelength is 30 cm. This gives

$$\beta = \frac{2\pi}{\lambda} = 0.2094 \text{ radians m}^{-1} \text{ and so } D_1 = 1.537 \text{ cm}.$$

Now Equation (2.22) gives the input impedance of a length of line, D_2, that is terminated in a short circuit as $Z_{IN} = \mathrm{j}Z_0 \tan(\beta D_2)$ so the normalised admittance will be

$$\frac{Z_0}{Z_{IN}} = \frac{-\mathrm{j}}{\tan(\beta D_2)}.$$

The only way this can give a value of j would be if $D_2 = 3\lambda/8$ or $\beta D_2 = 3\pi/4$. This would give a value of $D_2 = 11.25$ cm. In contrast to this, Equation (2.23) gives the input impedance of a length of line terminated in an open circuit as

$$\frac{Z_{IN}}{Z_0} = -\mathrm{j}\cot(\beta D)$$

and so the normalised admittance will be

$$\frac{Z_0}{Z_{IN}} = \mathrm{j}\tan(\beta D).$$

So for this stub to give a value of j the length need only be $\lambda/8$ or $D = 3.75$ cm.

By writing out the equations for the input impedance and then calculating the reflection coefficient, it is possible to plot the frequency response of these two solutions. In Figure 2.7 is shown the frequency response for the

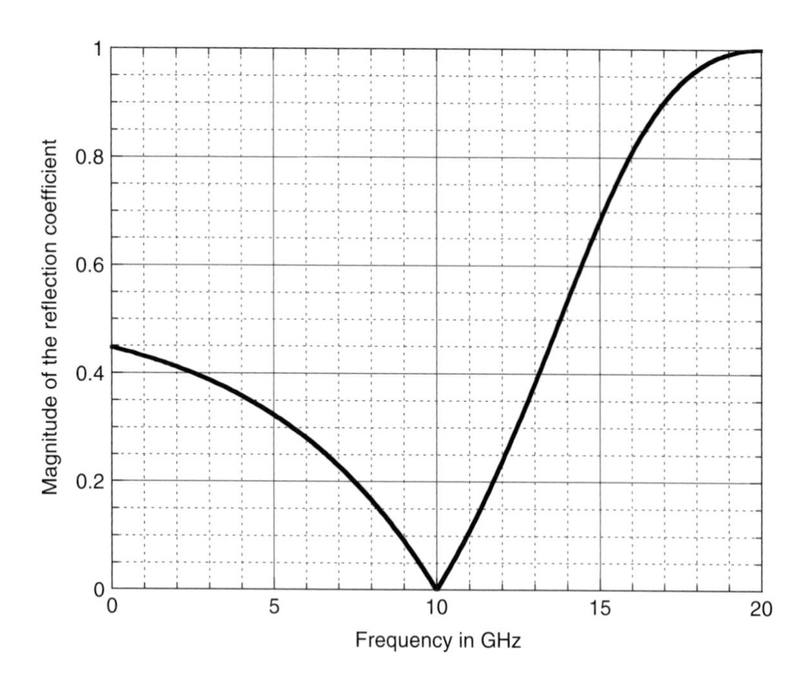

Figure 2.8 The frequency response of a single open-circuit stub matching at 10 GHz.

short-circuit stub matching at 10 GHz. It can be seen that the reflection coefficient amplitude is zero at 10 GHz, but away from this frequency it is far from matched. The reason for the points where the amplitude reaches unity, i.e. at 0 Hz and 13.333 GHz is that the short circuit is effectively zero or a half wavelength away from the main line at these frequencies, respectively. In other words it shorts the line at that point and hence the reflection coefficient has an amplitude of unity.

The frequency response of the open-circuit stub is plotted in Figure 2.8. Here the stub's frequency response is less invasive. At 20 GHz the open circuit is just a quarter of a wavelength away and so it appears as a short circuit to the main line. Hence the amplitude of the reflection coefficient is unity at that point. Clearly, the shorter the stubs, the broader the range of frequencies where the impedance is reasonably matched. For a match better than $|\rho| \leq 0.2$ the open-circuit stub in this case gives over twice the bandwidth. However, there may be applications where a narrow band is required and this topic is thoroughly discussed in the further reading on circuit design at the end of the chapter.

The solution via the Smith chart

In Figure 2.9 is a Smith chart without the full grid of the classic chart in Figure 2.5. The first step is to superimpose on the chart a circle representing the reflection coefficient. Now the reflection coefficient for the impedance will be given by

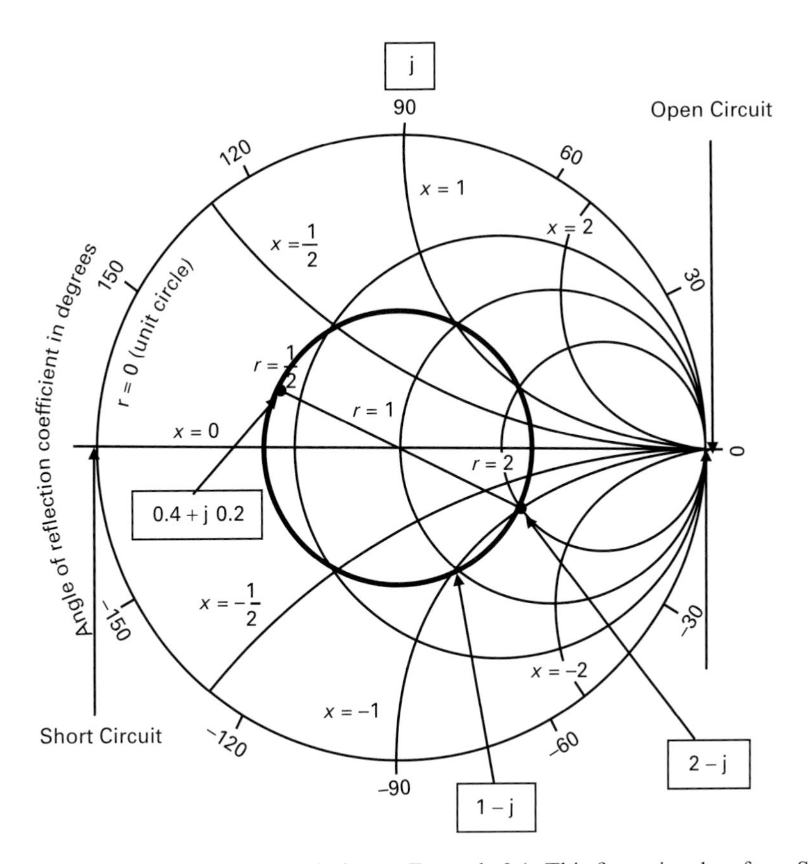

Figure 2.9 The Smith chart solution to Example 2.1. This figure is taken from S. Ramo, J. R. Whinnery and T. Van Duzer, *Fields and Waves in Communication Electronics*, Third edition, 1994, John Wiley and Sons. This material is reproduced with permission of John Wiley and Sons.

$$\rho = \frac{0.4 + \mathrm{j}0.2 - 1}{0.4 + \mathrm{j}0.2 + 1} = \frac{-6 + \mathrm{j}2}{14 + \mathrm{j}2} \text{ and } |\rho|^2 = \frac{40}{200} = \frac{1}{5} \text{ and so } |\rho| = \frac{1}{\sqrt{5}} = 0.447.$$

So the radius of the circle is 0.447, assuming the edge of the chart has a radius of unity. It can be seen that on the chart this circle passes through the point 0.4 + j0.2. The next step is to find the normalised impedance. This is the point 180° round this circle, since

$$\rho = \frac{z-1}{z+1} = \frac{1 - \frac{1}{z}}{1 + \frac{1}{z}} \text{ and } \frac{\frac{1}{z} - 1}{\frac{1}{z} + 1} = -\rho = \rho \exp(-\mathrm{j}\pi).$$

The normalised impedance from the chart is 2 − j. The distance down the line, D_1, is found by moving clockwise along the circle until the normalised admittance is 1 − j. Using a full sized chart this is about 37°. Now rotation on the chart is equivalent to a phase angle of $2\beta D_1$, so this gives $\beta D_1 = 18.5°$, and using $\beta = 0.20943$ radians m^{-1}, this gives a value of $D_1 = 1.54$ cm. Finally, the length of the stub is found in a similar way. For the short-circuit stub, starting at the short-circuit point, its normalised admittance is 180° round the chart at the open-circuit point. To obtain the value of j the length of the stub, D_2, is equivalent to a rotation clockwise round the chart of 270°, or in wavelengths $3\lambda/8$, which gives $D_2 = 11.25$ cm. Similarly, for the open circuit, the admittance is at the short-circuit point and so only 90° is required to reach j in this case and $D_2 = 3.75$ cm.

It will be observed that the second solution using the Smith chart is not quite so accurate but does have the merit of speed and a visual check on the method. However, it is not easy to use for anything other than a few frequencies. The frequency responses in Figures 2.7 and 2.8 are much simpler to compute from the appropriate equations. It should also be noted that the Smith chart was available well before the widespread use of computers and is still an excellent way of checking complex computations.

Example 2.2 A transmission line is terminated in an impedance equivalent to $50 + \mathrm{j}100\ \Omega$. A 2 GHz sine wave, with an amplitude of 10 V, is incident on this impedance. Assume the source is matched to the line. If the transmission line has characteristic impedance of 50 Ω and a velocity of propagation equal to 2.10^8 ms^{-1}, what is the power dissipated in the termination? Also find the distance from the termination to the first voltage minimum on the line. What is the value of this minimum? The circuit is shown in Figure 2.10.

Solution to Example 2.2

First of all, the reflection coefficient is

$$\rho = \frac{50 + \mathrm{j}100 - 50}{50 + \mathrm{j}100 + 50} = \frac{\mathrm{j}100}{100 + \mathrm{j}100} \text{ and } |\rho|^2 = \frac{1}{2},$$

so half of the incident power is reflected and the other half is dissipated in the termination. The incident power is given by Equation (2.9) as

$$\frac{10^2}{2 \times 50} = 1\ \mathrm{W},$$

so the power dissipated in the termination is 0.5 W.

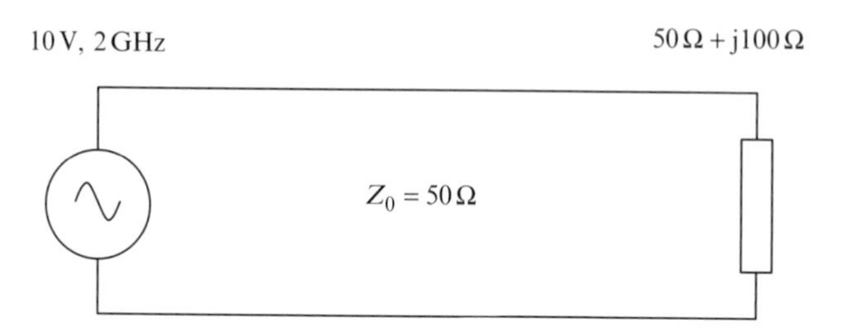

Figure 2.10 The circuit diagram for Example 2.2.

To check this answer, find the voltage across the termination:

$$V_{\mathrm{L}} = V_+ + \rho V_+ = \frac{10(1+2\mathrm{j})}{1+\mathrm{j}}.$$

Now the fraction of this voltage across the 50 Ω resistor in the termination is given by

$$\frac{50}{50+\mathrm{j}100},$$

so the voltage across the resistor is:

$$\frac{10V(1+2\mathrm{j})}{1+\mathrm{j}} \times \frac{1}{1+2\mathrm{j}} = \frac{10}{1+\mathrm{j}}.$$

Now the amplitude of this voltage is $5\sqrt{2}$V so the power in the resistor is

$$\frac{V_{\mathrm{R}}^2}{2R} = \frac{50}{2\times 50} = \frac{1}{2}\mathrm{W}.$$

This second method is longer, but it serves to check the first method.

Now at 2 GHz and a velocity of propagation of $2.10^8\,\mathrm{ms}^{-1}$, the wavelength is 10 cm and $\beta = 2\pi/10$ radians per centimetre. The phase or argument of ρ is $\pi/4$ so the position of the first minimum is given by Equation (2.16) as

$$2\beta D - \angle\rho = \pi, \text{ that is } \frac{4\pi D}{10} - \frac{\pi}{4} = \pi \text{ and } D = \frac{50}{16} = 3.125 \text{ cm}.$$

Finally, from Equation (2.17):

$$V_{\mathrm{MIN}} = V_+(1-|\rho|) = 10\left(1-\frac{1}{\sqrt{2}}\right) = 2.928 \text{ V}.$$

These results are shown in Figure 2.11.

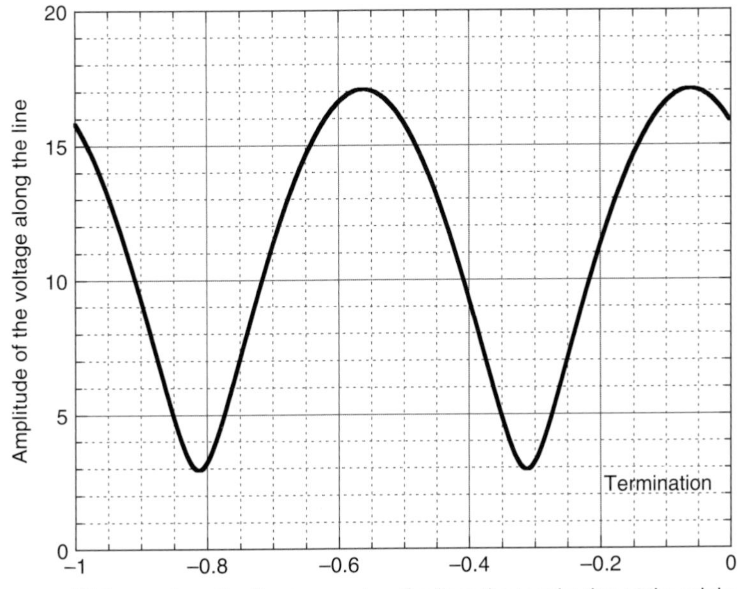

Figure 2.11 Voltage standing waves in Example 2.2.

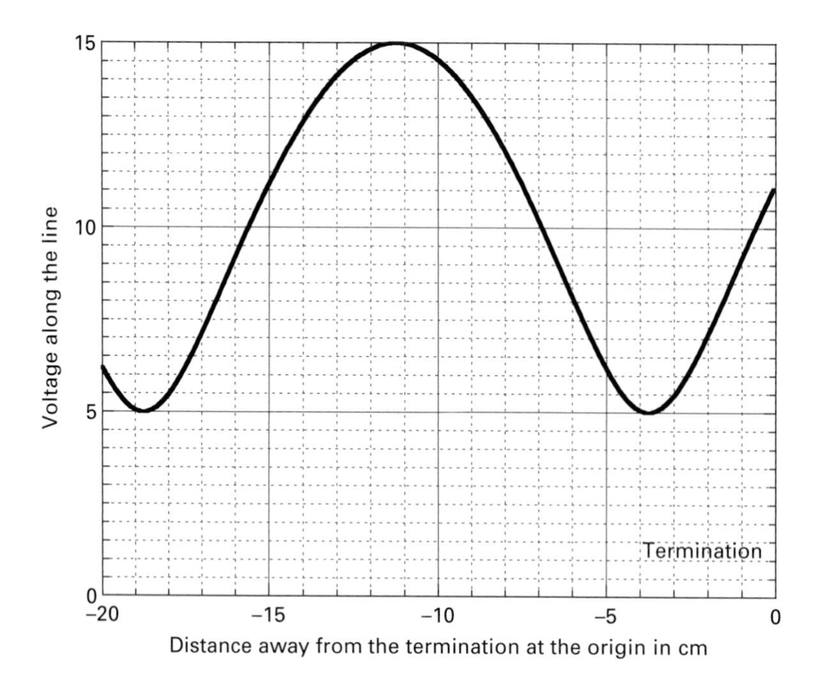

Figure 2.12 The standing waves in Example 2.3.

Example 2.3 An unknown impedance is connected to a transmission line. Using a 1 GHz oscillator, the *VSWR* was found to be 3 and the position of the first minimum was 3.75 cm away from the impedance. If the characteristic impedance of the transmission line is 50 Ω and the velocity of propagation is $3.10^8\,\mathrm{ms}^{-1}$, find this unknown impedance.

Solution to Example 2.3

First use Equation (2.19) to find the value of $|\rho|$:

$$|\rho| = \frac{3-1}{3+1} = 0.5.$$

Next the value of the wavelength and β:

$$\lambda = \frac{3.10^8}{10^9} = 0.3 \text{ m}, \quad \beta = \frac{2\pi}{0.3} \text{ radians m}^{-1}.$$

Then using Equation (2.16) to find the phase of ρ:

$$2\beta D - \angle\rho = \pi,$$

$$\angle\rho = 2 \times \frac{2\pi}{0.3} \times \frac{15}{400} - \pi = -\frac{\pi}{2},$$

so $\rho = -j/2$. Now by rewriting Equation (1.13):

$$\frac{Z_L}{Z_0} = \frac{1-\rho}{1+\rho} = \frac{2-j}{2+j} = \frac{3}{5} - j\frac{4}{5}.$$

Since $Z_0 = 50\,\Omega$, $Z_L = 30\,\Omega - j40\,\Omega$.

Figure 2.12 shows the result of a 10 V sine wave incident on this impedance. The $VSWR = 3$ can be seen as $V_{MAX} = 15\,V$ and $V_{MIN} = 5\,V$. The position of the first minimum is also at 3.75 cm from the termination.

Example 2.4 A quarter wavelength of transmission line is used to match an impedance of 200 Ω to a transmission line with a characteristic impedance of 50 Ω at a frequency of 1 GHz. If the velocity of propagation is $10^8\,\text{ms}^{-1}$ in both lines, find the length and characteristic impedance of the quarter wavelength of line. Plot the fraction of the incident power in the 200 Ω resistor from 0 to 4 GHz.

Solution to Example 2.4

The wavelength is 10 cm, so a quarter wavelength is 2.5 cm. Using Equation (2.24), the input impedance must be 50 Ω so

$$50 = \frac{Z_0^2}{200} \text{ so } Z_0 = 100\ \Omega.$$

Now for the harder part! Using Equation (2.21) and putting in the values:

$$Z_{IN} = 100\left(\frac{200 + j100\tan(\beta D)}{100 + j200\tan(\beta D)}\right) = 100\left(\frac{2 + j\tan(\beta D)}{1 + 2j\tan(\beta D)}\right).$$

Now the reflection coefficient will be

$$\rho = \frac{Z_{IN} - 50}{Z_{IN} + 50} = \frac{3}{5 + 4j\tan(\beta D)}.$$

Hence the fraction of the incident power in the 200 Ω impedance is given by Equation (2.12):

$$1 - |\rho|^2 = \frac{1}{1 + \frac{9}{16}\cos^2(\beta D)}.$$

This is plotted in Figure 2.13, where it can be seen that when the length of the 100 Ω line is $\lambda/4 + n\lambda/2$ for $n = 0, 1, 2, \ldots$ all the incident power is absorbed in the 200 Ω termination. However, when the length is $n\lambda/2$ for $n = 0, 1, 2, \ldots$ the

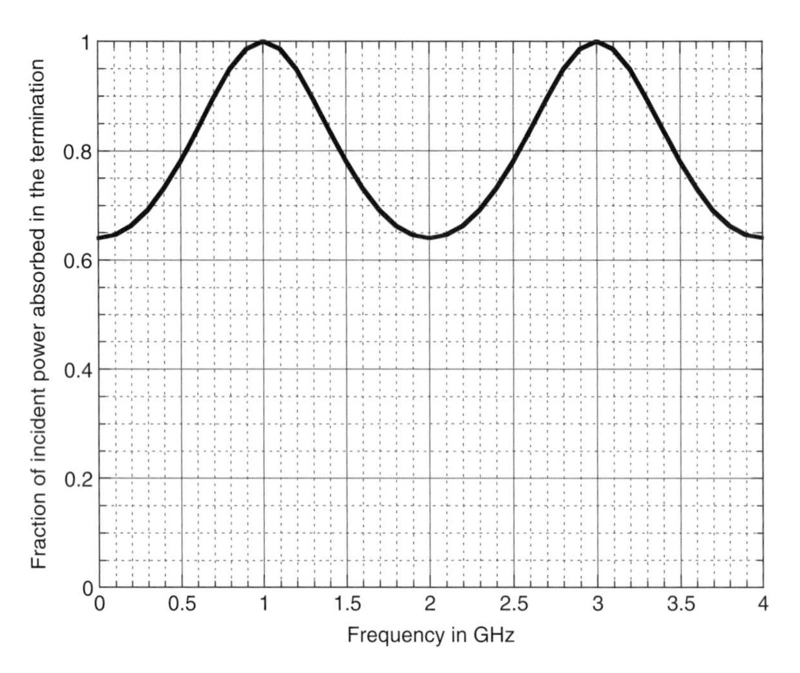

Figure 2.13 This shows the fraction of the incident power absorbed in the 200 Ω termination against frequency.

mismatch is worse and only 64 % of the incident power is absorbed. In practice this means that all odd harmonics of the main frequency, i.e. 1 GHz, are also matched to the termination, whereas even harmonics are partially rejected. More sophisticated techniques are required for either broad-band or selective matching.

2.7 The transmission coefficient

The next examples concern the transmission of waves and the transmission coefficient, τ. The definition of τ is the ratio of the transmitted voltage to the incident voltage:

$$\tau = \frac{V_t}{V_+} \text{ and } |\tau|^2 = \left|\frac{V_t}{V_+}\right|^2 = \frac{\text{transmitted power}}{\text{incident power}}.$$

If there are no resistors involved, then

$$|\rho|^2 + |\tau|^2 = 1.$$

However, in general, where there are losses,

$$|\rho|^2 + |\tau|^2 \leq 1. \tag{2.30}$$

In simple circuits, there is a relationship between ρ and τ, but this is not generally the case. This is illustrated in the following example.

Example 2.5 Find the reflection coefficient and the transmission coefficient for each of the following circuits, a, b, c and d, shown in Figures 2.14, 2.15, 2.16 and 2.17.

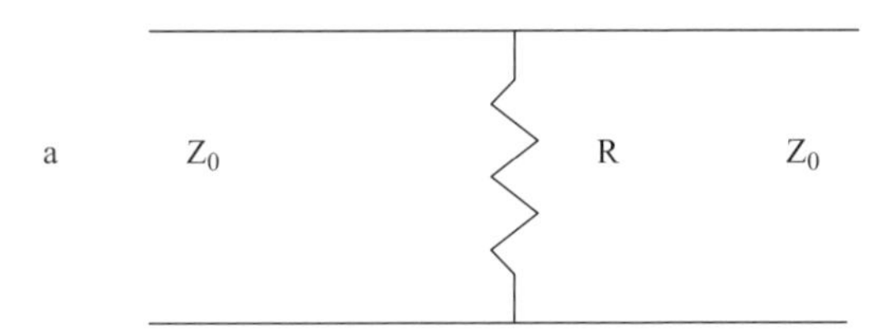

Figure 2.14 Circuit a for Example 2.5 – a parallel resistor.

Solution to Example 2.5 circuit a

First of all consider a wave approaching from the left and assume that to the right the line goes on to a matched load. In other words no signal comes in from the right. When the wave arrives at the resistor, which is assumed to be much smaller than the wavelength, it 'sees' not only the resistor, but the line in parallel beyond. So the impedance it 'sees' is given by

$$\frac{RZ_0}{R+Z_0}$$

and so the reflection coefficient, ρ_a, is

$$\rho_a = \frac{-Z_0}{2R+Z_0}. \tag{2.31}$$

It is negative because this impedance is less than Z_0. The transmitted wave in this case will be the same as the voltage across the resistor. So τ_a is given by

$$\tau_a = 1 + \rho_a = \frac{2R}{2R+Z_0}. \tag{2.32}$$

It is worth noting that the fraction of the incident power reflected is

$$\frac{Z_0^2}{(2R+Z_0)^2}.$$

The fraction of power transmitted is

$$\frac{4R^2}{(2R+Z_0)^2}.$$

Since the voltage is common across the resistor and the transmitted wave, the power in the resistor is

$$\frac{4RZ_0}{(2R+Z_0)^2}.$$

The three fractions add up to give unity, which verifies these results.

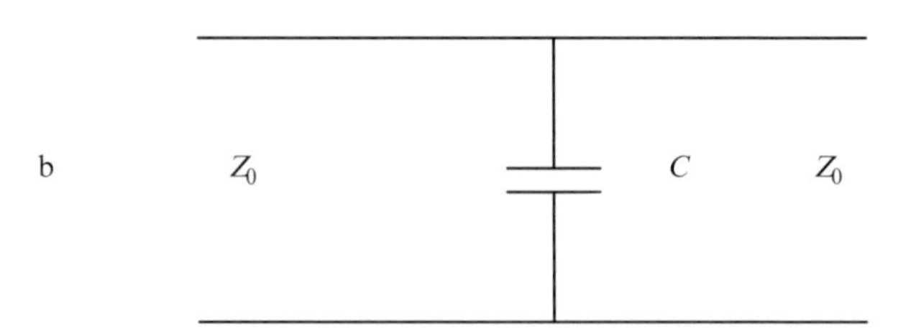

Figure 2.15 Circuit b for Example 2.5 – a parallel capacitor.

Solution to Example 2.5 circuit b

Using similar assumptions to circuit a, the capacitor 'appears in parallel' with the 'Z_0' of the line on the right-hand side. So the impedance 'seen' by a wave approaching from the left-hand side is

$$\frac{Z_0}{1 + j\omega C Z_0}$$

and the reflection coefficient, ρ_b, is

$$\rho_b = \frac{-j\omega C Z_0}{2 + j\omega C Z_0} \tag{2.33}$$

and, by a similar argument,

$$\tau_b = 1 + \rho_b = \frac{2}{2 + j\omega C Z_0}. \tag{2.34}$$

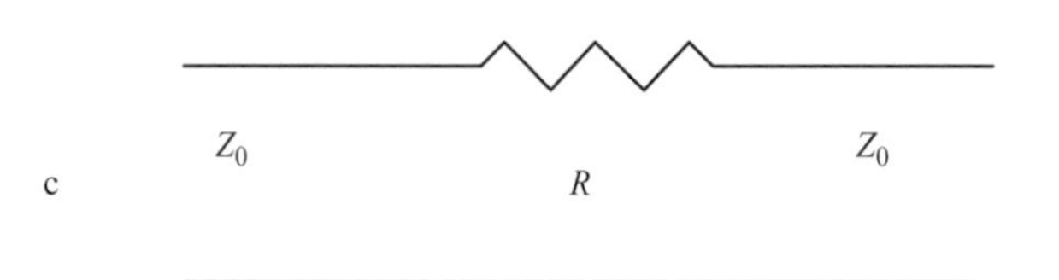

Figure 2.16 Circuit c for Example 2.5 – a series resistor.

Solution to Example 2.5 circuit c

In this case the impedance 'seen' by a wave approaching from the left-hand side is the sum of R and Z_0. So the reflection coefficient is now

$$\rho_c = \frac{R}{2Z_0 + R}. \tag{2.35}$$

The transmission coefficient is not the same as in the above examples. The fraction of the incident voltage on the left-hand side of the resistor is $1 + \rho$ but only a fraction of this goes on to the line on the right-hand side, given by the potentiometer effect:

$$\tau_c = (1 + \rho_c)\frac{Z_0}{R + Z_0} = \frac{2Z_0}{2Z_0 + R}. \tag{2.36}$$

d Z_0 R_1 R_2 Z_0

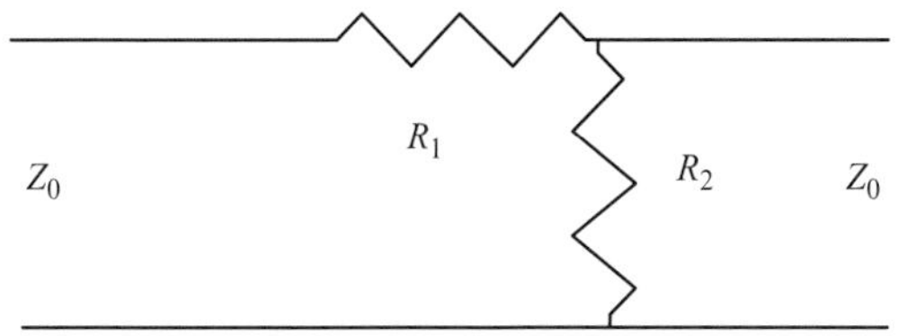

Figure 2.17 Circuit d for Example 2.5 – a potentiometer.

Solution to Example 2.5 circuit d

This is a combination of circuits a and c. Consider waves approaching from the left-hand side, the impedance 'seen' is

$$R_1 + \frac{R_2 Z_0}{R_2 + Z_0}.$$

So the reflection coefficient is

$$\rho_{dl} = \frac{-Z_0^2 + R_1(R_2 + Z_0)}{2R_2 Z_0 + Z_0^2 + R_1(R_2 + Z_0)} \tag{2.37}$$

and the transmission coefficient is

$$\tau_{dl} = \frac{2R_2 Z_0}{2R_2 Z_0 + Z_0^2 + R_1(R_2 + Z_0)}. \tag{2.38}$$

However, if a wave approaches from the right-hand side, the impedance 'seen' is now

$$\frac{R_2(R_1 + Z_0)}{R_1 + R_2 + Z_0},$$

which gives a reflection coefficient of

$$\rho_{dr} = \frac{-Z_0^2 + R_1(R_2 - Z_0)}{2R_2 Z_0 + Z_0^2 + R_1(R_2 + Z_0)} \tag{2.39}$$

and a transmission coefficient of

$$\tau_{dr} = \frac{2R_2 Z_0}{2R_2 Z_0 + Z_0^2 + R_1(R_2 + Z_0)}. \tag{2.40}$$

The first thing to notice is the increase in complexity of these equations. The second thing to notice is that the transmission one way is the same as the other direction, i.e. Equations (2.38) and (2.40) are identical. Finally, there is just a single change of sign distinguishing Equations (2.37) and (2.39).

2.8 Scattering parameters

This is the moment to introduce scattering parameters, which have become well established in both microwave and millimetre wave measurements. The wave voltages are normalised in a rather arbitrary way:

$$a_n = \frac{V_{n+}}{\sqrt{Z_{0n}}} \text{ and } b_n = \frac{V_{n-}}{\sqrt{Z_{0n}}},$$

where n represents the nth port of a circuit. So far in this chapter, the circuits have only had one or two ports. The characteristic impedance is now related to the line connecting to the particular port. Now if all the characteristic impedances are the same and equal to 50 Ω, then this reduces the voltages by a common factor of $10/\sqrt{2}$ or 7.07107. The divergence between real voltages and the new wave parameters only occurs when there is a change of characteristic impedance. So the various quantities relate as follows:

$$V_{1L} = V_{1+} + V_{1-} = \sqrt{Z_{01}}(a_1 + b_1), \tag{2.41}$$

$$I_{1L} = I_{1+} + I_{1-} = \frac{1}{\sqrt{Z_{01}}}(a_1 - b_1), \tag{2.42}$$

where the suffix 1 refers to the port 1. Clearly the 'a' waves are going towards the port, and the 'b' waves are reflected – this replaces the + and – signs used so far. The power in a wave is also related:

$$\frac{|V_+^2|}{2Z_0} = \frac{1}{2}\left|a_1^2\right|, \quad \frac{|V_-^2|}{2Z_0} = \frac{1}{2}\left|b_1^2\right|. \tag{2.43}$$

It might have been more useful if the definition had included a factor of two to make the modulus squared equal to the power, but it is too late to change it now. For two ports these new waves form a scattering matrix, which gives the scattered waves, i.e. those leaving the ports, in terms of the waves arriving at the ports as

$$\begin{bmatrix} b_1 \\ b_2 \end{bmatrix} = \begin{bmatrix} S_{11} & S_{12} \\ S_{21} & S_{22} \end{bmatrix} \begin{bmatrix} a_1 \\ a_2 \end{bmatrix}. \tag{2.44}$$

Now if $a_2 = 0$, then $b_1 = S_{11}a_1$ and by analogy $S_{11} = \rho_1$ and similarly $S_{22} = \rho_r$. As far as transmission goes, $b_2 = S_{21}a_1$ and again by analogy $\tau_{l \to r} = S_{21}$, $\tau_{r \to l} = S_{12}$. The suffixes 'l' and 'r' are from circuit d above and refer to the left or port 1 and the right or port 2. So to finish this transition to scattering matrices, the four circuits a to d in Example 2.5 have the following scattering matrices:

$$\text{(a)}\ \frac{1}{2R+Z_0}\begin{bmatrix} -Z_0 & 2R \\ 2R & -Z_0 \end{bmatrix}, \tag{2.45}$$

$$\text{(b)}\ \frac{1}{2+\mathrm{j}\omega CZ_0}\begin{bmatrix} -\mathrm{j}\omega CZ_0 & 2 \\ 2 & -\mathrm{j}\omega CZ_0 \end{bmatrix}, \tag{2.46}$$

$$\text{(c)}\ \frac{1}{2Z_0+R}\begin{bmatrix} R & 2Z_0 \\ 2Z_0 & R \end{bmatrix}, \tag{2.47}$$

$$\text{(d)}\ \frac{1}{2R_2Z_0+Z_0^2+R_1(R_2+Z_0)}\begin{bmatrix} -Z_0^2+R_1(R_2+Z_0) & 2R_2Z_0 \\ 2R_2Z_0 & -Z_0^2+R_1(R_2-Z_0) \end{bmatrix}. \tag{2.48}$$

2.9 Transmission parameters

Since these scattering matrices relate waves on both sides of a network, it is often useful to use a transmission matrix, which relates waves on one side to waves on the other side. In order to use these matrices in a chain for solving networks, it is also useful to alternate the variables as follows:

$$\begin{bmatrix} a_1 \\ b_1 \end{bmatrix} = \begin{bmatrix} T_{11} & T_{12} \\ T_{21} & T_{22} \end{bmatrix}\begin{bmatrix} b_2 \\ a_2 \end{bmatrix}. \tag{2.49}$$

These matrices are limited to those with four elements, i.e. just networks with two ports. Using Equations (2.44) and (2.49), the following relationships can be found:

$$T_{11} = \frac{1}{S_{21}}, \quad T_{12} = -\frac{S_{22}}{S_{21}}, \quad T_{21} = \frac{S_{11}}{S_{21}}, \quad T_{22} = \frac{S_{12}S_{21}-S_{11}S_{22}}{S_{21}} = \frac{-\Delta S}{S_{21}},$$

$$[T] = \frac{1}{S_{21}}\begin{bmatrix} 1 & -S_{22} \\ S_{11} & -\Delta S \end{bmatrix}. \tag{2.50}$$

In reverse:

$$S_{11} = \frac{T_{21}}{T_{11}}, \quad S_{12} = \frac{T_{11}T_{22}-T_{12}T_{21}}{T_{11}} = \frac{\Delta T}{T_{11}}, \quad S_{21} = \frac{1}{T_{11}}, \quad S_{22} = -\frac{T_{12}}{T_{11}},$$

$$[S] = \frac{1}{T_{11}}\begin{bmatrix} T_{21} & \Delta T \\ 1 & -T_{12} \end{bmatrix}. \tag{2.51}$$

Example 2.6 Transmission matrices: Find the transmission matrices for circuits a, c and d and show that the one from d can be derived from a and c.

Solution to Example 2.6

Using Equations (2.45), (2.47) and (2.50) find the transmission matrices for circuits a and c in Figures 2.14 and 2.16. For circuit a the transmission matrix is

$$\text{(a)} \begin{bmatrix} 1+\dfrac{Z_0}{2R_2} & \dfrac{Z_0}{2R_2} \\ -\dfrac{Z_0}{2R_2} & 1-\dfrac{Z_0}{2R_2} \end{bmatrix}, \tag{2.52}$$

where the suffix 2 has been added ready for the final part of the example. For circuit c the transmission matrix is

$$\text{(c)} \begin{bmatrix} 1+\dfrac{R_1}{2Z_0} & -\dfrac{R_1}{2Z_0} \\ \dfrac{R_1}{2Z_0} & 1-\dfrac{R_1}{2Z_0} \end{bmatrix}, \tag{2.53}$$

where again the suffix 1 has been added in readiness for the last section. For the circuit d in Figure 2.17 and Equation (2.48), the transmission matrix is

$$\text{(d)} \begin{bmatrix} 1+\dfrac{Z_0}{2R_2}+\dfrac{R_1(R_2+Z_0)}{2R_2Z_0} & \dfrac{Z_0}{2R_2}-\dfrac{R_1(R_2-Z_0)}{2R_2Z_0} \\ -\dfrac{Z_0}{2R_2}+\dfrac{R_1(R_2+Z_0)}{2R_2Z_0} & 1-\dfrac{Z_0}{2R_2}-\dfrac{R_1(R_2-Z_0)}{2R_2Z_0} \end{bmatrix}. \tag{2.54}$$

To show that the matrix given in (2.54) is the product of the matrices given in (2.52) and (2.53) requires some manipulation and this seems a lot of hard work for just two resistors. For many networks, the impedance method, used initially to find the scattering matrix, is more efficient than the transmission matrix method. For systems where there are more than two ports, using a scattering matrix is the only method. The transmission matrices can be useful when there is a long chain of similar networks. The Cayley Hamilton theorem can greatly simplify manipulations in these circumstances.

Example 2.7 Find the scattering matrices for the following three tee junctions shown in Figures 2.18, 2.19 and 2.20.

Solution to Example 2.7, tee number 1

For this tee junction, which is a parallel tee junction, assume ports 2 and 3 are terminated in matched loads. The input impedance at port 1 is then the parallel combination of two Z_0 impedances in parallel. So the reflection coefficient at port 1 is given by

$$\rho_1 = \frac{\dfrac{Z_0}{2}-Z_0}{\dfrac{Z_0}{2}+Z_0} = -\frac{1}{3}.$$

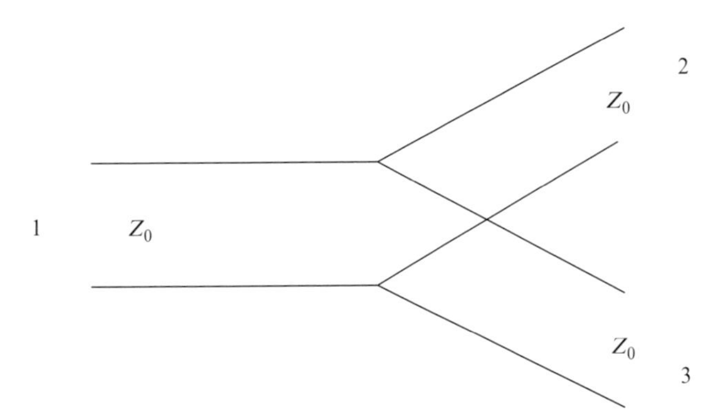

Figure 2.18 Example 2.7, tee number 1: a parallel tee junction.

Now the fraction of the incident voltage at the junction will be

$$1 + \rho_1 = \frac{2}{3} = \tau.$$

So two thirds of the incident voltage is transmitted to both ports 2 and 3. Checking the power, by squaring the values,

fraction of incident power reflected at port 1 is $\frac{1}{9}$,
fraction of power transmitted to both ports 2 and 3 is $\frac{4}{9}$.

So the tee junction does divide the signal but with a loss which is reflected back to the input. Since the circuit is symmetrical the scattering matrix is given by

$$\text{(a)} \begin{bmatrix} \rho & \tau & \tau \\ \tau & \rho & \tau \\ \tau & \tau & \rho \end{bmatrix} = \frac{1}{3} \begin{bmatrix} -1 & 2 & 2 \\ 2 & -1 & 2 \\ 2 & 2 & -1 \end{bmatrix}. \tag{2.55}$$

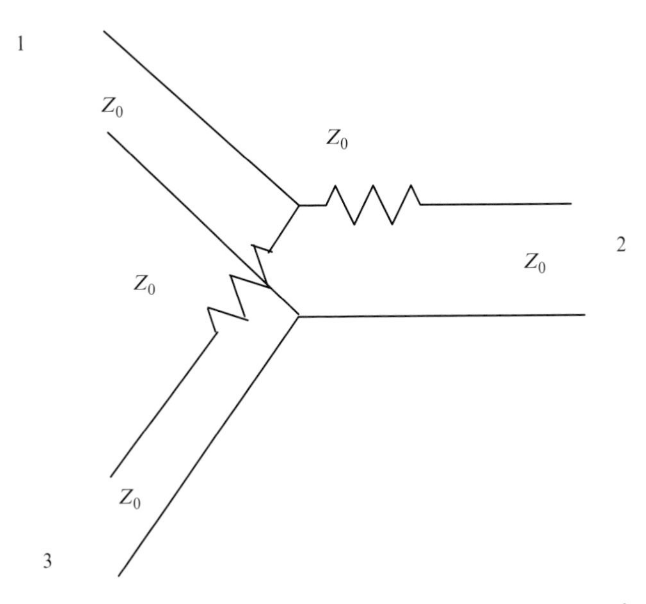

Figure 2.19 Example 2.7, tee number 2: a partially compensated parallel tee.

Solution to Example 2.7, tee number 2

For tee number 2, some resistors have been added to prevent the reflection of power and the possibility of standing waves in the line leading to port 1. This circuit is often used in the comparison of power detectors. Clearly, as far as port 1 goes, the input impedance is Z_0 and it is therefore matched and all of the input voltage appears at the junction. However, only half of this voltage is transmitted on to ports 2 and 3. Checking the power reveals that a quarter of the available power is lost in each of the two resistors. In this case there is not the symmetry of circuit (a) and, by matching ports 1 and 3, the input can now be made at port 2. The input impedance is

$$50 + \frac{100 \times 50}{100 + 50} = \frac{300}{6} + \frac{200}{6} = \frac{500}{6}\,\Omega,$$

which gives a reflection coefficient of

$$\rho_2 = \frac{\frac{500}{6} - \frac{300}{6}}{\frac{500}{6} + \frac{300}{6}} = \frac{200}{800} = \frac{1}{4},$$

so the fraction of the input voltage at the junction is given by

$$(1 + \rho_2) \times \frac{\frac{200}{6}}{\frac{500}{6}} = \frac{5}{4} \times \frac{2}{5} = \frac{1}{2}.$$

This is the fraction of the incident voltage that is transmitted to port 1. However, because of the resistor leading to port 3, only half of this goes on to port 3. So the scattering matrix is given by

$$\frac{1}{4}\begin{bmatrix} 0 & 2 & 2 \\ 2 & 1 & 1 \\ 2 & 1 & 1 \end{bmatrix}. \tag{2.56}$$

Checking the powers for an input at port 2:

the fraction of incident power reflected at port 2 $= \frac{1}{16}$,
the fraction of incident power transmitted to port 1 $= \frac{1}{4}$,
the fraction of incident power transmitted to port 3 $= \frac{1}{16}$,
the fraction of the incident power dissipated in the resistor leading to port 2 $= \frac{9}{16}$,
finally, the fraction in the resistor leading to port 3 $= \frac{1}{16}$.

The powers in the resistors were calculated from the voltages across them and as the sum of all these fractions is 1, this accounts for all of the power.

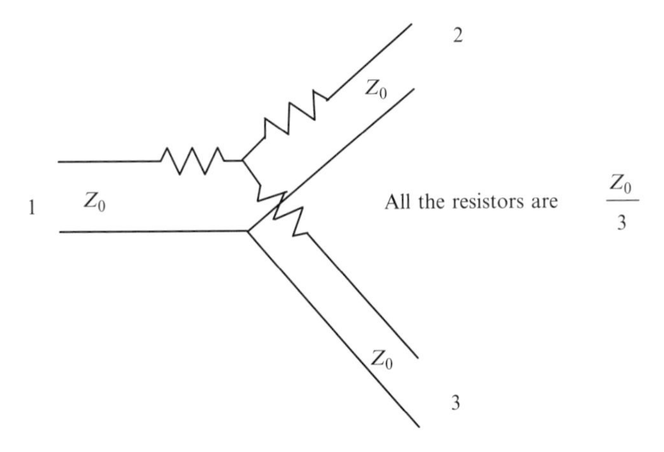

Figure 2.20 Example 2.7, tee number 3: a fully compensated parallel tee.

Solution to Example 2.7, tee number 3

Finally, for tee number 3, the input resistance at any of the ports, assuming the other two are terminated in matched impedances, is

$$\frac{Z_0}{3} + \frac{\frac{4Z_0}{3} \times \frac{4Z_0}{3}}{\frac{4Z_0}{3} + \frac{4Z_0}{3}} = Z_0,$$

so the reflection coefficient at all three ports is zero. The transmission coefficient is thus

$$\frac{\frac{2Z_0}{3}}{Z_0} \times \frac{Z_0}{\frac{4Z_0}{3}} = \frac{1}{2}.$$

So the scattering matrix is given by

$$\frac{1}{2}\begin{bmatrix} 0 & 1 & 1 \\ 1 & 0 & 1 \\ 1 & 1 & 0 \end{bmatrix}. \tag{2.57}$$

This is called the compensated tee because it is matched at every port, i.e. the diagonal elements of the scattering matrix are zero. Checking the power again shows for an input at port 1:

the fraction of incident power transmitted to each of port 2 and port 3 $= \frac{1}{4}$,
the fraction of incident power dissipated in the resistor leading to port 1 $= \frac{1}{3}$,
the fraction of incident power in each of the resistors leading to ports 2 and 3 $= \frac{1}{12}$.

Again the total is unity, and half the available power is dissipated in the resistors.

Example 2.8 Calibration of power meters: The three tee junctions shown in Example 2.7 can be used to calibrate power detectors or meters. As an example of the use of scattering matrices, the error in the measurement can be found for each circuit. In order to make the analysis simple, it is assumed that a matched source of power is connected to port 1 in each case and that the elements of the scattering matrix are correct. At port 2 a reference power detector with a reflection coefficient, ρ_2, is connected, and finally at port 3 is connected an uncalibrated power detector with a reflection coefficient, ρ_3. These examples can easily be extended, if the source is not matched and the elements of the scattering matrix are different from those described above.

Solution to Example 2.8, tee number 1

The wave reflected from the reference power meter is $a_2 = \rho_2 b_2$, and similarly for the uncalibrated power meter $a_3 = \rho_3 b_3$, so the scattering matrix for tee number 1 using Equation (2.55) becomes

$$\begin{bmatrix} b_1 \\ b_2 \\ b_3 \end{bmatrix} = \frac{1}{3}\begin{bmatrix} -1 & 2 & 2 \\ 2 & -1 & 2 \\ 2 & 2 & -1 \end{bmatrix}\begin{bmatrix} 1 \\ \rho_2 b_2 \\ \rho_3 b_3 \end{bmatrix},$$

where $a_1 = 1$, that is a unit input from the oscillator. Now if the oscillator is matched, any reflection back to the oscillator will not disturb the main power flow into the circuit. So by solving the second two equations, it is possible to obtain

$$b_2 = \frac{2(1+\rho_3)}{(3+\rho_2+\rho_3-\rho_2\rho_3)}, \quad b_3 = \frac{2(1+\rho_2)}{(3+\rho_2+\rho_3-\rho_2\rho_3)}.$$

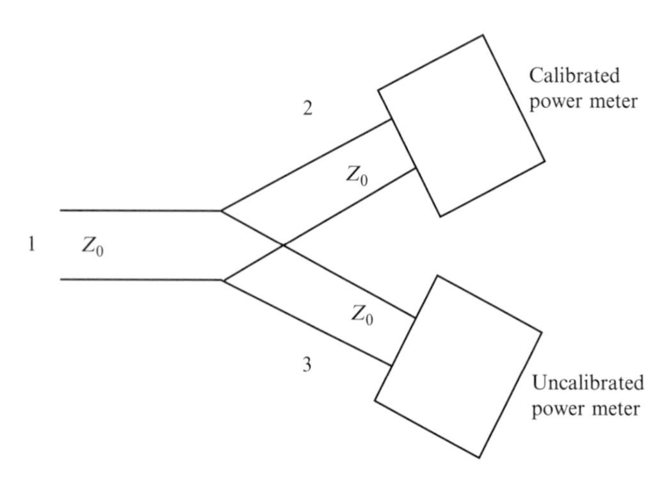

Figure 2.21 Calibration of power meters using a parallel tee.

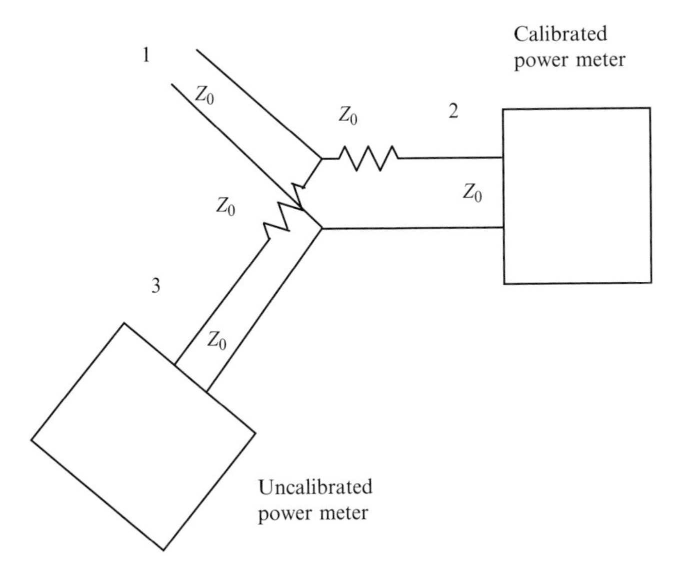

Figure 2.22 Calibration of power meters using a partially compensated tee.

So the power entering the reference power detector is $|b_2|^2(1-|\rho_2|^2)/2$ and the power entering the uncalibrated detector is $|b_3|^2(1-|\rho_3|^2)/2$. Clearly, a complex correction is required to obtain a correct calibration.

Solution to Example 2.8, tee number 2

In a similar fashion to circuit (a), the scattering matrix from Equation (2.56) becomes

$$\begin{bmatrix} b_1 \\ b_2 \\ b_3 \end{bmatrix} = \frac{1}{4}\begin{bmatrix} 0 & 2 & 2 \\ 2 & 1 & 1 \\ 2 & 1 & 1 \end{bmatrix}\begin{bmatrix} 1 \\ \rho_2 b_2 \\ \rho_3 b_3 \end{bmatrix}$$

and this time the equations for b_2 and b_3 become

$$b_2 = b_3 = \frac{2}{(4-\rho_2-\rho_3)}.$$

Solution to Example 2.8, tee number 3

Finally, the scattering matrix from Equation (2.57) becomes

$$\begin{bmatrix} b_1 \\ b_2 \\ b_3 \end{bmatrix} = \frac{1}{2}\begin{bmatrix} 0 & 1 & 1 \\ 1 & 0 & 1 \\ 1 & 1 & 0 \end{bmatrix}\begin{bmatrix} 1 \\ \rho_2 b_2 \\ \rho_3 b_3 \end{bmatrix},$$

which gives equations for b_2 and b_3:

$$b_2 = \frac{2+\rho_3}{4-\rho_2\rho_3} \text{ and } b_3 = \frac{2+\rho_2}{4-\rho_2\rho_3}.$$

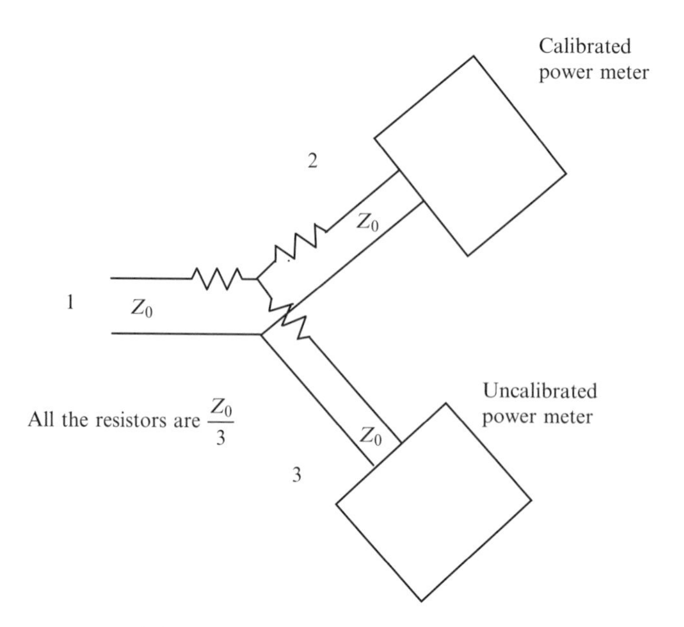

Figure 2.23 Calibration of power meters using a compensated parallel tee.

So, not surprisingly, the arrangement using tee number 2 is usually used for calibrating power detectors, and although it has the simplest corrections, it still needs the further correction described in the solution to tee number 1 for the power reflected from each detector. For further practical details see the reference at the end of this chapter.

Example 2.9 Design a 6 dB attenuator involving a π network of three resistors. The attenuator must have S_{11} and S_{22} equal to zero, i.e. be a matched attenuator.

Solution to Example 2.9

Assuming a symmetrical network as shown in Figure 2.24, as $S_{11} = 0$ the attenuator is matched, so then the full input voltage appears across (say) the left-hand resistor R_2. For a 6 dB attenuator, the reduction in voltage must be a factor of 0.5. So

$$R_1 = \frac{R_2 Z_0}{R_2 + Z_0}$$

and to be matched:

$$Z_0 = \frac{R_2\left(R_1 + \dfrac{R_2 Z_0}{R_2 + Z_0}\right)}{R_2 + \left(R_1 + \dfrac{R_2 Z_0}{R_2 + Z_0}\right)},$$

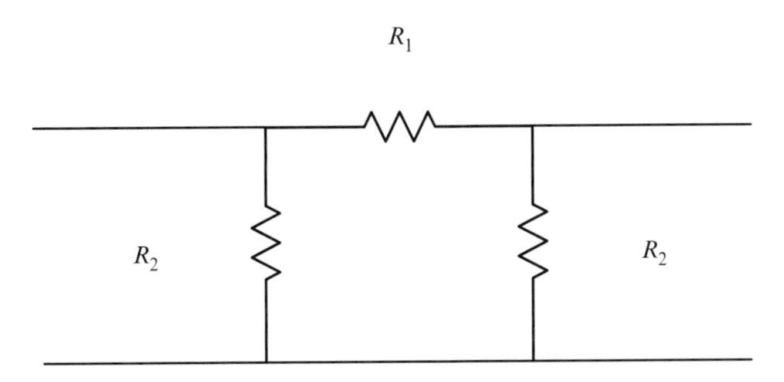

Figure 2.24 A π network attenuator for Example 2.9.

which gives $R_1 = 3Z_0/4$ and $R_2 = 3Z_0$ and the scattering matrix is simply

$$\frac{1}{2}\begin{bmatrix} 0 & 1 \\ 1 & 0 \end{bmatrix}. \tag{2.58}$$

This result could also be found by first finding the scattering matrix in terms of R_1, R_2 and Z_0 and equating the elements to either 0 or 0.5. The resulting equations can then be solved for R_1 and R_2. However, this route to the solution is much longer than the above method.

Example 2.10 A transmission line is terminated in a short circuit with a resistor in parallel placed a distance of 3 mm in front of the short circuit. The resistor has the same value as the characteristic impedance of the line, and the velocity of propagation is $3.10^8\,\text{ms}^{-1}$. Find the frequency response of the amplitude of the voltage across the resistor assuming the incident wave has an amplitude of 10 V. If the characteristic impedance is 50 Ω, find the frequency response of the amplitude of the current in the short circuit. The circuit diagram is shown in Figure 2.25.

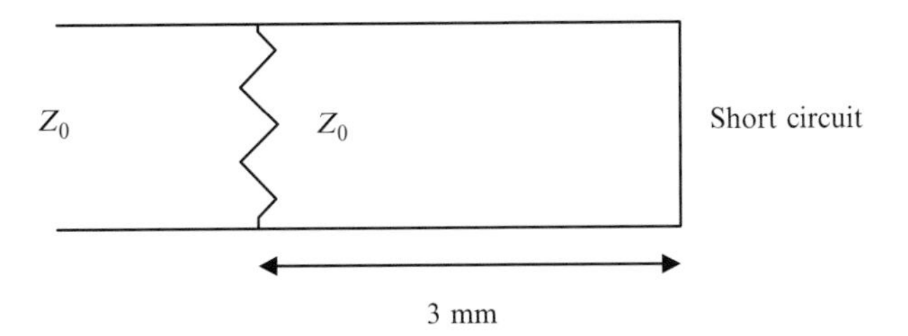

Figure 2.25 The circuit diagram for Example 2.10.

Solution to Example 2.10

Using Equation (2.22) for the impedance of a short circuit at the end of a line of length D:

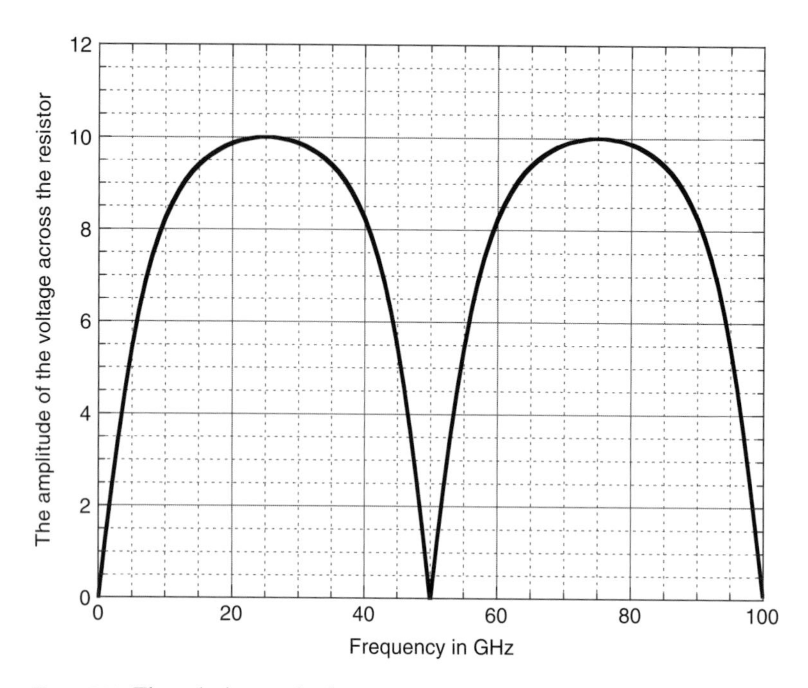

Figure 2.26 The solution to the first part of Example 2.10. The graph shows the amplitude of the voltage across the resistor against frequency.

$$Z_{IN} = jZ_0 \tan(\beta D)$$

and combining it in parallel with the resistor gives a new Z_{IN} as

$$Z_{IN} = \frac{jZ_0 \tan(\beta D)}{1 + j \tan(\beta D)}$$

and hence the reflection coefficient as

$$\rho = -\frac{1}{1 + 2j \tan(\beta D)},$$

so the amplitude of the voltage across the resistor is given by

$$|V| = 10|1 + \rho| = \frac{20 \tan(\beta D)}{(1 + 4 \tan^2(\beta D))^{\frac{1}{2}}}. \tag{2.59}$$

Now $D = 3$ mm, and this is equivalent to a quarter wavelength at 25 GHz, that is the first frequency at which $\beta D = \pi/2$. So by plotting Equation (2.59) from 0 to 100 GHz the frequency response of the voltage across the resistor can be obtained, and this is shown in Figure 2.26.

For the current, the voltage across Z_{IN} is

$$V = \frac{20j \tan(\beta D)}{1 + 2j \tan(\beta D)},$$

and so the amplitude of the current in that part of Z_{IN} which corresponds to the short circuit is

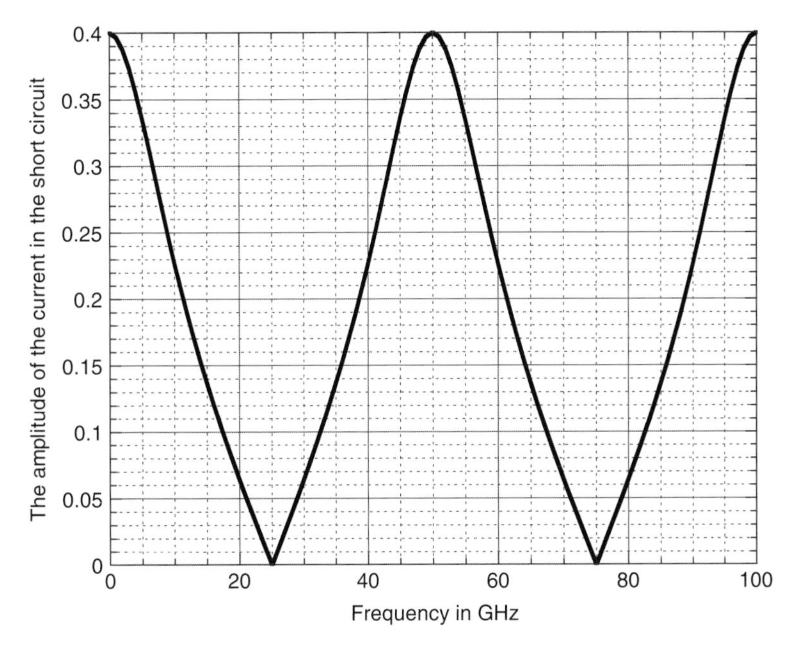

Figure 2.27 The solution to the second part of Example 2.10. The graph shows the amplitude of the current in the short circuit against frequency.

$$|I| = \frac{20}{Z_0(1 + 4\tan^2(\beta D))^{\frac{1}{2}}} = \frac{1}{2.5(1 + 4\tan^2(\beta D))^{\frac{1}{2}}}. \tag{2.60}$$

This is shown in Figure 2.27.

So at 25 GHz, the short circuit is a quarter of a wavelength away and, using Equation (2.24), it appears as an 'open circuit' in parallel with the resistor. Hence the resistor acts like a matched termination with the full 10 V across it. However, at twice the frequency, 50 GHz, Equation (2.25) applies and the resistor is effectively short-circuited and so no voltage appears across the resistor. This pattern then repeats for 75 and 100 GHz and so on. It is also significant to notice that between 10 and 40 GHz there is at least 8 V across the resistor. So this circuit is a reasonable match over quite a wide bandwidth, with only sharp rejections at the half wavelength points, i.e. 50 and 100 GHz, etc.

In contrast to Figure 2.26, the current peaks at 50 GHz where the distance between the short circuit and the resistor is half a wavelength and hence is forming a resonant structure. The resonant energy has a peak of 20 V halfway between the resistor and the short circuit.

2.10 Sine waves in the time domain

To complete this chapter on sine waves, a few examples are discussed concerning what happens when a sinusoidal source is switched on. This has similarities with examples in Chapter 1, where mainly step waves were discussed.

Here the waves are sine waves that begin at a point moving along the transmission lines, just like the step previously.

Example 2.11 Using the same circuit as in Example 2.10, but this time examine what happens in the time domain when a sine wave first arrives at the resistor. Analyse a 25 GHz sine wave first, and then a 50 GHz wave.

Solution to Example 2.11 part 1, 25 GHz

In circuit (a) of Example 2.5 there is a parallel resistor similar to the one in this problem. The reflection and transmission coefficients derived in that example are applicable in both the time and frequency domains. So when a wave approaches the resistor it will initially be partly reflected with a reflection coefficient of $-1/3$ as in Equation (2.27) with $R = Z_0$. The wave will also be partly transmitted with a transmission coefficient of 2/3 derived from Equation (2.32). The fraction of the wave that goes on to the short circuit will be delayed in time by 10 ps, which is the equivalent of a quarter wavelength of phase delay at 25 GHz. At the short circuit, the wave will not be delayed but the waveform will be inverted or multiplied by -1. On returning to the resistor, there will be a further delay in time of 10 ps. The repeated reflections are shown in the wave diagram in Figure 2.28.

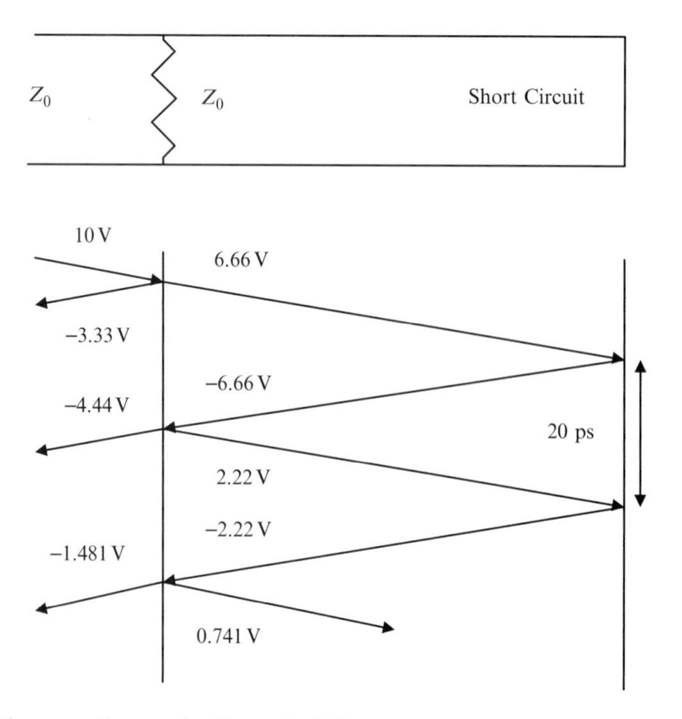

Figure 2.28 The wave diagram for Example 2.11.

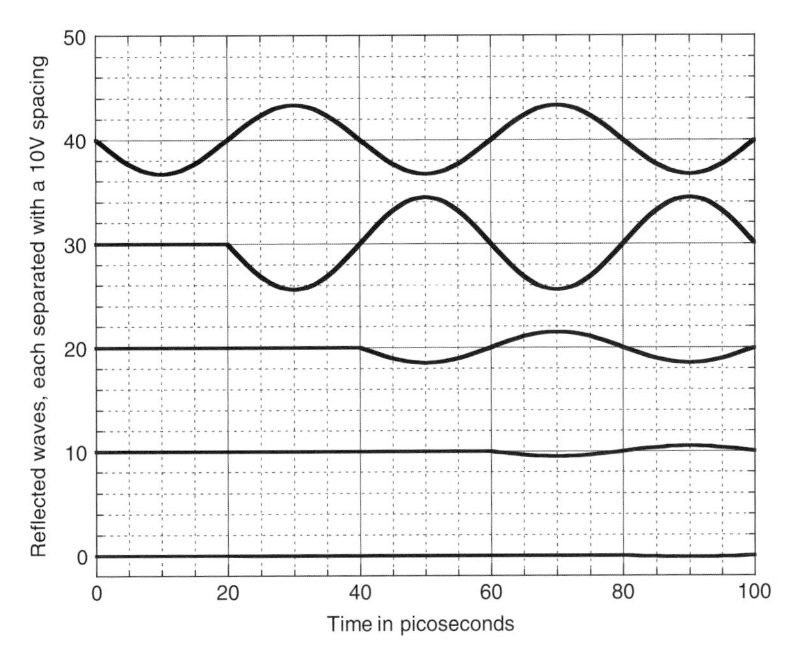

Figure 2.29 The reflected waves for Example 2.11. The vertical axis is in volts and each wave has been separated by a 10 V step. The waves correspond to those on the left of the wave diagram in Figure 2.28.

These various reflections can now be added to give the time domain response of the circuit. In Figure 2.29 the reflected waves are shown separately. It can be seen that in addition to the inversion, caused by the short circuit, there is also the shift in time.

Adding together all these reflected waves gives the waveform shown in Figure 2.30. It can be seen that the waves decay away quite rapidly to zero leaving the voltage across the resistor as 10 V. Figure 2.31 shows the waveform across the resistor. Adding the waves across the resistor gives

$$V = \frac{20}{3} + \frac{40}{9}\left(1 - \frac{1}{3} + \frac{1}{9} - \frac{1}{27} + \cdots\right) = \frac{20}{3} + \frac{10}{3} = 10.$$

Solution to Example 2.11 part 2, 50 GHz

The waveforms for 50 GHz are very similar, except that the reflected wave now builds up to 10 V as shown in Figure 2.32 and the voltage across the resistor fades to zero as shown in Figure 2.33. The delay is the same, but at twice the frequency this is equivalent to a half wavelength of phase delay.

Example 2.12 Using the same circuit as in Example 2.11, but decreasing the value of the resistor so that $R = 0.1\,Z_0$, find waves leaving the resistor on the way to the short circuit at 50 GHz in the time domain and by returning to the frequency domain find the current in the short circuit. Find a relationship between these two results.

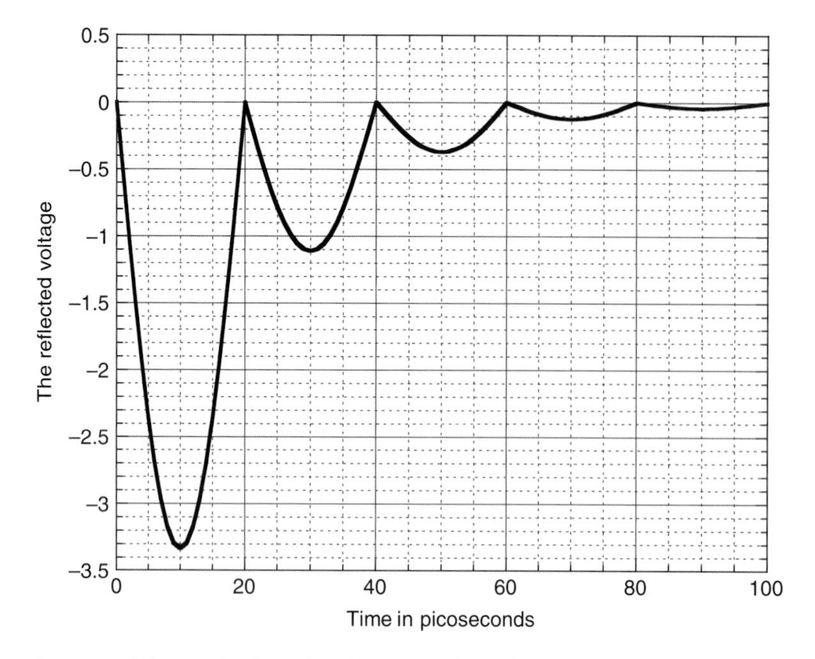

Figure 2.30 The total reflected voltage waveform for Example 2.11.

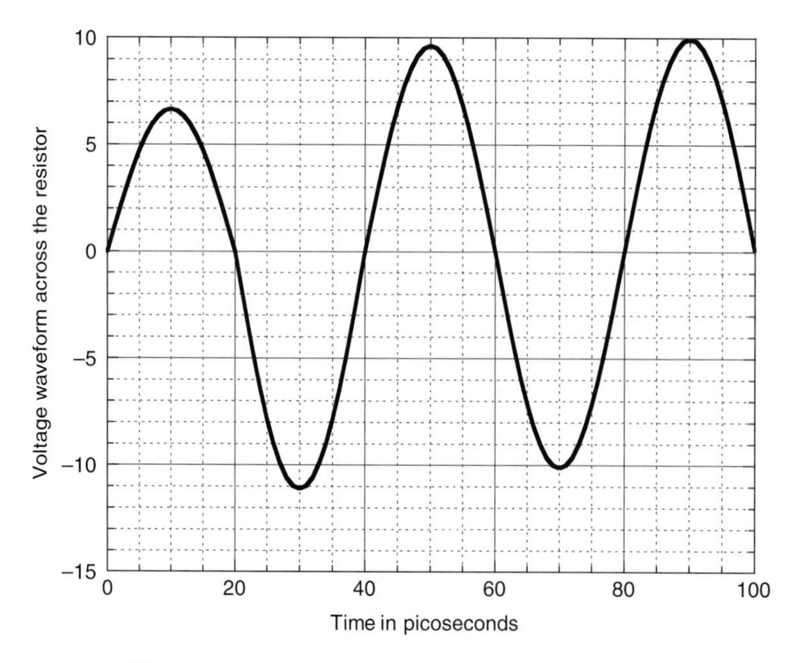

Figure 2.31 The voltage across the resistor in Example 2.11 against time.

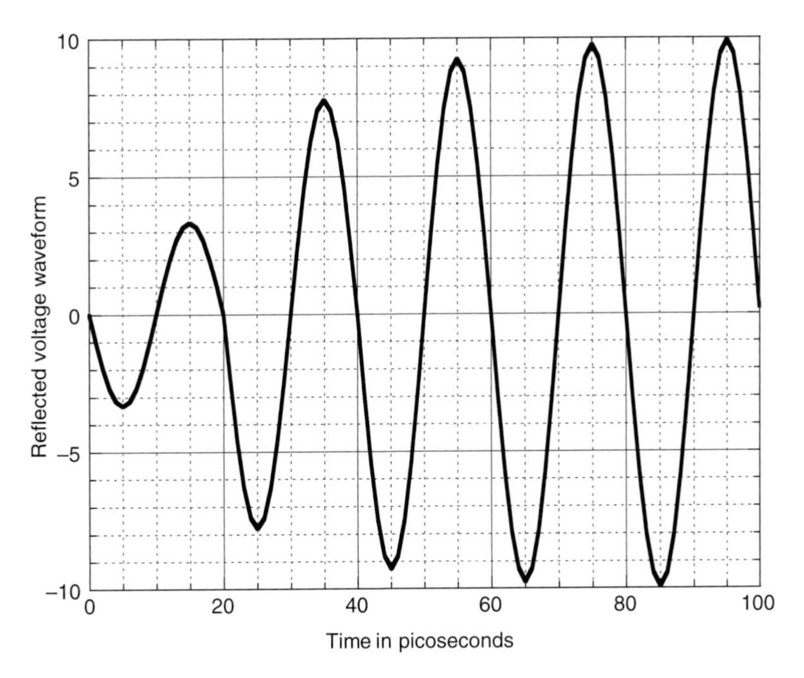

Figure 2.32 The reflected voltage waveform at 50 GHz.

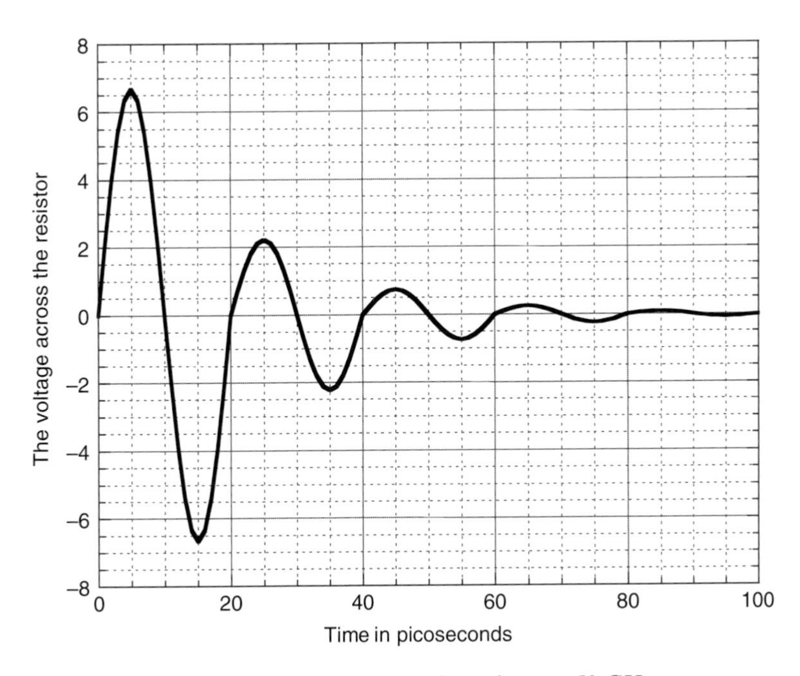

Figure 2.33 The voltage waveform across the resistor at 50 GHz.

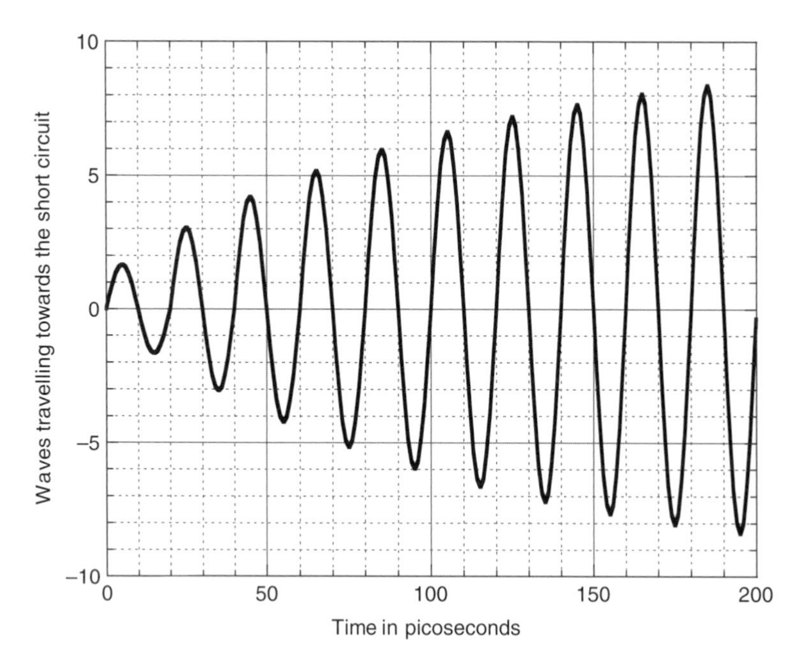

Figure 2.34 The sum of the waves transmitted past the resistor towards the short circuit at 50 GHz.

Solution to Example 2.12

Reducing the value of the resistor increases the amplitude of the reflection coefficient and reduces the transmission coefficient. In this case,

$$\rho = -\frac{5}{6} \text{ and } \tau = \frac{1}{6}.$$

So the waves leaving the resistor on the way to the short circuit will be eventually the sum of

$$V = 10\left(\tau - \rho + \rho^2 - \rho^3 + \cdots\right) = 10\frac{\tau}{1+\rho} = 10.$$

This is because at 50 GHz the waves add together because the phase delay from the resistor and back again is 2π and there are two inversions, one at the short circuit and one at the resistor due to the negative sign of the reflection coefficient. The results of adding the waves are shown in Figure 2.34.

Using the same technique as Example 2.10, the current in the short circuit can be shown to be

$$|I| = \frac{20}{\left(1 + 121\,\tan^2(\beta D)\right)^{\frac{1}{2}}}. \qquad (2.61)$$

The result is shown in Figure 2.35.

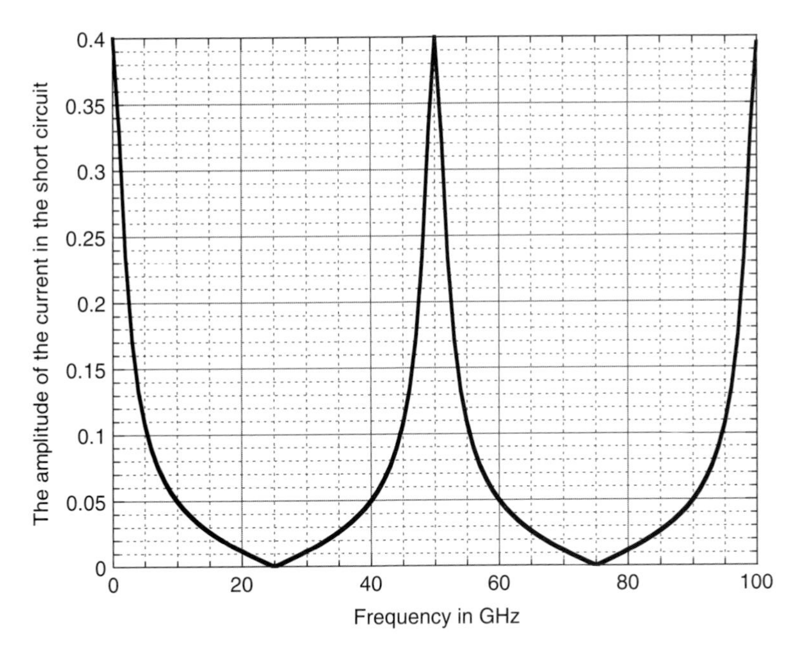

Figure 2.35 The amplitude of the current in the short circuit for $R = 0.1\ Z_0$. This is the solution for the second part of Example 2.12.

Linking time and frequency domains in Example 2.12

Now Figure 2.35 shows a resonance in the current at 50 GHz and the 'Q' factor is 16.66. The relationship between the time domain and the frequency domain can be developed in terms of the ringing time of a resonant circuit. Since the 'Q' factor is the ratio of the stored energy to the energy lost per cycle, for a Q of 16.66 this suggests that the energy should be at half its value after about 11 cycles. Figure 2.34 shows that the energy reaches its half value after about seven cycles or 140 ps and considering it did not start from zero this shows the link between these two results, one from the frequency domain and one from the time domain.

2.11 Modulation of sinusoidal waves

It can be seen in figures like 2.31 and 2.32 that it takes at least ten times the transit time before the circuit settles down to its steady state in the time domain. For the circuit described in Figure 2.34 the 'settling time' is over twenty times the transit time. So if this circuit were modulated, each change in the waveform would have a similar 'settling time'. This suggests that the modulating frequency should be much less than the main or carrier frequency to avoid these switching transients. A rough guide would be for straight pulse modulation of the carrier wave that the pulse width should be much greater than any settling or ringing time. From the above examples, this might be as long as 500 ps. Translating this into a frequency modulation this would mean that a 25 GHz carrier should not be modulated at a frequency higher than 2 GHz. This is not far removed from the 10 % maximum modulation 'rule of thumb' in current usage. However, for higher Q circuits, some as high as 1000 or more, this should be reduced to only 0.1 % or 25 MHz. Clearly, in most circuits, compromises are made, but these switching transients are present and will cause a distortion of the signal when it is demodulated, even in the digital case.

This also has relevance when sweeping the frequency in order to measure the frequency response of a circuit. If the sweep rate is too high there is not enough time for the energy to enter a resonant circuit and reach equilibrium. In some cases the Q is not measured but a derivative shape is measured, which is due to the resonant energy escaping after the swept frequency has moved away. This effect can be detected by reversing the direction of the frequency sweep. In physics this effect is called an 'adiabatic fast passage' through a resonance and it is usually to be avoided in measurements. The best technique is to reduce the sweep rate until the frequency response does not change and then it is safe to assume that all the high Q resonances are being detected. In some crystal-controlled oscillators used in many network analysers, the frequency is swept in small jumps. So, for example, a complete sweep might take 10 ms to make it appear instantaneous on a display. If the frequency range was 1 to 100 GHz, and the crystal-controlled clock frequency was 30 MHz, this would involve about 3300 jumps in frequency. So the period at one frequency would be 3 μs and this would easily allow for a settling time of (say) 140 ps. Indeed, it would allow for the very much longer settling times which might occur at the lower frequencies.

2.12 Further reading

Smith chart

S. Ramo, J. R. Whinnery and T. Van Duzer *Fields and Waves in Communication Electronics*, Third edition, New York, Wiley, 1994. Section 5.9.

J. Everard *Fundamentals of RF Circuit Design with Low Noise Oscillators*, New York, Wiley, 2001. Section 3.7.

K. Chang *Microwave Solid State Circuits and Applications*, New York, Wiley, 1994. Chapter 2.

J. D. Krauss *Electromagnetics*, Fourth edition, New York, McGraw-Hill, 1991. Chapter 12, sections 12.7–12.9.

R. A. Chipman *Transmission Lines*, Schaum's Outline Series, New York, McGraw-Hill, 1968. Chapter 9.

R. L. Liboff and G. C. Dalman *Transmission Lines, Waveguides and Smith Charts*, New York, Macmillan Publishing Company, 1985. Chapter 5.

E. da Silva *High Frequency and Microwave Engineering*, Oxford, Butterworth Heinemann, 2011. Chapter 3.

T. H. Lee *Planar Microwave Engineering, a Practical Guide to Theory, Measurements and Circuits*, Cambridge, Cambridge University Press, 2004. Chapter 3, sections 3.2 and 3.3.

R. E. Collin *Foundations for Microwave Engineering*, New York, McGraw-Hill, 1992. Chapter 5, sections 5.1 to 5.4.

D. M. Pozar *Microwave Engineering*, Second edition, New York, Wiley, 1998. Chapter 5, sections 5.1 to 5.3.

Scattering and transmission parameters

S. Ramo, J. R. Whinnery and T. Van Duzer *Fields and Waves in Communication Electronics*, Third edition, New York, Wiley, 1994. Section 11.5.

J. Everard *Fundamentals of RF Circuit Design with Low Noise Oscillators*, New York, Wiley, 2001. Section 2.7.

T. Edwards *Foundations for Microstrip Circuit Design*, Second edition, New York, Wiley, 1992. Appendix C (his definition of power lacks a factor of two in C.7 onwards).

K. Chang *Microwave Solid State Circuits and Applications*, New York, Wiley, 1994. Chapter 3.

A. J. Baden-Fuller *Microwaves*, Oxford, Pergamon Press, 1969. Chapter 1.

J. D. Krauss *Electromagnetism*, Fourth edition, New York, McGraw-Hill, 1991. Chapter 12, section 12.3.

P. A. Rizzi *Microwave Engineering*, London, Prentice Hall, 1988. Chapter 4 and Appendix D.

R. J. Collier and A. D. Skinner *Microwave Measurements*, Third edition, London, IET, 2007. Chapter 2.

E. da Silva *High Frequency and Microwave Engineering*, Oxford, Butterworth Heinemann, 2011. Chapter 3, section 3.10.

R. E. Collin *Foundations for Microwave Engineering*, New York, McGraw-Hill, 1992. Chapter 4, sections 4.7 to 4.9.

D. M. Pozar *Microwave Engineering*, Second edition, New York, Wiley, 1998. Chapter 4, section 4.3.

Networks

D. M. Kerns and R. W. Beatty *Basic Theory of Waveguide Junctions and Introductory Microwave Network Analysis*, Oxford, Pergamon Press, 1967. Part 2.

B. J. Minnis *Designing Microwave Circuits by Exact Synthesis*, London, Artech House, 1996.

G. Matthaei, L. Young and E. M. T. Jones *Microwave Filters, Impedance-Matching Networks and Coupling Structures*, Dedham, MA, Artech House Books, 1980.

R. N. Simons *Coplanar Wave-guide Circuits*, New York, Wiley, 2001.

Calibration of power meters

R. J. Collier and A. D. Skinner *Microwave Measurements*, Third edition, London, IET, 2007. Chapter 15.

3 Coupled transmission lines and circuits

In many applications of transmission lines there is a degree of coupling to neighbouring lines. This can be a disadvantage when it is a source of unwanted signals or, as it is often called, 'cross-talk'. On the other hand, it can be a way of coupling a fixed quantity of a signal to another circuit for monitoring, measuring or signal processing. This chapter will use the equivalent circuit model to describe this coupling and, in the same format as in the previous chapters, the basic theory will be given first, before examples of its application are discussed.

3.1 Basic theory

For certain transmission lines there is very little coupling to other lines, for example, a solid coaxial cable (as opposed to the variety with a braided outer conductor) or a metallic rectangular waveguide. In others, like stripline, microstrip and dielectric waveguides, considerable coupling is possible. Details of these and other transmission lines will be discussed in later chapters. In this chapter, the equivalent circuit given in Figure 1.1 will be analysed to show the coupling to an adjacent line. There are two types of coupling: electric and magnetic. The first is caused by the charges on one line inducing charges on another and this is usually described by a mutual capacitance. The second is caused by magnetic flux coupling between the lines and this is usually described by a mutual inductance. The extended equivalent circuit is shown in Figure 3.1 for two identical lines adjacent to each other.

Here the new distributed variables are C_M and M, the mutual capacitance and inductance per metre of line. The mutual inductance has been taken as positive and the mutual capacitance as negative. The reasons for the negative sign of C_M will be discussed in Chapter 4, Example 4.5. Using a very similar analysis to that given in Chapter 1, the equations for the voltage and current will be developed first:

$$V_{10} = V_1 - L\Delta x\frac{\partial I_1}{\partial t} - M\Delta x\frac{\partial I_2}{\partial t} \text{ and } I_{10} = I_1 - C\Delta x\frac{\partial V_1}{\partial t} + C_M\Delta x\frac{\partial V_2}{\partial t},$$

$$V_{20} = V_2 - L\Delta x\frac{\partial I_2}{\partial t} - M\Delta x\frac{\partial I_1}{\partial t} \text{ and } I_{20} = I_2 - C\Delta x\frac{\partial V_2}{\partial t} + C_M\Delta x\frac{\partial V_1}{\partial t}, \quad (3.1)$$

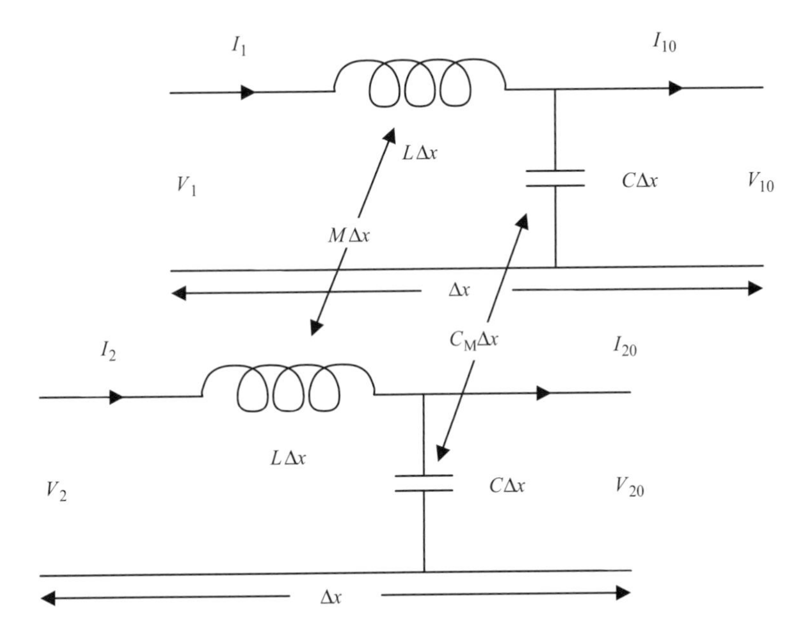

Figure 3.1 The equivalent circuit of two coupled transmission lines.

where the suffixes 10 and 20 are for the output on the right-hand side of the circuit diagram. So the changes in voltage and current are:

$$V_{10} - V_1 = \Delta V_1 = -L\Delta x\frac{\partial I_1}{\partial t} - M\Delta x\frac{\partial I_2}{\partial t} \text{ and } I_{10} - I_1 = \Delta I_1 = -C\Delta x\frac{\partial V_1}{\partial t} + C_M\Delta x\frac{\partial V_2}{\partial t},$$

$$V_{20} - V_1 = \Delta V_2 = -L\Delta x\frac{\partial I_2}{\partial t} - M\Delta x\frac{\partial I_1}{\partial t} \text{ and } I_{20} - I_2 = \Delta I_2 = -C\Delta x\frac{\partial V_2}{\partial t} + C_M\Delta x\frac{\partial V_1}{\partial t},$$

which gives

$$\frac{\Delta V_1}{\Delta x} = -L\frac{\partial I_1}{\partial t} - M\frac{\partial I_2}{\partial t} \text{ and } \frac{\Delta I_1}{\Delta x} = -C\frac{\partial V_1}{\partial t} + C_M\frac{\partial V_2}{\partial t},$$
$$\frac{\Delta V_2}{\Delta x} = -L\frac{\partial I_2}{\partial t} - M\frac{\partial I_1}{\partial t} \text{ and } \frac{\Delta I_2}{\Delta x} = -C\frac{\partial V_2}{\partial t} + C_M\frac{\partial V_1}{\partial t}, \tag{3.2}$$

and in the limit as $\Delta x \to 0$:

$$\frac{\partial V_1}{\partial x} = -L\frac{\partial I_1}{\partial t} - M\frac{\partial I_2}{\partial t} \text{ and } \frac{\partial I_1}{\partial x} = -C\frac{\partial V_1}{\partial t} + C_M\frac{\partial V_2}{\partial t},$$
$$\frac{\partial V_2}{\partial x} = -L\frac{\partial I_2}{\partial t} - M\frac{\partial I_1}{\partial t} \text{ and } \frac{\partial I_2}{\partial x} = -C\frac{\partial V_2}{\partial t} + C_M\frac{\partial V_1}{\partial t}. \tag{3.3}$$

These are the Telegraphists' equations for two coupled lines. They can be written more succinctly in matrix form as

$$\frac{\partial}{\partial x}\begin{bmatrix} V_1 \\ V_2 \end{bmatrix} = -\begin{bmatrix} L & M \\ M & L \end{bmatrix}\frac{\partial}{\partial t}\begin{bmatrix} I_1 \\ I_2 \end{bmatrix} \text{ and } \frac{\partial}{\partial x}\begin{bmatrix} I_1 \\ I_2 \end{bmatrix} = -\begin{bmatrix} C & -C_M \\ -C_M & C \end{bmatrix}\frac{\partial}{\partial t}\begin{bmatrix} V_1 \\ V_2 \end{bmatrix}. \tag{3.4}$$

Using the same technique as in Chapter 1, the equations for the voltage and the current can be separated out by differentiating each equation with respect to both time and distance:

$$\frac{\partial^2}{\partial x^2}\begin{bmatrix} V_1 \\ V_2 \end{bmatrix} = -\begin{bmatrix} L & M \\ M & L \end{bmatrix}\frac{\partial^2}{\partial x \partial t}\begin{bmatrix} I_1 \\ I_2 \end{bmatrix} \text{ and } \frac{\partial^2}{\partial t \partial x}\begin{bmatrix} I_1 \\ I_2 \end{bmatrix} = -\begin{bmatrix} C & -C_M \\ -C_M & C \end{bmatrix}\frac{\partial^2}{\partial t^2}\begin{bmatrix} V_1 \\ V_2 \end{bmatrix}.$$

So by equating the mixed differentials, the wave equation for the voltages is

$$\frac{\partial^2}{\partial x^2}\begin{bmatrix} V_1 \\ V_2 \end{bmatrix} = \begin{bmatrix} (LC - MC_M) & (MC - LC_M) \\ (MC - LC_M) & (LC - MC_M) \end{bmatrix}\frac{\partial^2}{\partial t^2}\begin{bmatrix} V_1 \\ V_2 \end{bmatrix}. \tag{3.5}$$

In a very similar fashion, the wave equation for the currents can be found to be

$$\frac{\partial^2}{\partial x^2}\begin{bmatrix} I_1 \\ I_2 \end{bmatrix} = \begin{bmatrix} (CL - C_M M) & (CM - C_M L) \\ (CM - C_M L) & (CL - C_M M) \end{bmatrix}\frac{\partial^2}{\partial t^2}\begin{bmatrix} I_1 \\ I_2 \end{bmatrix}. \tag{3.6}$$

Both equations are equivalent as the capacitances and inductances are scalar quantities. The equations still have two variables each and to solve them it is usual to change these variables. This is achieved by finding the eigenvalues of the above matrices and their associated eigenvectors. First of all, the eigenvalues are solutions of

$$\begin{vmatrix} (A - \lambda) & B \\ B & (A - \lambda) \end{vmatrix} = 0, \text{ where } A = (LC - MC_M) \text{ and } B = (MC - LC_M),$$

which gives $\lambda = A \pm B = (C - C_M)(L + M)$ or $= (C + C_M)(L - M)$, and the eigenvectors are found from

$$\begin{bmatrix} (A - \lambda) & B \\ B & (A - \lambda) \end{bmatrix}\begin{bmatrix} U_1 \\ U_2 \end{bmatrix} = 0.$$

For $\lambda = A + B$, $U_1 = U_2$ and for $\lambda = A - B$, $U_1 = -U_2$. So the eigenvectors are

$$\begin{bmatrix} 1 \\ 1 \end{bmatrix} \text{ for } \lambda = (C - C_M)(L + M) \text{ and } \begin{bmatrix} 1 \\ -1 \end{bmatrix} \text{ for } \lambda = (C + C_M)(L - M). \tag{3.7}$$

The first of these is often called the even mode and it occurs when $V_1 = V_2 = V_e$ and $I_1 = I_2 = I_e$, where the suffix e denotes this mode. Equations (3.5) and (3.6) become:

$$\frac{\partial^2 V_e}{\partial x^2} = (C - C_M)(L + M)\frac{\partial^2 V_e}{\partial t^2} \text{ and } \frac{\partial^2 I_e}{\partial x^2} = (C - C_M)(L + M)\frac{\partial^2 I_e}{\partial t^2}. \tag{3.8}$$

Similarly, the second is often called the odd mode and it occurs when $V_1 = -V_2 = V_o$ and $I_1 = -I_2 = I_o$, where the suffix o denotes this mode. These are the new variables which will be used instead of V_1, V_2, I_1 and I_2. They are the fundamental modes for the coupled lines, and the energy propagates in either the even mode or the odd mode or combinations of these two modes. Equations (3.5) and (3.6) become:

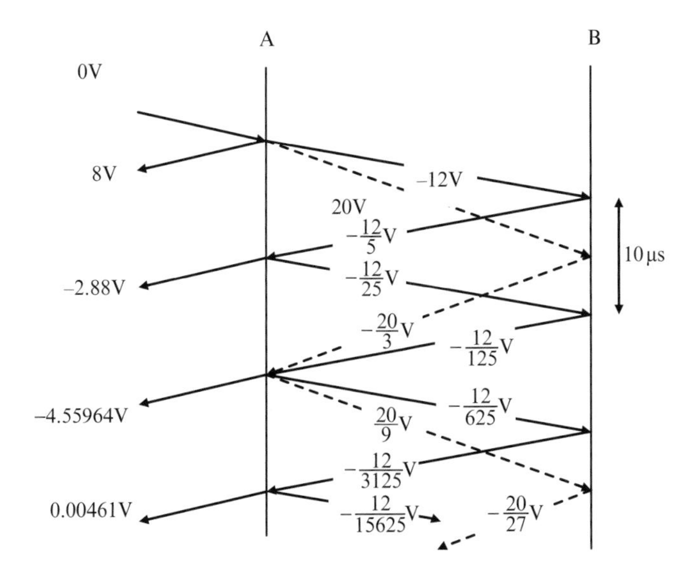

Figure 3.6 The wave diagram for Example 3.4 showing the waves on line 2. The solid lines between A and B are the odd mode pulses and the dotted lines are the even ones.

Solution to Example 3.4

The problem still involves odd and even modes, even if they are ignored. So instead of a matched termination there will be multiple reflections. These are shown in a wave diagram in Figure 3.6. The reflection coefficients at the termination are

$$\rho_e = -\frac{1}{3} \text{ and } \rho_o = \frac{1}{5},$$

and when the waves return to A, the reflection coefficients are the same and the transmission coefficients are

$$\tau_e = \frac{2}{3} \text{ and } \tau_o = \frac{6}{5}.$$

The even mode takes longer to die away as it has half the number of reflections and a larger reflection coefficient. The odd mode has practically died away in 30 μs. The results are shown graphically in Figure 3.7. Clearly the effects of the coupled lines are demonstrated by the unexpected waveform on line 2, which is not 'connected' to the circuit of line 1.

Example 3.5 A pair of coupled lines have capacitance and inductance matrices given by

$$60\begin{bmatrix} 1 & -0.2 \\ -0.2 & 1 \end{bmatrix} \text{pFm}^{-1} \text{ and } 2\begin{bmatrix} 1 & 0.2 \\ 0.2 & 1 \end{bmatrix} \mu\text{Hm}^{-1}.$$

Find the characteristic impedances and velocities of the odd and even modes.

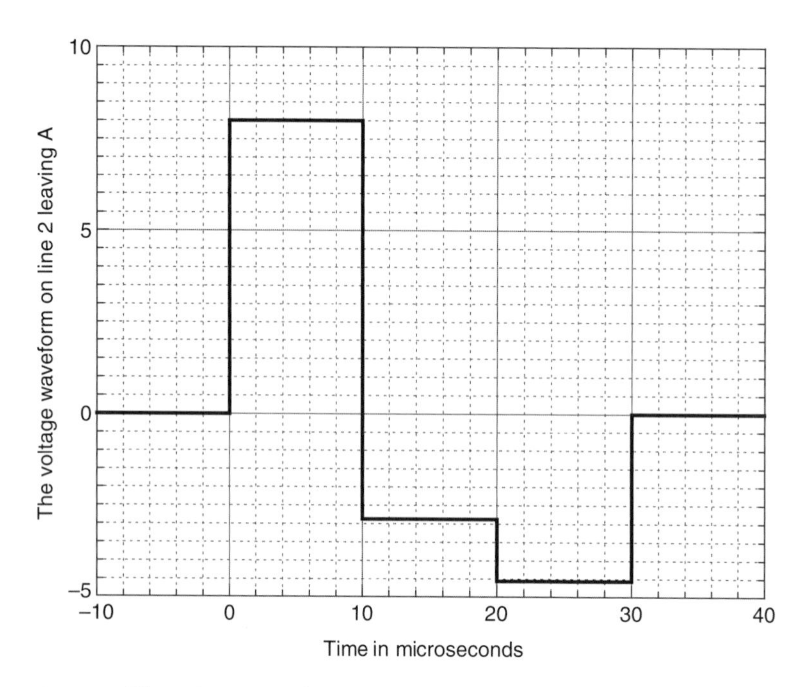

Figure 3.7 The voltage waveform leaving point A on line 2.

Solution to Example 3.5

Using Equation (3.11) to obtain the two mode impedances:

$$Z_{0e} = \sqrt{\frac{2.4 \times 10^{-6}}{48 \times 10^{-12}}} = 223.6\ \Omega \text{ and } Z_{0o} = \sqrt{\frac{1.6 \times 10^{-6}}{72 \times 10^{-12}}} = 149.1\ \Omega.$$

Using Equation (3.10) to obtain the two mode velocities:

$$v_e = \frac{1}{\sqrt{(2.4 \times 10^{-6})\ (48 \times 10^{-12})}}$$

$$= v_o = \frac{1}{\sqrt{(1.6 \times 10^{-6})\ (72 \times 10^{-12})}} = 0.93169 \times 10^8\ \text{ms}^{-1}.$$

Since the change in capacitance due to the coupling is equal to the change in inductance, both the velocities are equal. In general, it is easier to obtain different characteristic impedances rather than different velocities.

Example 3.6 A pair of coupled lines have even and odd mode characteristic impedances equal to 120 Ω and 40 Ω respectively. At the end of these lines, there is a short uncoupled region where each line has a characteristic impedance equal to 120 Ω. At the end of line 1 is a termination of 60 Ω and the end of line 2 is terminated with 40 Ω, as shown in Figure 3.8. If an even mode of amplitude 10 V is travelling towards the terminations, what are the amplitudes of the reflected even mode and odd modes on the coupled lines? Check the answers by working out the powers in the problem.

Solution to Example 3.6

Starting the problem with the termination, the reflection coefficients for line 1 and line 2 are

$$\rho_1 = \frac{60 - 120}{60 + 120} = -\frac{1}{3} \text{ and } \rho_2 = \frac{40 - 120}{40 + 120} = -\frac{1}{2}.$$

Now the reflected waves will not be equal and the average reflection will form the even mode reflected wave and half the difference will form the odd mode reflection. So for the termination:

$$\begin{bmatrix} V_{e-} \\ V_{o-} \end{bmatrix} = \frac{1}{2} \begin{bmatrix} (\rho_1 + \rho_2) & (\rho_1 - \rho_2) \\ (\rho_1 - \rho_2) & (\rho_1 + \rho_2) \end{bmatrix} \begin{bmatrix} V_{e+} \\ V_{o+} \end{bmatrix}. \tag{3.14}$$

Using the numbers for ρ_1 and ρ_2, Equation (3.14) becomes:

$$\begin{bmatrix} V_{e-} \\ V_{o-} \end{bmatrix} = \frac{1}{12} \begin{bmatrix} -5 & 1 \\ 1 & -5 \end{bmatrix} \begin{bmatrix} V_{e+} \\ V_{o+} \end{bmatrix}.$$

Now considering the transition from the coupled lines to the short uncoupled section, the even mode travels through this without reflection. However, the odd

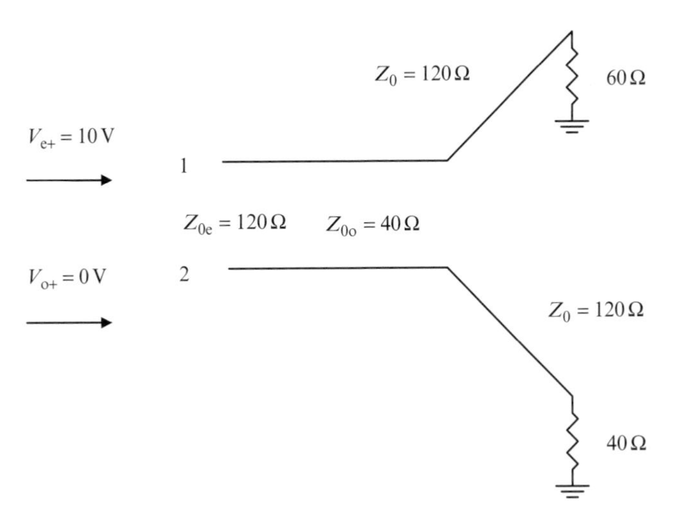

Figure 3.8 The circuit diagram from Example 3.6. The lines are coupled in the left-hand section and uncoupled in the right-hand section.

mode is partly reflected and partly transmitted. When an odd mode goes from the coupled to the uncoupled region the reflection and transmission coefficients are

$$\rho_{oc} = \frac{Z_{0e} - Z_{0o}}{Z_{0e} + Z_{0o}} = \frac{1}{2} \text{ and } \tau_{oc} = 1 + \rho_{oc} = \frac{3}{2}.$$

For the reverse direction the coefficients are

$$\rho_{ou} = \frac{Z_{0o} - Z_{0e}}{Z_{0o} + Z_{0e}} = -\rho_{oc} = -\frac{1}{2} \text{ and } \tau_{ou} = 1 + \rho_{ou} = \frac{1}{2},$$

so putting these results together in the form of a scattering matrix for the transition between the coupled and the uncoupled regions as follows:

$$\begin{bmatrix} V_{ec-} \\ V_{eu-} \\ V_{oc-} \\ V_{ou-} \end{bmatrix} = \frac{1}{2}\begin{bmatrix} 0 & 2 & 0 & 0 \\ 2 & 0 & 0 & 0 \\ 0 & 0 & 1 & 1 \\ 0 & 0 & 3 & -1 \end{bmatrix}\begin{bmatrix} V_{ec+} \\ V_{eu+} \\ V_{oc+} \\ V_{ou+} \end{bmatrix},$$

where the suffixes c and u are used to indicate the coupled and uncoupled regions. Now in the example $V_{ec+} = 10\,\text{V}$ and $V_{oc+} = 0\,\text{V}$. Using all the above equations gives the following results:

amplitude of the reflected even mode, V_{ec-}, as $-\frac{80}{19}\,\text{V}$,
amplitude of the reflected odd mode, V_{oc-}, as $\frac{10}{19}\,\text{V}$,
amplitude of the voltage across the $60\,\Omega$ resistor as $\frac{120}{19}\,\text{V}$,
amplitude of the voltage across the $40\,\Omega$ resistor as $\frac{100}{19}\,\text{V}$.

The powers are as follows:

input power in the even mode $= \frac{5}{3}\,\text{W}$,
reflected power in the even mode $= \frac{64}{361} \times \frac{5}{3}\,\text{W}$,
reflected power in the odd mode $= \frac{3}{361} \times \frac{5}{3}\,\text{W}$,
power in the $60\,\Omega$ resistor $= \frac{144}{361} \times \frac{5}{3}\,\text{W}$,
power in the $40\,\Omega$ resistor $= \frac{150}{361} \times \frac{5}{3}\,\text{W}$.

Fortunately $64 + 3 + 144 + 150 = 361$, so the powers equate to the input power. This problem shows how mode mixing can take place and that it is not a straightforward matter to estimate the answers from a casual examination of the circuit diagram.

3.2 Coupled transmission line circuits in the frequency domain

---

Example 3.7 A directional coupler using backward waves: Design a directional coupler using a quarter wavelength of coupled lines as shown in Figure 3.9. If the four uncoupled lines leading to the coupled region all have a characteristic impedance of $50\,\Omega$, find the even and odd mode characteristic impedances needed to achieve a coupling of 12 dB. Assuming the coupled modes both travel at a

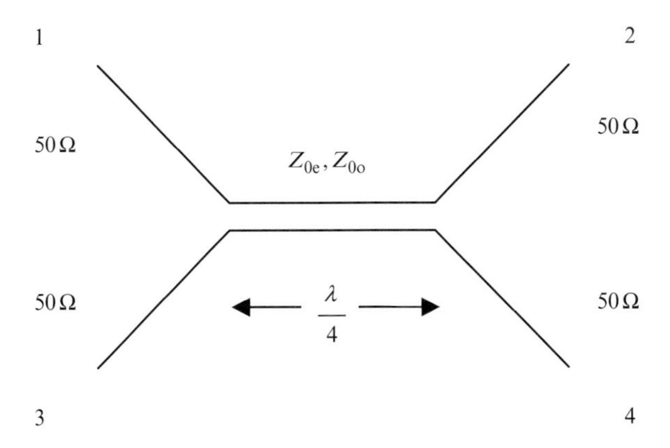

Figure 3.9 The circuit diagram of the directional coupler.

velocity of $2.10^8\,\mathrm{ms}^{-1}$, find the length of the coupled region for operation at 10 GHz. Finally, find the frequency response of the coupler.

Solution to Example 3.7

A directional coupler is a device which allows the coupling in one direction only of a fixed amount out of the main circuit. So if the main circuit is from port 1 to port 2, then the coupled output will be at port 3 only. If the waves travel in the reverse direction, that is port 2 to port 1, then the output will be in port 4 only. The analysis is in terms of odd and even modes as follows. If ports 2 and 4 are terminated with a matched load, then an even mode travelling towards the coupler from ports 1 and 3 will 'see' an impedance at the input given by Equation (2.24), i.e.

$$Z_{\mathrm{INe}} = \frac{Z_{0e}^2}{Z_0} \quad \text{and } \rho_e = \frac{Z_{0e}^2 - Z_0^2}{Z_{0e}^2 + Z_0^2};$$

similarly for the odd mode:

$$Z_{\mathrm{INo}} = \frac{Z_{0o}^2}{Z_0} \quad \text{and } \rho_o = \frac{Z_{0o}^2 - Z_0^2}{Z_{0o}^2 + Z_0^2}.$$

Now if there is an input of 1 V on just line 1, this can be split into even and odd modes each of amplitude $\frac{1}{2}$, then the reflected even and odd modes will be given by

$$\tfrac{1}{2}(\rho_e + \rho_o) \text{ on line 1 and } \tfrac{1}{2}(\rho_e - \rho_o) \text{ on line 2.} \tag{3.15}$$

In terms of the characteristic impedances, these reflections become

$$\frac{Z_{0e}^2 Z_{0o}^2 - Z_0^4}{\left(Z_{0e}^2 + Z_0^2\right)\left(Z_{0o}^2 + Z_0^2\right)} \quad \text{and} \quad \frac{Z_0^2\left(Z_{0e}^2 - Z_{0o}^2\right)}{\left(Z_{0e}^2 + Z_0^2\right)\left(Z_{0o}^2 + Z_0^2\right)}.$$

Now the above equations can be simplified by making

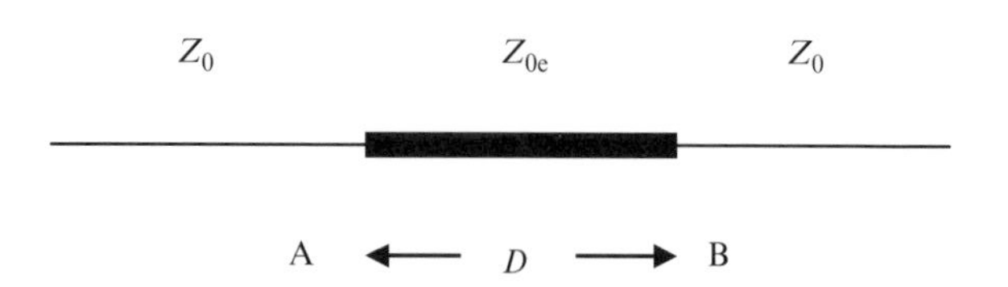

Figure 3.10 The circuit diagram for the frequency dependent solution for the even mode.

$$Z_0^2 = Z_{0e}Z_{0o}. \tag{3.16}$$

This is a widely used design criterion for these circuits, and in this case there is no reflection at port 1 and the output at port 2 is simply

$$\frac{Z_{0e} - Z_{0o}}{Z_{0e} + Z_{0o}}, \tag{3.17}$$

which is called the coupling coefficient, k.

Now the amount of coupling is 12 dB, which is a factor of 16 in power or 4 in amplitude, so $k = \frac{1}{4}$. This gives the values of Z_{0e} and Z_{0o} as

$$Z_{0e} = \sqrt{\frac{12500}{3}} = 64.54\ \Omega,\ Z_{0o} = \sqrt{1500} = 38.73\ \Omega.$$

The length of the coupled region is given by

$$\frac{2 \times 10^8}{4 \times 10^{10}} = 0.005\ \text{m} = 5\ \text{mm}.$$

The frequency dependence involves an equation for the fraction of the incident voltage which is transmitted through the coupled region. Let the coupled region have a length D, then in Figure 3.10 is shown the circuit for the even mode. Now Equation (2.21) applying to Figure 3.10 gives

$$Z_{\text{INe}} = Z_{0e}\left(\frac{Z_0 + \text{j}Z_{0e}\tan(\beta D)}{Z_{0e} + \text{j}Z_0\tan(\beta D)}\right). \tag{3.18}$$

So the reflection coefficient for waves arriving at A is found from Equation (1.13) as

$$\rho_{\text{Ae}} = \frac{\text{j}(Z_{0e}^2 - Z_0^2)\tan(\beta D)}{2Z_{0e}Z_0 + \text{j}(Z_{0e}^2 + Z_0^2)\tan(\beta D)}. \tag{3.19}$$

Now, using the condition given in Equation (3.16), this simplifies to:

$$\rho_{\text{Ae}} = \frac{\text{j}(Z_{0e} - Z_{0o})\tan(\beta D)}{2Z_0 + \text{j}(Z_{0e} + Z_{0o})\tan(\beta D)}. \tag{3.20}$$

In a similar manner, the odd mode reflection coefficient can be shown to be

$$\rho_{\text{Ao}} = \frac{\text{j}(Z_{0o} - Z_{0e})\tan(\beta D)}{2Z_0 + \text{j}(Z_{0e} + Z_{0o})\tan(\beta D)} = -\rho_{\text{Ae}}. \tag{3.21}$$

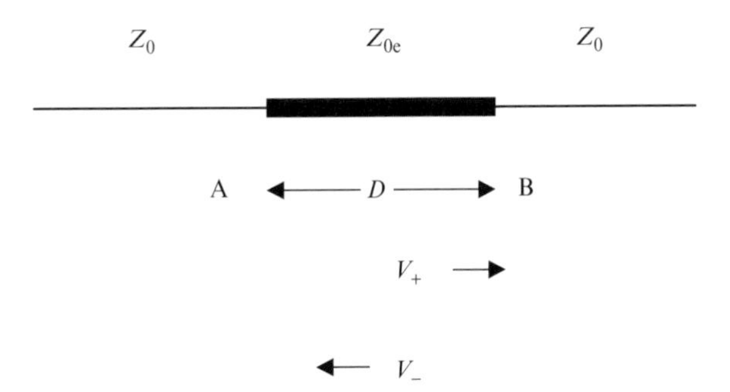

Figure 3.11 This figure shows the voltage waves at B, for Example 3.7.

For all frequencies, Equation (3.15) becomes

$$\frac{1}{2}(\rho_{Ae} + \rho_{Ao}) = 0 \text{ and } \frac{1}{2}(\rho_{Ae} - \rho_{Ao}) = \rho_{Ae}. \tag{3.22}$$

So the power reflected on line 1 is zero for all frequencies and on line 2 the fraction of the incident power on line 1 coupled to line 2 is found from Equation (3.22) as

$$|\rho_{Ae}|^2 = \frac{(Z_{0e} - Z_{0o})^2 \tan^2(\beta D)}{4Z_{0e}Z_{0o} + (Z_{0e} + Z_{0o})^2 \tan^2(\beta D)}. \tag{3.23}$$

Putting the numbers into Equation (3.23) gives

$$|\rho_{Ae}|^2 = \frac{\tan^2(\beta D)}{15 + 16 \tan^2(\beta D)}. \tag{3.24}$$

In order to derive the transmission coefficients, a useful equation linking A and B needs to be derived first.

In Figure 3.11, the voltage wave arriving at B is V_+ and the reflected wave is V_-. The sum of these waves is the voltage at B, V_B, and also the wave which is transmitted onwards from B. So if ρ_B is the reflection coefficient at B then

$$\rho_B = \frac{Z_0 - Z_{0e}}{Z_0 + Z_{0e}} = \frac{V_-}{V_+} \text{ and } V_B = V_+ + V_- = V_+(1 + \rho_B). \tag{3.25}$$

Moving back towards A, both the waves will have a phase change. For V_+ it will be a phase advance and the opposite for V_-. If V_A is the voltage at A, then

$$V_A = V_+ \exp(j\beta D) + V_- \exp(-j\beta D) = V_+(\exp(j\beta D) + \rho_B \exp(-j\beta D)). \tag{3.26}$$

Hence the useful equation linking V_A and V_B can be found from Equations (3.25) and (3.26) as follows:

$$\frac{V_B}{V_A} = \frac{V_+(1 + \rho_B)}{V_+(\exp(j\beta D) + \rho_B \exp(-j\beta D))}.$$

Expanding the exponentials and using the first part of Equation (3.25) gives:

$$\frac{V_B}{V_A} = \frac{2Z_0}{(Z_0 + Z_{0e})(\cos(\beta D) + \mathrm{j}\sin(\beta D)) + (Z_0 - Z_{0e})(\cos(\beta D) - \mathrm{j}\sin(\beta D))}$$

$$\text{or } \frac{V_B}{V_A} = \frac{Z_0 \sec(\beta D)}{Z_0 + \mathrm{j}Z_{0e}\tan(\beta D)}. \tag{3.27}$$

Now V_A is also given by using Equation (3.20):

$$\frac{V_A}{V_{INAe}} = 1 + \rho_{Ae} = \frac{2(Z_0 + \mathrm{j}Z_{0e}\tan(\beta D))}{2Z_0 + \mathrm{j}(Z_{0e} + Z_{0o})\tan(\beta D)}. \tag{3.28}$$

Finally, multiplying Equations (3.27) and (3.28) gives the transmission coefficient for the even mode:

$$\frac{V_A}{V_{INAe}} \times \frac{V_B}{V_A} = \tau_e = \frac{2Z_0 \sec(\beta D)}{2Z_0 + \mathrm{j}(Z_{0e} + Z_{0o})\tan(\beta D)} = \tau_o. \tag{3.29}$$

Repeating the analysis for the odd mode yields the same equation. Now the output on line 2 is $\frac{1}{2}(\tau_e + \tau_o) = \tau_e$ and for line 4 is $\frac{1}{2}(\tau_e - \tau_o) = 0$, so the output power on line 2 is given by

$$|\tau_e|^2 = \frac{4Z_{0e}Z_{0o}\sec^2(\beta D)}{4Z_{0e}Z_{0o} + (Z_{0e} + Z_{0o})^2\tan^2(\beta D)}. \tag{3.30}$$

Putting in the numbers gives

$$|\tau_e|^2 = \frac{15\sec^2(\beta D)}{15 + 16\tan^2(\beta D)} = \frac{15 + 15\tan^2(\beta D)}{15 + 16\tan^2(\beta D)}. \tag{3.31}$$

By considering Equations (3.24) and (3.31), it can be seen that

$$|\rho_e|^2 + |\tau_e|^2 = 1. \tag{3.32}$$

The scattering matrix for this coupler can now be constructed as follows:

$$\begin{bmatrix} b_1 \\ b_2 \\ b_3 \\ b_4 \end{bmatrix} = \begin{bmatrix} 0 & \tau_e & \rho_e & 0 \\ \tau_e & 0 & 0 & \rho_e \\ \rho_e & 0 & 0 & \tau_e \\ 0 & \rho_e & \tau_e & 0 \end{bmatrix} \begin{bmatrix} a_1 \\ a_2 \\ a_3 \\ a_4 \end{bmatrix}. \tag{3.33}$$

It is also worth noting that the phases of ρ_e and τ_e differ by $\pi/2$ independently of frequency. In Figure 3.12 are shown the frequency responses given in Equations (3.24) and (3.31). Because this directional coupler appears to send out its coupled power backwards, it is often called a *backward-wave* coupler. It is a common circuit at microwave frequencies as it can be easily manufactured. However, at optical frequencies a *forward-coupler* is often used and this will be described in the next example. It is also worth noting that the zeros in the scattering matrix given in Equation (3.33) are a direct result of the assumption that the two modes have the same velocities. Over a quarter of a wavelength this is normally a reasonable assumption. In the forward-wave coupler, however, the distances are much longer and the difference in the velocities is crucial to the operation.

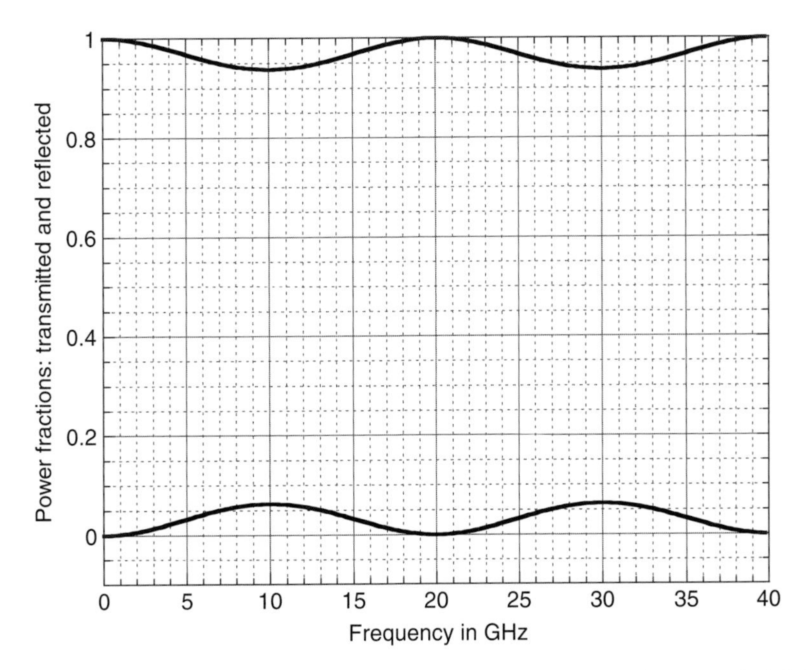

Figure 3.12 The frequency responses of the directional coupler in Example 3.7. The upper trace shows the fraction of transmitted power $|\tau_e|^2$ and the lower trace the fraction of reflected power $|\rho_e|^2$. The coupler is designed to operate at 10 GHz.

Example 3.8 A directional coupler using forward waves: Design a forward-wave directional coupler using two weakly coupled transmission lines. Assume the characteristic impedances of the odd and even modes are approximately equal to the characteristic impedances of the individual lines. The even mode velocity is $2.10^8\,\mathrm{ms}^{-1}$ and the odd mode velocity is $2.03.10^8\,\mathrm{ms}^{-1}$ in the coupled region. Find the length of the coupler which will just couple 3 dB of the wave across from one line to another. Assume the frequency is 10^{14} Hz.

Solution to Example 3.8

The solution to this problem is similar to the other problems in that it is assumed that odd and even modes propagate in the coupled region. So if a wave of amplitude V_0 is introduced into line 1, this immediately splits into two modes each of amplitude $\frac{1}{2}V_0$. Now, as the propagation velocities are different, it is possible to write the waves on the two lines as follows:

$$\text{on line 1,}\quad V_1 = \tfrac{1}{2}V_0 \exp(-\mathrm{j}\beta_e D) + \tfrac{1}{2}V_0 \exp(-\mathrm{j}\beta_o D),$$

$$\text{on line 2,}\quad V_2 = \tfrac{1}{2}V_0 \exp(-\mathrm{j}\beta_e D) - \tfrac{1}{2}V_0 \exp(-\mathrm{j}\beta_o D), \qquad (3.34)$$

Now these equations can be recast by using two different phase constants:

$$\beta_a = \tfrac{1}{2}(\beta_e + \beta_o) \text{ and } \beta_b = \tfrac{1}{2}(\beta_e - \beta_o). \tag{3.35}$$

This gives

$$\text{on line 1,} \quad V_1 = \tfrac{1}{2}V_0 \exp(-j(\beta_a + \beta_b)D) + \tfrac{1}{2}V_0 \exp(-j(\beta_a - \beta_b)D),$$

$$\text{on line 2,} \quad V_2 = \tfrac{1}{2}V_0 \exp(-j(\beta_a + \beta_b)D) - \tfrac{1}{2}V_0 \exp(-j(\beta_a - \beta_b)D),$$

Taking out the common factors and expanding the exponentials gives

$$\text{on line 1,} \quad V_1 = V_0 \exp(-j\beta_a D)\cos(\beta_b D),$$

$$\text{on line 2,} \quad V_2 = jV_0 \exp(-j\beta_a D)\sin(\beta_b D). \tag{3.36}$$

So the voltage on line 1 falls to zero when $\beta_b D = \pi/2$, and the voltage on line 2 reaches a peak. The pattern repeats with the energy wandering from one line to another, just like it does between a pair of coupled pendulums. This is the basic mechanism for 'cross-talk' between adjacent lines in telephony. So for a 3 dB coupling the length is given by

$$\beta_b D = \frac{\pi}{4} \text{ so } \tfrac{1}{2}(\beta_e - \beta_o)D = \frac{\pi}{4} \text{ and } D = \frac{1}{4f\left(\dfrac{1}{v_e} - \dfrac{1}{v_o}\right)}, \tag{3.37}$$

which gives $D = 33.8\,\mu\text{m}$. If this had been at microwave frequencies, say 10 GHz, then $D = 33.8$ cm. Clearly far too long for this 'hand-held' age!

Example 3.9 A band-stop filter: Find the scattering matrix and the frequency characteristic for the circuit shown in Figure 3.13. For certain values of the coupling coefficients, comment on possible applications.

Solution to Example 3.9

The solution involves the scattering matrix given in Equation (3.33):

$$\begin{bmatrix} b_1 \\ b_2 \\ b_3 \\ b_4 \end{bmatrix} = \begin{bmatrix} 0 & \tau & \rho & 0 \\ \tau & 0 & 0 & \rho \\ \rho & 0 & 0 & \tau \\ 0 & \rho & \tau & 0 \end{bmatrix} \begin{bmatrix} a_1 \\ a_2 \\ a_3 \\ a_4 \end{bmatrix}, \tag{3.38}$$

where the suffix e has been dropped for clarity.

First assume an input at port 1 and a matched load at port 2. So the boundary conditions are as follows:

$a_2 = 0$, $b_3 = -a_3$ (short circuit at port 3) and $b_4 = a_4$ (open circuit at port 4).

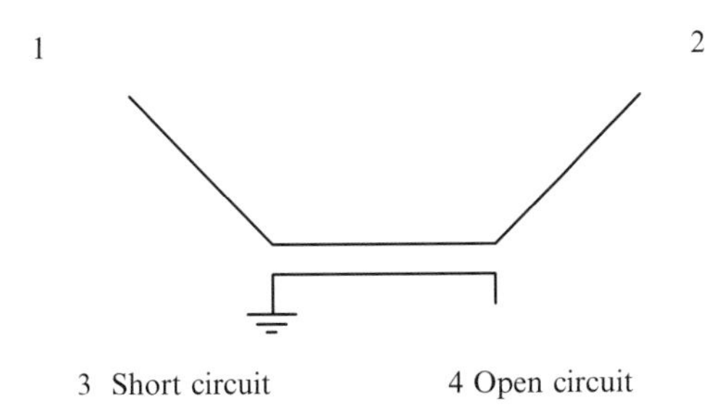

Figure 3.13 The circuit for Example 3.9 – a band-stop filter.

Using these conditions and the equations in (3.38) gives

$$b_1 = -\frac{\rho^2}{1+\tau^2}\,a_1 \text{ and } b_2 = \frac{\tau\left(1+\tau^2-\rho^2\right)}{1+\tau^2}\,a_1.$$

Then, by assuming the input is at 2 and a matched load is at 1, the equations become

$$b_1 = \frac{\tau\left(1+\tau^2-\rho^2\right)}{1+\tau^2}\,a_2 \text{ and } b_2 = \frac{\rho^2}{1+\tau^2}\,a_2.$$

So the scattering matrix for the circuit is

$$\begin{bmatrix} b_1 \\ b_2 \end{bmatrix} = \frac{1}{1+\tau^2}\begin{bmatrix} -\rho^2 & \tau\left(1+\tau^2-\rho^2\right) \\ \tau\left(1+\tau^2-\rho^2\right) & \rho^2 \end{bmatrix}\begin{bmatrix} a_1 \\ a_2 \end{bmatrix}. \tag{3.39}$$

For the centre frequency, when the length of the coupled region is equal to a quarter of a wavelength (or three quarters, etc.), using Equations (3.20) and (3.29):

$$\rho = k = \frac{Z_{0e}-Z_{0o}}{Z_{0e}+Z_{0o}} \text{ and } \tau = -\frac{2\mathrm{j}\sqrt{Z_{0e}Z_{0o}}}{Z_{0e}+Z_{0o}}.$$

At this frequency, $1+\tau^2 = \rho^2$ and so the matrix given in Equation (3.39) becomes

$$\begin{bmatrix} b_1 \\ b_2 \end{bmatrix} = \begin{bmatrix} -1 & 0 \\ 0 & 1 \end{bmatrix}\begin{bmatrix} a_1 \\ a_2 \end{bmatrix}, \tag{3.40}$$

which is a band-stop filter, that is nothing is transmitted from ports 1 to 2 at the centre frequency. At double the centre frequency, $\rho = 0$ and $\tau = -1$, so the matrix becomes

$$\begin{bmatrix} b_1 \\ b_2 \end{bmatrix} = \begin{bmatrix} 0 & -1 \\ -1 & 0 \end{bmatrix}\begin{bmatrix} a_1 \\ a_2 \end{bmatrix}, \tag{3.41}$$

which allows total transmission between the ports, with a delay of π as expected.

The physical explanation for this response is that the line joining ports 3 and 4 becomes a resonant structure at the centre frequency and the build up of energy in that line prevents the transmission of energy in the other line. This build up of energy depends on the coupling to the circuit. In Figures 3.14 and 3.15 are shown the frequency responses of this circuit for two separate coupling coefficients.

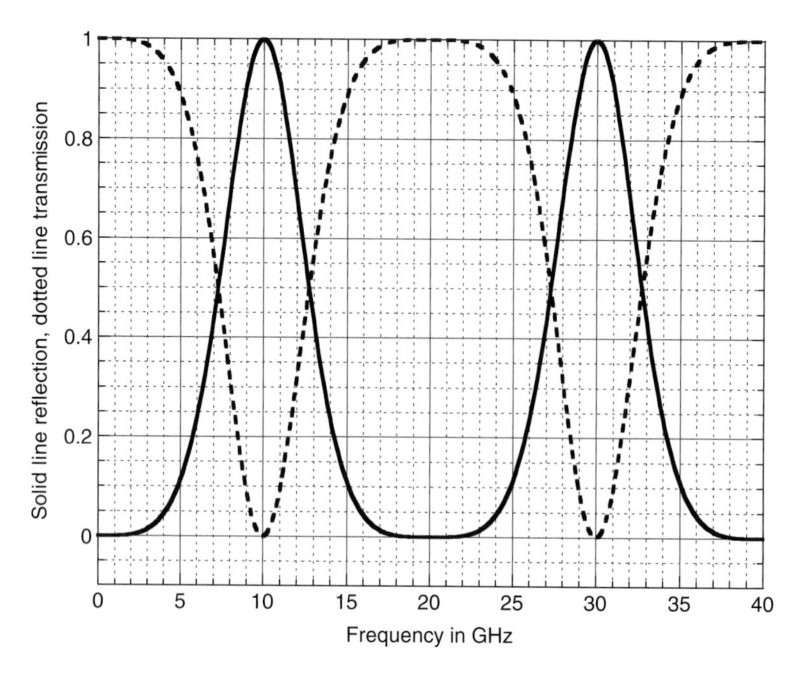

Figure 3.14 The fraction of reflected (solid line) and transmitted power (dotted line) in a band-stop filter with $k = \frac{1}{\sqrt{2}}$. The centre frequency is 10 GHz and the 'bandwidth' is 5.4 GHz.

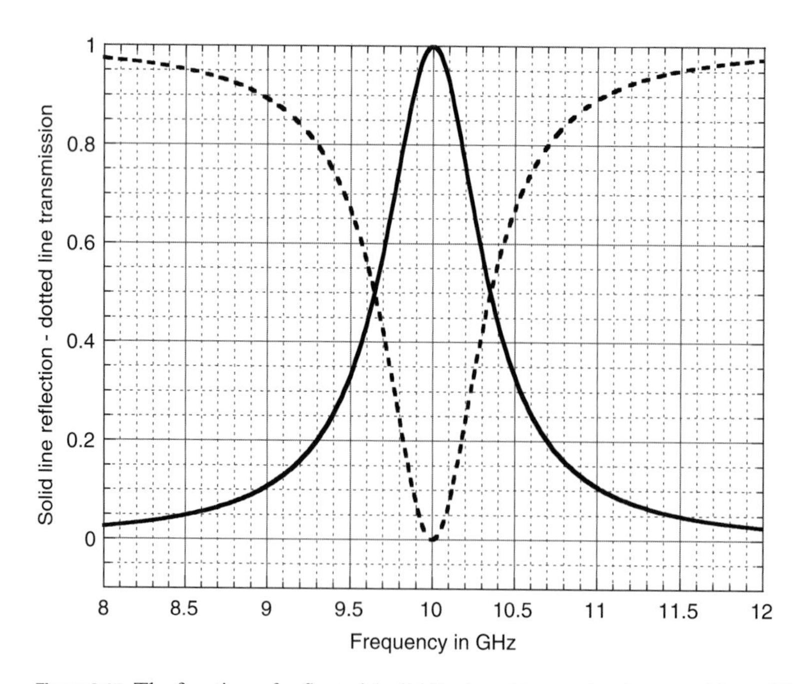

Figure 3.15 The fraction of reflected (solid line) and transmitted power (dotted line) in a band-stop filter with $k = \frac{1}{\sqrt{10}}$. The centre frequency is 10 GHz and the 'bandwidth' is 0.7 GHz.

The reflection is the fraction of the incident power reflected and similarly the transmission is the fraction of the incident power transmitted. The bandwidth for a coupling of $\frac{1}{\sqrt{2}}$ is 5.4 GHz, whereas the bandwidth for the smaller coupling of $\frac{1}{\sqrt{10}}$ is only 0.7 GHz. The amplitude at the open circuit at port 4 at the centre frequency is

$$\frac{2\tau a_1}{\rho} \text{ for input at port 1 and } \frac{2a_2}{\rho} \text{ for input at port 2.}$$

So the build up of voltage at port 2 is as high as $2\sqrt{2}a_2$ or $2\sqrt{10}a_2$ depending on the coupling. For very weak coupling there will be an even bigger build up of energy and a narrower stop band. This is characteristic of a parallel resonant circuit with an increasing 'Q' factor. It is worthy of note that a very weakly coupled resonant circuit can produce a narrow stop band in an adjacent circuit and that the energy in the resonant circuit may have an amplitude much higher than in the adjacent circuit. Since this build up of energy will take many times the transit time of the quarter-wavelength of the resonant circuit, this band-stop filter will only operate after this has happened. So an initial sine wave will partially pass through this filter until the energy has built up sufficiently in the resonant circuit to prevent any further transmission. In other words, the time domain response of this filter is dependent on the inverse of the bandwidth.

Example 3.10 A band-pass filter: Find the scattering matrix and the frequency response for the circuit shown in Figure 3.16. For certain values of the coupling coefficient, discuss the applications.

Solution to Example 3.10

Assume the input is at port 1 and that there is a matched load at port 4. Then the following boundary conditions apply:

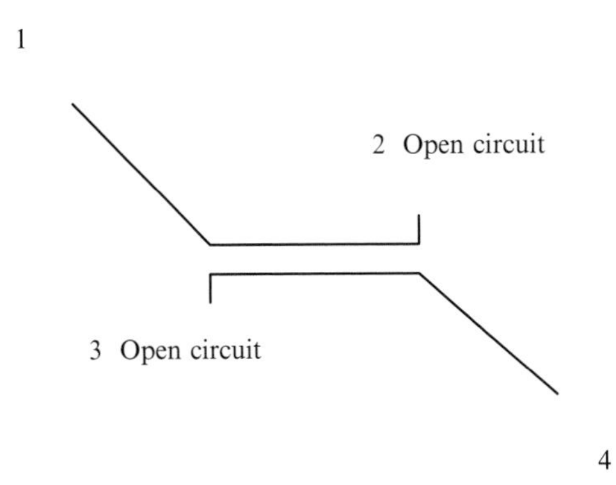

Figure 3.16 The circuit for Example 3.10.

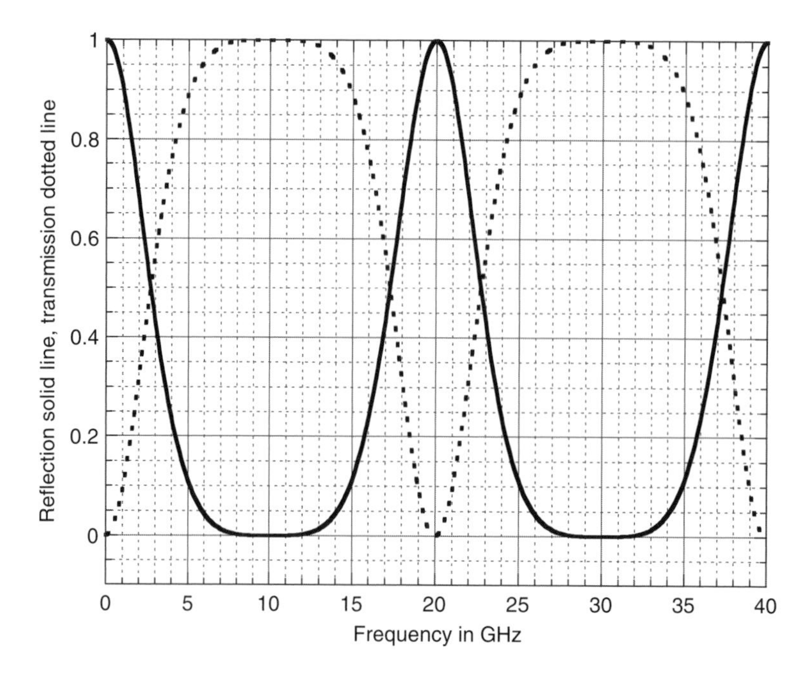

Figure 3.17 The frequency response of a band-pass filter. The solid lines are the fraction of the incident power which is reflected and the dotted lines are the same but for the transmitted power. Around the centre frequency of 10 GHz, this circuit has excellent transmission and over a bandwidth of nearly 14 GHz. The main use of the circuit is as a 'DC break'.

$b_2 = a_2$ (open circuit at port 2), $b_3 = a_3$ (open circuit at port 3) and $a_4 = 0$.

Using these equations in (3.38) gives

$$b_1 = \left(\rho^2 + \tau^2\right) a_1 \text{ and } b_4 = 2\rho\tau.$$

From the symmetry of the circuit, the scattering matrix becomes

$$\begin{bmatrix} b_1 \\ b_4 \end{bmatrix} = \begin{bmatrix} \rho^2 + \tau^2 & 2\rho\tau \\ 2\rho\tau & \rho^2 + \tau^2 \end{bmatrix} \begin{bmatrix} a_1 \\ a_4 \end{bmatrix}. \tag{3.42}$$

Now, at the centre frequency, $1 + \tau^2 = \rho^2$ and so $\rho^2 + \tau^2 = 2\rho^2 - 1$. So for just one value of $\rho = \frac{1}{\sqrt{2}}$ there will be no reflection and the scattering matrix becomes

$$\begin{bmatrix} b_1 \\ b_4 \end{bmatrix} = \begin{bmatrix} 0 & -\mathrm{j} \\ -\mathrm{j} & 0 \end{bmatrix} \begin{bmatrix} a_1 \\ a_4 \end{bmatrix}. \tag{3.43}$$

So this circuit allows total transmission at the centre frequency, and at double the centre frequency it becomes a band-stop with a scattering matrix given by

$$\begin{bmatrix} b_1 \\ b_4 \end{bmatrix} = \begin{bmatrix} 1 & 0 \\ 0 & 1 \end{bmatrix} \begin{bmatrix} a_1 \\ a_4 \end{bmatrix}. \tag{3.44}$$

This is particularly important at 0 Hz where the obvious break in the circuit means that no direct currents can pass through. The frequency response is shown in Figure 3.17.

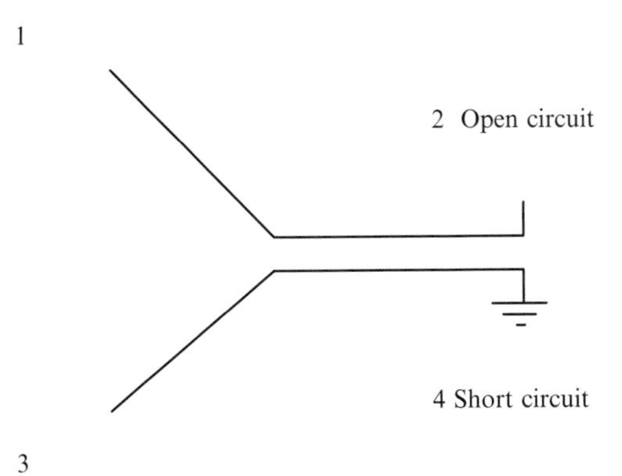

Figure 3.18 The circuit for Example 3.11.

Example 3.11 A transformer: Find the scattering matrix and frequency response for the circuit shown in Figure 3.18.

For certain values of the coupling coefficient, discuss the applications.

Solution to Example 3.11

First assume an input at port 1 and a matched load at port 3. The boundary conditions become

$b_2 = a_2$ (open circuit at port 2), $b_4 = -a_4$ (short circuit at port 4) and $a_3 = 0$

Using these equations in (3.38) gives

$$b_1 = \frac{\tau^2 a_1}{1+\rho^2} \text{ and } b_3 = \frac{\rho(1+\rho^2-\tau^2)a_1}{1+\rho^2}.$$

Now repeating the process with an input at port 3 and a matched load at port 1 gives

$$b_1 = \frac{\rho(1+\rho^2-\tau^2)a_3}{1+\rho^2} \text{ and } b_3 = -\frac{\tau^2 a_3}{1+\rho^2}.$$

These can be combined to give the scattering matrix as

$$\begin{bmatrix} b_1 \\ b_3 \end{bmatrix} = \frac{1}{1+\rho^2}\begin{bmatrix} \tau^2 & \rho(1+\rho^2-\tau^2) \\ \rho(1+\rho^2-\tau^2) & -\tau^2 \end{bmatrix}\begin{bmatrix} a_1 \\ a_3 \end{bmatrix}. \tag{3.45}$$

Now at the centre frequency, when $1 + \tau^2 = \rho^2$, this matrix becomes

$$\begin{bmatrix} b_1 \\ b_3 \end{bmatrix} = \frac{1}{1+\rho^2}\begin{bmatrix} \rho^2-1 & 2\rho \\ 2\rho & 1-\rho^2 \end{bmatrix}\begin{bmatrix} a_1 \\ a_3 \end{bmatrix}, \tag{3.46}$$

whereas at double the centre frequency the matrix becomes

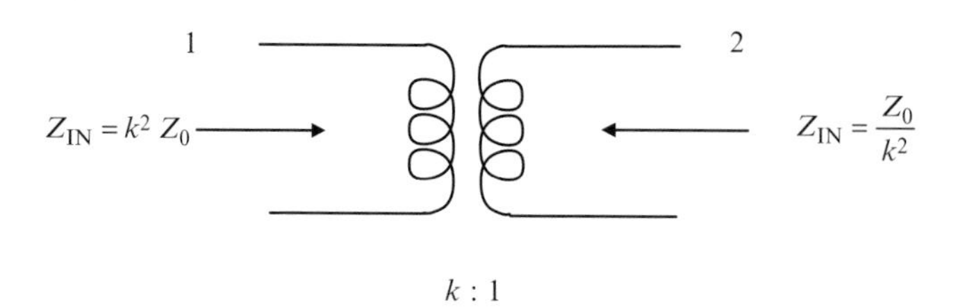

Figure 3.19 The transformer in Example 3.11.

$$\begin{bmatrix} b_1 \\ b_3 \end{bmatrix} = \begin{bmatrix} -1 & 0 \\ 0 & 1 \end{bmatrix} \begin{bmatrix} a_1 \\ a_3 \end{bmatrix}. \tag{3.47}$$

The significance of this circuit is that at the centre frequency it acts like a transformer with a turns ratio k:1, as shown in Figure 3.19.

Using the usual impedance transformation for transformers, the input impedance at port 1 is k^2Z_0 and so the reflection coefficient is given by

$$\rho_{11} = \frac{k^2Z_0 - Z_0}{k^2Z_0 + Z_0} = \frac{k^2 - 1}{k^2 + 1} = \frac{\rho^2 - 1}{\rho^2 + 1}$$

and the transmission coefficient is given by

$$\tau_{21} = (1 + \rho_{11}) \times \frac{1}{k} = \frac{2k}{1 + k^2} = \frac{2\rho}{1 + \rho^2}.$$

Moving to port 3, the input impedance is now Z_0/k^2 and so

$$\rho_{22} = \frac{\dfrac{Z_0}{k^2} - Z_0}{\dfrac{Z_0}{k^2} + Z_0} = \frac{1 - k^2}{1 + k^2} = \frac{1 - \rho^2}{1 + \rho^2}$$

$$\text{and } \tau_{12} = (1 + \rho_{22}) \times k = \frac{2k}{1 + k^2} = \frac{2\rho}{1 + \rho^2}.$$

So the circuit can be used to change impedances by choosing the appropriate value of the coupling coefficient. The frequency response is shown in Figure 3.20 for a value of $k = 1/2$. At the centre frequency the power ratios are 0.36 for the reflection and 0.64 for the transmission and these are only valid over a limited frequency range.

Example 3.12 The Wilkinson Power Divider: Use even and odd mode analysis to find the scattering matrices and the frequency response of the circuit shown in Figure 3.21. The first circuit is called the Compensated Wilkinson Power Divider and the second the hybrid or mixer ring. The Wilkinson Divider, as its name suggests, is used to divide power entering port 1 into ports 2 and 3. In its compensated form it also isolates the ports 2 and 3.

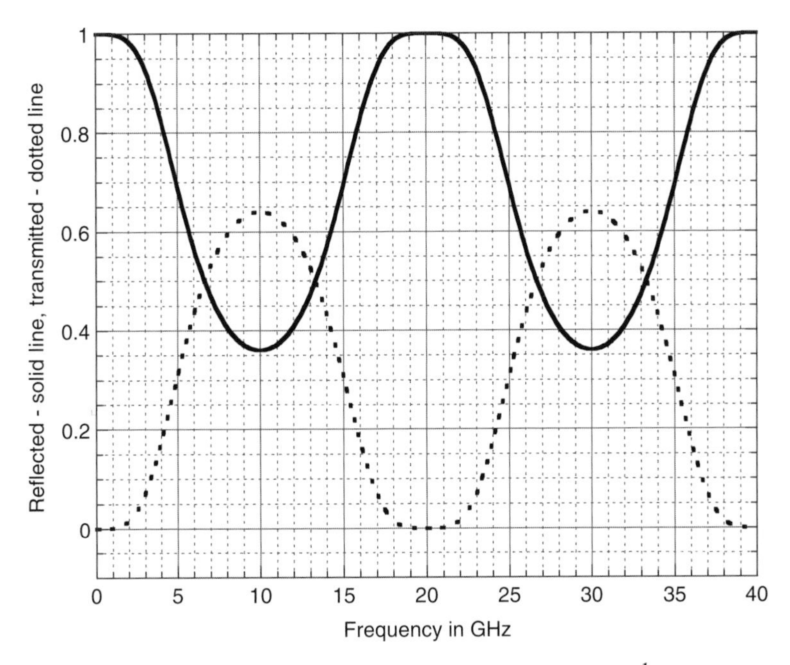

Figure 3.20 The frequency response for a transformer with $k = \frac{1}{2}$. The solid line shows the fraction of the incident power which is reflected and the dotted line shows the ratio for transmission.

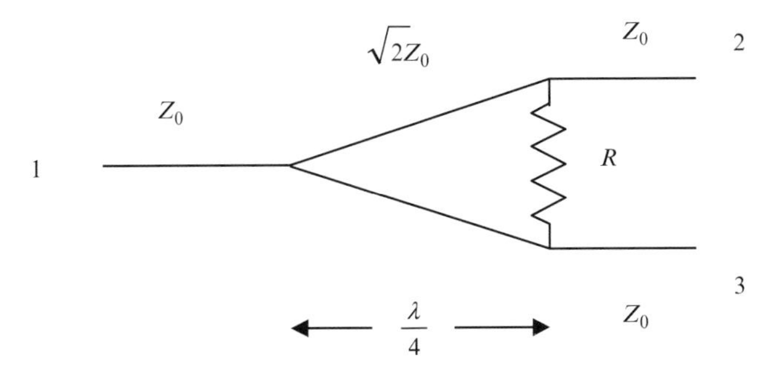

Figure 3.21 The circuit diagram for Example 3.12, the Compensated Wilkinson Power Divider (or Combiner).

Solution to Example 3.12

Odd and even modes are used to analyse this circuit and this analysis begins with the even mode. Bisecting the circuit horizontally, as shown in Figure 3.22, the transmission line at port 1 effectively doubles its characteristic impedance. The resistor, R, is omitted from the equivalent circuit as it will not affect the even mode. This is because there will be no voltage difference across its terminals. The quarter-wave sections are assumed to be uncoupled to each other in this analysis.

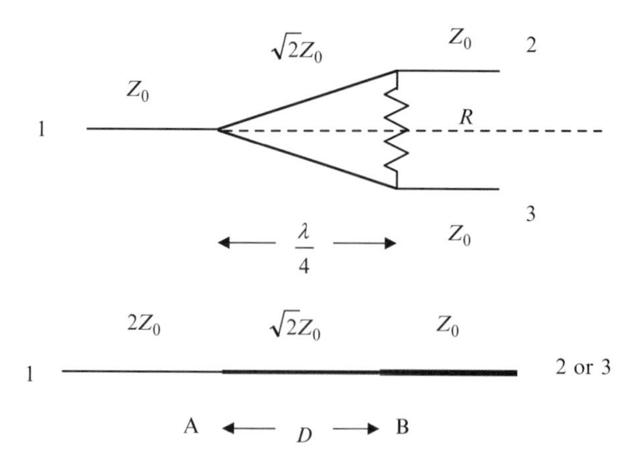

Figure 3.22 The circuit of Figure 3.21 cut along its horizontal axis of symmetry for the even mode. Below is the equivalent circuit of either the top or the bottom of this circuit.

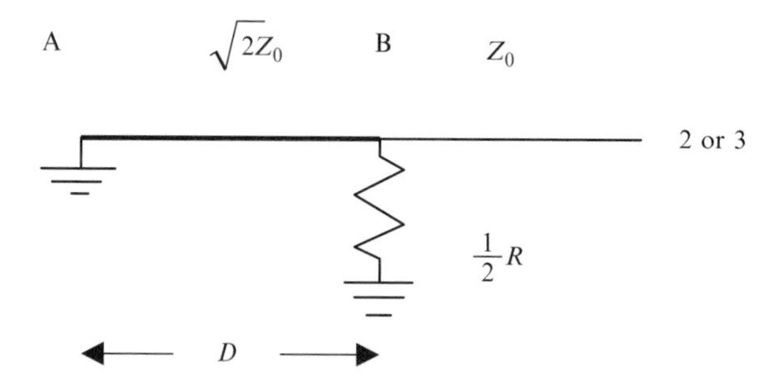

Figure 3.23 The circuit of Figure 3.21 cut along its horizontal axis of symmetry for the odd mode.

Using a similar analysis as in Example 3.7, for an even mode input at A, the reflection and transmission coefficients are

$$\rho_{eA} = \frac{-1}{3 + j2\sqrt{2}\tan(\beta D)} \text{ and } \tau_{eA\to B} = \frac{2\sec(\beta D)}{3 + j2\sqrt{2}\tan(\beta D)}. \tag{3.48}$$

For an input at either port 2 or 3:

$$\rho_{eB} = \frac{1}{3 + j2\sqrt{2}\tan(\beta D)} \text{ and } \tau_{eB\to A} = \frac{4\sec(\beta D)}{3 + j2\sqrt{2}\tan(\beta D)}. \tag{3.49}$$

Clearly the odd mode cannot go beyond point A where it 'sees' a short circuit. Figure 3.23 shows the circuit again cut along its axis of symmetry but this time for the odd mode. Half the resistor is now included as there will be a zero of voltage at the middle of the resistor. Again using the usual analysis, the reflection coefficient is

$$\rho_{\mathrm{oB}} = \frac{\mathrm{j}\sqrt{2}(R - 2Z_0)\tan(\beta D) - R}{\mathrm{j}\sqrt{2}(R + 2Z_0)\tan(\beta D) + R}. \tag{3.50}$$

Now the purpose of the resistor R is to reduce the coupling between ports 2 and 3, i.e. to compensate for this coupling. Clearly, if $R = 2Z_0$ this will be achieved at the centre frequency. Applying this condition, which is used on all Compensated Wilkinson Dividers, gives

$$\rho_{\mathrm{oB}} = \frac{-1}{1 + 2\sqrt{2}\tan(\beta D)}. \tag{3.51}$$

Now the scattering matrix can be constructed as follows:

$$\begin{bmatrix} b_1 \\ b_2 \\ b_3 \end{bmatrix} = \begin{bmatrix} \rho_{\mathrm{eA}} & \frac{1}{2}\tau_{\mathrm{eB}\to\mathrm{A}} & \frac{1}{2}\tau_{\mathrm{eB}\to\mathrm{A}} \\ \tau_{\mathrm{eA}\to\mathrm{B}} & \frac{1}{2}(\rho_{\mathrm{eB}} + \rho_{\mathrm{oB}}) & \frac{1}{2}(\rho_{\mathrm{eB}} - \rho_{\mathrm{oB}}) \\ \tau_{\mathrm{eA}\to\mathrm{B}} & \frac{1}{2}(\rho_{\mathrm{eB}} - \rho_{\mathrm{oB}}) & \frac{1}{2}(\rho_{\mathrm{eB}} + \rho_{\mathrm{oB}}) \end{bmatrix} \begin{bmatrix} a_1 \\ a_2 \\ a_3 \end{bmatrix}. \tag{3.52}$$

The input at port 2 has been resolved into an even and an odd mode in phase and both of an amplitude 1/2. At the centre frequency the matrix becomes

$$\begin{bmatrix} b_1 \\ b_2 \\ b_3 \end{bmatrix} = -\frac{\mathrm{j}}{\sqrt{2}} \begin{bmatrix} 0 & 1 & 1 \\ 1 & 0 & 0 \\ 1 & 0 & 0 \end{bmatrix} \begin{bmatrix} a_1 \\ a_2 \\ a_3 \end{bmatrix}, \tag{3.53}$$

which shows it is a perfect power combiner or divider. However, at double the centre frequency, the worst case, the matrix becomes

$$\begin{bmatrix} b_1 \\ b_2 \\ b_3 \end{bmatrix} = \frac{1}{3} \begin{bmatrix} -1 & -2 & -2 \\ -2 & -1 & -2 \\ -2 & -2 & -1 \end{bmatrix} \begin{bmatrix} a_1 \\ a_2 \\ a_3 \end{bmatrix}, \tag{3.54}$$

which is similar to the matrix for the tee section in Example 2.6 circuit (a). The extra minus signs appear because of the half wavelength between A and B at this frequency. For the frequency response, the following are displayed in Figure 3.24:

$$\begin{aligned} A &= |S_{11}|^2, \\ B &= |S_{22}|^2 = |S_{33}|^2, \\ C &= |S_{12}|^2 = |S_{13}|^2 = |S_{21}|^2 = |S_{31}|^2, \\ D &= |S_{23}|^2 = |S_{32}|^2. \end{aligned}$$

These four cover all the amplitude characteristics of the scattering matrix given in Equation (3.52). The power division, represented by $|S_{12}|^2$, is nearly 50 % over most of the bandwidth, dropping to 44.4 % at double the centre frequency of 10 GHz.

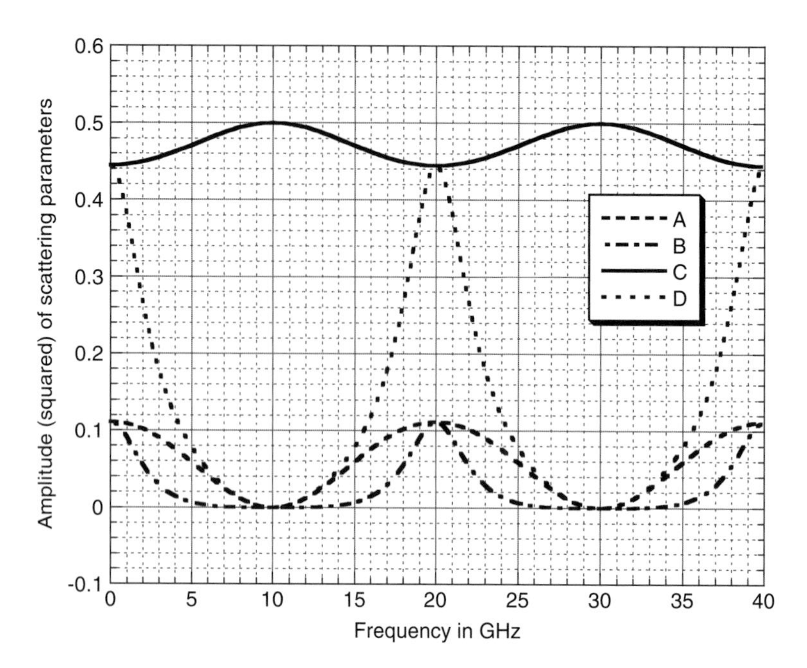

Figure 3.24 The frequency response of some of the scattering parameters for the Compensated Wilkinson Power Divider given in Equation (3.52). The square of the modulus of the elements is plotted against frequency. The centre frequency is 10 GHz.

Example 3.13 The hybrid or mixer ring: The hybrid or mixer ring shown in Figure 3.25 is commonly used at the front end of many microwave receivers, for example in mobile phones. It is designed to add an incoming signal at port 1 to the local oscillator signal at port 4, prior to their detection by mixer diodes at ports 2 and 3. At the centre frequency, both the ports 1 and 4 and the ports 2 and 3 are isolated from each other. These circuits are commonly used in printed circuits like microstrip and stripline. Using even and odd mode analysis, find the scattering matrices and the frequency response of the circuit.

Solution to Example 3.13

This circuit can be cut horizontally in a similar fashion to the circuit of Figure 3.21, as shown in Figure 3.26. Even and odd modes can be applied to ports 1 and 4 as well as ports 2 and 3. When an even mode is applied, there will be no current across the 'cut' so at that point it will be equivalent to an open circuit. On the other hand, when the odd mode is applied there will be current but no voltage at the 'cut'. So this point will be equivalent to a short circuit. Starting with the even mode, only the top part of Figure 3.26 needs to be analysed as the bottom part will be identical. In Figure 3.27 just this top half is shown.

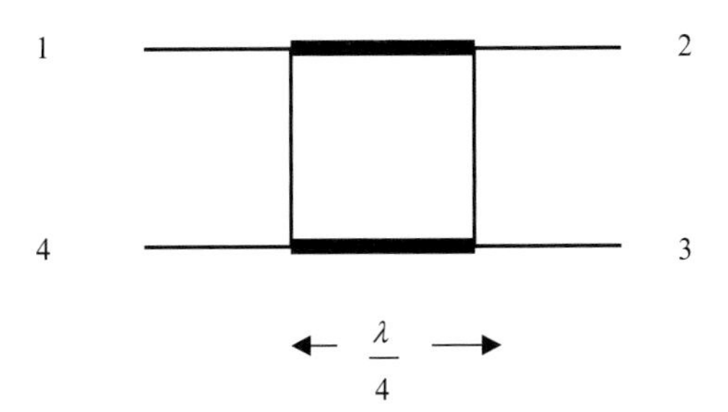

Figure 3.25 The hybrid or mixer ring. The thicker lines represent transmission lines with a characteristic impedance of $Z_0/\sqrt{2}$.

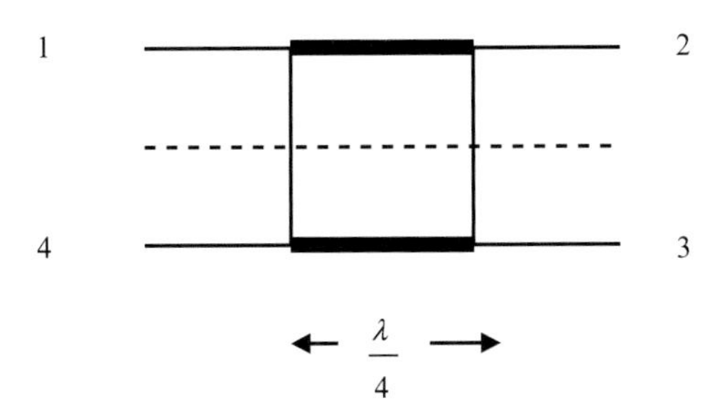

Figure 3.26 The hybrid or mixer ring. The dotted line shows the cut made to analyse the circuit.

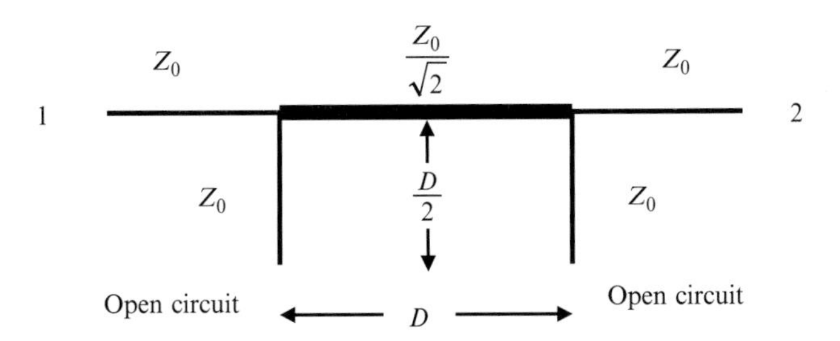

Figure 3.27 The top half of the mixer ring when even modes are applied.

Now the analysis is the same as in previous examples. By assuming the input is at port 1 in Figure 3.27 and port 2 is terminated with a matched termination, it is possible to find the input impedance at port 1. The input impedance of each of the side arms is given by

$$-\mathrm{j}Z_0 \cot\left(\tfrac{1}{2}\beta D\right)$$

and the input impedance at port 1 is given by

$$Z_{\mathrm{INe}} = \frac{Z_0\left(T + \mathrm{j}\left(tT - \sqrt{2}\right)\right)}{\left(T(2 - t^2) + 2\sqrt{2}t\right) + \mathrm{j}\left(tT - \sqrt{2}\right)}, \tag{3.55}$$

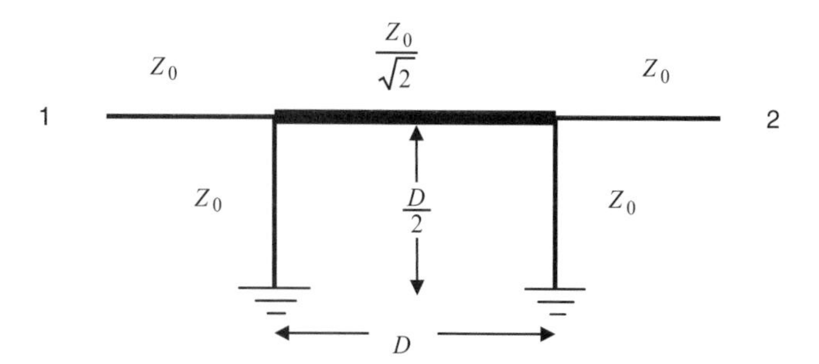

Figure 3.28 The top section of the mixer ring when odd modes are applied.

where $T = \tan(\beta D)$ and $t = \tan(\frac{1}{2}\beta D)$. From this can be derived the even mode reflection coefficient ρ_e:

$$\rho_e = \frac{T(t^2 - 1) - 2\sqrt{2}t}{\left(T(3 - t^2) + 2\sqrt{2}t\right) + 2j\left(tT - \sqrt{2}\right)}, \tag{3.56}$$

and modifying Equation (3.27) to include the side arm, the transmission coefficient can be found as

$$\tau_e = \frac{-2j\sqrt{2}\sec(\beta D)}{\left(T(3 - t^2) + 2\sqrt{2}t\right) + 2j\left(tT - \sqrt{2}\right)}. \tag{3.57}$$

For the odd mode, the top section is given in Figure 3.28. The side arms now have an impedance given by

$$jZ_0 \tan\left(\frac{1}{2}\beta D\right) = jZ_0 t.$$

The input impedance at port 1 with port 2 matched is

$$Z_{INo} = \frac{Z_0\left(\left(\sqrt{2}t^2 + tT\right) + jt^2T\right)}{\left(\sqrt{2}t^2 + tT\right) - j\left(2\sqrt{2}t + T(1 - 2t^2)\right)}. \tag{3.58}$$

As before, the reflection coefficient can now be found as

$$\rho_o = \frac{T\left(1 - t^2\right) + 2\sqrt{2}t}{\left(T(3t^2 - 1) - 2\sqrt{2}t\right) - 2jt\left(\sqrt{2}t + T\right)}. \tag{3.59}$$

The transmission coefficient becomes

$$\tau_o = \frac{-2\sqrt{2}jt^2\sec(\beta D)}{\left(T(3t^2 - 1) - 2\sqrt{2}t\right) - 2jt\left(\sqrt{2}t + T\right)}. \tag{3.60}$$

At the centre frequency, $t = 1$ and $T \to \infty$, so the coefficients become

$$\rho_e = 0 \text{ as } Z_{INe} = Z_0 \text{ and similarly } \rho_o = 0 \text{ as } Z_{INo} = Z_0,$$

and the transmission coefficients are given by

$$\tau_e = \frac{-\sqrt{2}\mathrm{j}}{1+\mathrm{j}} \text{ and } \tau_o = \frac{-\sqrt{2}\mathrm{j}}{1-\mathrm{j}}.$$

Combining all these results into a scattering matrix gives

$$\begin{bmatrix} b_1 \\ b_2 \\ b_3 \\ b_4 \end{bmatrix} = \frac{1}{2} \begin{bmatrix} \rho_e + \rho_o & \tau_e + \tau_o & \tau_e - \tau_o & \rho_e - \rho_o \\ \tau_e + \tau_o & \rho_e + \rho_o & \rho_e - \rho_o & \tau_e - \tau_o \\ \tau_e - \tau_o & \rho_e - \rho_o & \rho_e + \rho_o & \tau_e + \tau_o \\ \rho_e - \rho_o & \tau_e - \tau_o & \tau_e + \tau_o & \rho_e + \rho_o \end{bmatrix} \begin{bmatrix} a_1 \\ a_2 \\ a_3 \\ a_4 \end{bmatrix}. \tag{3.61}$$

So the scattering matrix for the centre frequency is

$$\begin{bmatrix} b_1 \\ b_2 \\ b_3 \\ b_4 \end{bmatrix} = -\frac{1}{\sqrt{2}} \begin{bmatrix} 0 & \mathrm{j} & 1 & 0 \\ \mathrm{j} & 0 & 0 & 1 \\ 1 & 0 & 0 & \mathrm{j} \\ 0 & 1 & \mathrm{j} & 0 \end{bmatrix} \begin{bmatrix} a_1 \\ a_2 \\ a_3 \\ a_4 \end{bmatrix}. \tag{3.62}$$

This shows that when a signal is introduced at port 1 it is divided equally between ports 2 and 3 with the appropriate phase delays. If the circuit were used to form a balanced mixer, the local oscillator would be introduced at port 4. Thus both ports 2 and 3 would have both signal and local oscillator and the mixer diodes would be connected there. There is isolation between ports 1 and 4, as well as ports 2 and 3, which are the characteristics of a good balanced mixer.

At double the centre frequency the coefficients become

$$\rho_e = -1,\ \rho_o = 0,\ \tau_e = 0 \text{ and } \tau_o = -1,$$

so the scattering matrix becomes

$$\begin{bmatrix} b_1 \\ b_2 \\ b_3 \\ b_4 \end{bmatrix} = -\frac{1}{2} \begin{bmatrix} 1 & 1 & -1 & 1 \\ 1 & 1 & 1 & -1 \\ -1 & 1 & 1 & 1 \\ 1 & -1 & 1 & 1 \end{bmatrix} \begin{bmatrix} a_1 \\ a_2 \\ a_3 \\ a_4 \end{bmatrix}. \tag{3.63}$$

Finally, the frequency responses will be displayed assuming the centre frequency is 10 GHz, as in the previous examples. In order to cover the response the following are plotted:

$$|S_{11}|^2,\ |S_{12}|^2,\ |S_{13}|^2,\ |S_{14}|^2.$$

The matrix in Equation (3.61) has all its other elements equal to one of the above. The frequency response is shown in Figure 3.29. It can be seen that the bandwidth of the circuit is less than 5 GHz, but that is adequate for most applications. This is the last example in this chapter and it shows in particular how a seemingly simple circuit has a complex characteristic. The only part that can be predicted from simple circuit theory is the solution for 0 Hz where all the parameters have a modulus squared equal to 1/4.

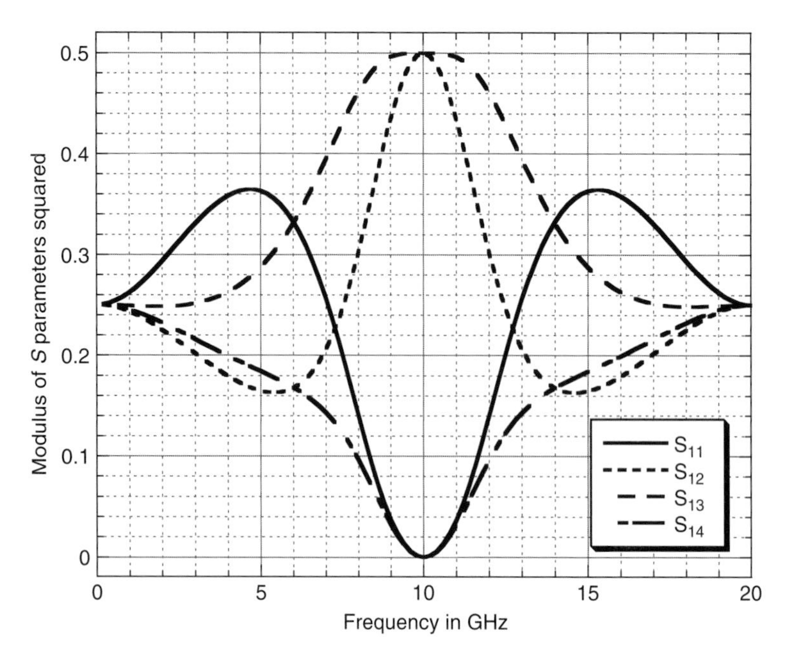

Figure 3.29 The frequency response of the hybrid or mixer ring. The squares of the moduli of the elements of the first row of the scattering matrix are displayed.

3.3 Conclusion

This chapter brings to an end the main discussion about the use of equivalent circuits. These can be used very efficiently to solve many of the problems in transmission lines. Basically they are concerned with propagation in only one direction and this is the most common type of propagation in transmission lines. However, the equivalent circuit method does not give the basic values of parameters such as the capacitance m^{-1} and the inductance m^{-1}. In order to obtain these and an insight into other types of guided wave, the theory of electromagnetism and electromagnetic waves is needed. Hence the discussion in the next three chapters is concerned with these theories applied to transmission lines. The section ends in Chapter 6 with a discussion about attenuation, which has been omitted so far. For many applications the assumption that there is no attenuation is valid, particularly for circuits smaller than a wavelength. However, as the frequency increases well above 100 GHz, this is no longer reasonable for transmission lines incorporating conductors. In Chapter 6, the discussion returns to the use of the equivalent circuit method to analyse the attenuation in two conductor lines before using electromagnetic theory for the attenuation in other types of transmission line.

3.4 Further reading

S. Ramo, J. R. Whinnery and T. Van Duzer *Fields and Waves in Communication Electronics*, Third edition, New York, Wiley, 1993. Chapter 11, section 11.10.

G. Matthaei, L. Young and E. M. T. Jones *Microwave Filters, Impedance-Matching Networks, and Coupling Structures*, New York, Artech House, 1980.

K. F. Sander *Microwave Components and Systems*, New York, Addison-Wesley, 1987. Chapter 5.

K. D. Granzow *Digital Transmission Lines*, Oxford, Oxford University Press, 1998.

C. R. Paul *Transmission Lines in Digital and Analogue Electronic Systems*, New York, Wiley, 2010.

R. E. Collin *Foundations for Microwave Engineering*, New York, McGraw-Hill, 1992. Chapter 6, sections 6.4 to 6.6.

D. M. Pozar *Microwave Engineering*, Second edition, New York, Wiley, 1998. Chapter 7.

A. K. Goel *High-Speed VLSI Interconnections*, New York, Wiley, 1994. Chapter 4.

Part 2

Transmission lines using electromagnetic theory

4 Transmission lines and electromagnetism

In this chapter the equations for the capacitance and inductance of various transmission lines will be derived using the principles of electromagnetism. Much of this material is extensively discussed in many textbooks and for some readers it may seem unnecessary. However, the main reason for including this topic here is to show the links between capacitance and inductance when applied to transmission lines and this chapter will form a precursor to the later chapter on electromagnetic waves. The treatment mainly involves static rather than dynamic fields and its relevance to transmission lines will be discussed at the end of the chapter.

4.1 The capacitance of transmission lines with one dielectric

Capacitance is arguably more basic than inductance since it is linked directly with charge rather than with the movement of charge. In the derivation of capacitance the starting point is the distribution of charge in the capacitor. Using Gauss' Theorem, the total charge, Q (coulombs), is equal to the total electric flux, Φ_e (C), leaving the charge. The next part is to find the electric flux density, D (C m^{-2}), and relate it to the electric field, E (V m^{-1}), using the permittivity, ε (F m^{-1}), as

$$D = \varepsilon E. \tag{4.1}$$

In these examples, ε has a single value. Finally, the voltage across the capacitor can be found by integrating the electric field between the plates of the capacitor:

$$V = \int E.\mathrm{d}x, \tag{4.2}$$

and then the capacitance is given by the definition

$$C = \frac{Q}{V}. \tag{4.3}$$

In some of the examples that follow, the route $Q \rightarrow \Phi_e \rightarrow D \rightarrow E \rightarrow V \rightarrow C$ will be followed and in others the capacitance will be found using conformal mapping.

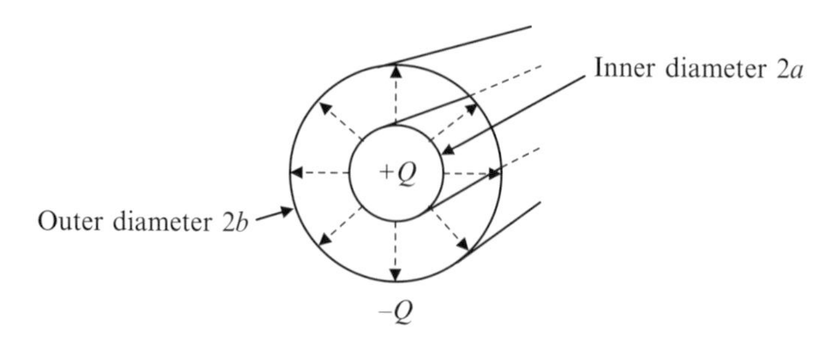

Figure 4.1 The coaxial cable. Both conductors have the same axis, hence the name 'co-axial cable'. The dotted lines show the direction of the electric flux, the electric flux density and the electric field.

Example 4.1 The coaxial cable capacitance: The coaxial cable is shown in Figure 4.1. It is normal to assume that equal and opposite charges are present on the two conductors. Since like charges repel, these are usually distributed uniformly round the surfaces of the conductors. These charges can be introduced from a source which adds positive charges to one conductor and negative charges to the other. In practice, these moveable charges are normally electrons and so a battery would add electrons to one conductor and remove them from another. When one of the conductors is 'earthed', that is it is in good electrical contact with the Earth, it is assumed to have a voltage of zero. The Earth potential acts as a reference for voltage. The Earth is also assumed to have exactly equal numbers of positive and negative charges, indeed all the heavenly bodies have the same assumption to allow for the observed domination by the gravitational force, which is so much weaker than electric force. So if a positive charge was introduced onto one conductor, an equal and opposite charge would flow into the other conductor, if it is earthed.

Starting with a charge $+Q$ on the inner conductor and a charge $-Q$ on the outer conductor, the electric flux will flow between the conductors. Outside the outer conductor there will be no electric flux as the charge enclosed is zero and Gauss' Theorem states that the flux there will also be zero. Similarly, inside the inner conductor, since the charges are only on the surface, there will be no flux. If the radius from the central axis is r and the length of the cable is l, then the electric flux density is given by

$$D = \frac{\Phi_e}{2\pi r l} = \frac{Q}{2\pi r l} \ \text{Cm}^{-2}(\text{radially outwards}) \text{ as } Q = \Phi_e. \tag{4.4}$$

Using Equation (4.1), the electric field is

$$E = \frac{Q}{2\pi r l \varepsilon} \ \text{Vm}^{-1}(\text{radially outwards}) \tag{4.5}$$

and from Equation (4.2), the voltage between the conductors is

$$V = \int_a^b \frac{Q}{2\pi r l \varepsilon} \, \mathrm{d}r = \frac{Q}{2\pi l \varepsilon} \ln\left(\frac{b}{a}\right) \text{ V}. \tag{4.6}$$

Finally, from Definition (4.3):

$$C = \frac{2\pi l\varepsilon}{\ln\left(\frac{b}{a}\right)} \text{ F.} \tag{4.7}$$

So for one metre of cable, i.e. $l = 1$, the distributed capacitance used in transmission line theory is

$$C = \frac{2\pi\varepsilon}{\ln\left(\frac{b}{a}\right)} \text{ Fm}^{-1}. \tag{4.8}$$

The coefficient of Q in Equation (4.6) is called the elastance, s, given by

$$s = \frac{\ln\left(\frac{b}{a}\right)}{2\pi l\varepsilon} = \frac{1}{C} \text{ F}^{-1}. \tag{4.9}$$

The elastance is a useful concept when deriving the equations for mutual capacitance, see Examples 4.8 and 4.9. The inductance of the coaxial cable is derived in Example 4.11. The above equations are derived again in Example 4.4 as an illustration of conformal mapping.

Example 4.2 The parallel plate line capacitance: This line is shown in Figure 4.2. If equal and opposite charges are placed on the plates then the electric flux is able to go between the plates. A simple solution can be found if two assumptions are made: firstly that the distribution of charge is uniform and secondly that the electric flux only goes straight between the plates. These assumptions are valid if the plates are close together and particularly if there is a dielectric between the plates. However, in practice there will be flux going outside the region of the plates and particularly at the edges of the plates there will be a concentration of charge. The exact solution to this problem including fringing fields requires conformal mapping and is given in several of the books listed at the end of the chapter.

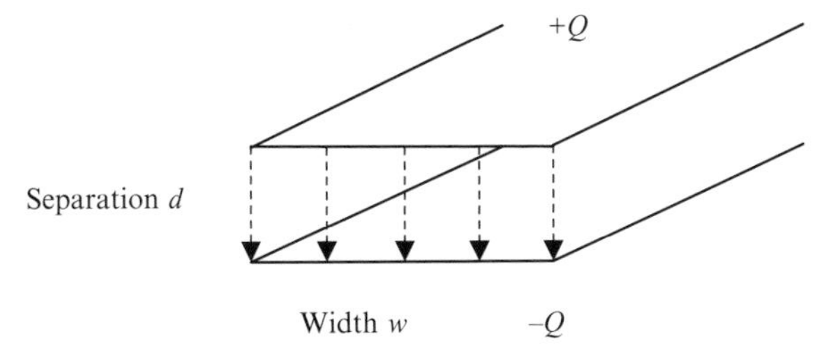

Figure 4.2 The parallel plate line. The dotted lines show the direction of the electric flux, the electric flux density and the electric field.

The purpose of this derivation is to demonstrate the simplicity of the solution under these assumptions.

The electric flux has been constrained to flow only in the direction shown in Figure 4.2. In other words, no fringing fields are being considered outside the width of the plates. If the length of the line is l, then the electric flux density and the electric field are given as

$$D = \frac{\Phi_e}{wl} = \frac{Q}{wl}\ \text{Cm}^{-2} \text{ and } E = \frac{Q}{wl\varepsilon}\ \text{Vm}^{-1}. \tag{4.10}$$

Since the electric field is constant throughout, the voltage is given by

$$V = \int_0^d \frac{Q}{wl\varepsilon}\, \mathrm{d}x = \frac{Qd}{wl\varepsilon}\ \text{V}. \tag{4.11}$$

So the capacitance is given by

$$C = \frac{wl\varepsilon}{d}\ \text{F}. \tag{4.12}$$

If $wl = A$, where A is the area of the plates, then

$$C = \frac{\varepsilon A}{d}\ \text{F}. \tag{4.13}$$

This is the most well-known equation of capacitance, and for many students, the only known equation for capacitance! By taking just one metre as the length, the capacitance per metre becomes

$$C = \frac{\varepsilon w}{d}\,\text{Fm}^{-1}. \tag{4.14}$$

In the practical case, the fringe fields add to this value and when $w = d$ this can be over twice this capacitance. The inductance of the parallel plate line is discussed in Example 4.12.

Example 4.3 Two parallel wires – capacitance: These wires are shown in Figure 4.3. When the wires are charged with equal and opposite charges, the electric flux leaves the wires in all directions. In order to find the electric flux density at any point, the electric flux density for each wire must be added vectorially. Now assuming the charges are unifomly distributed round the wires, the flux density for each wire will be given by

$$D = \frac{Q}{2\pi rl}\ \text{Cm}^{-2}(\text{radially outwards}). \tag{4.15}$$

This is identical to Equation (4.4) and it is the situation when the outer conductor of a coaxial cable has a very large radius. If the origin is taken at the centre of this

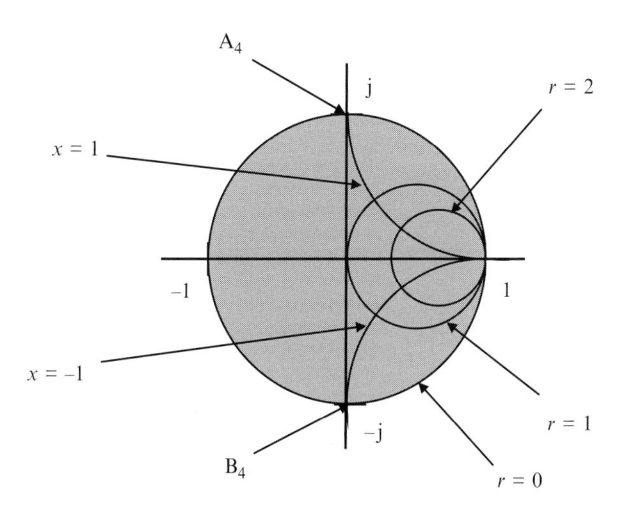

Figure 4.9(e) The w_4 or ρ plane for Equation (4.38). The shift by one to the right makes the Smith chart complete. For the full chart see Figure 2.5 in Chapter 2. The values of A_4 and B_4 are j and –j respectively, in other words, the same values they had in the z plane in Figure 4.9(a).

Finally, the last of the four conformal transformations completes the Smith chart. This is simply to shift the chart to the right, in the same way as the first transformation. This is shown in Equation (4.38) and Figure 4.9(e):

$$w_4 = w_3 + 1 = -2w_2 + 1 = \frac{-2}{w_1} + 1 = \frac{-2}{z+1} + 1 = \frac{z-1}{z+1} = \rho. \tag{4.38}$$

The four consecutive transformations thus combine to form the familiar equation for the reflection coefficient:

$$\rho = \frac{z-1}{z+1},$$

which is sometimes called the bilinear transformation. The simplicity of the equation does not reveal the final shape of the Smith chart. Figures 4.9(a) to 4.9(e) do, however, show more clearly the consequences of these transformations. This chart has all the possible values of z within the unit circle as well as being the Argand diagram for the reflection coefficient.

Example 4.7 A single wire above a conducting ground plane – capacitance: When a charged wire is above a conducting ground plane, it induces in that plane an equal and opposite charge. The electric fields at the ground plane are normal to the surface as it is assumed the conductor is sufficiently good to reduce all tangential electric fields to zero. The solution to the capacitance between the wire and the ground plane is usually carried out by the method of images. This involves

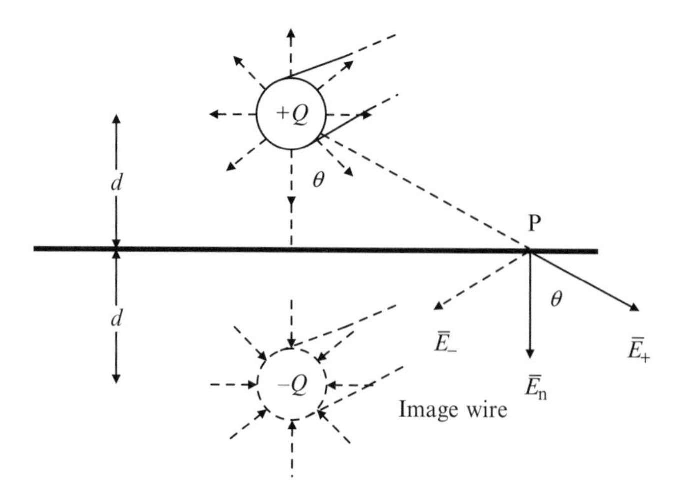

Figure 4.10 A single wire above a conducting ground plane. The equivalent image wire is shown in the lower part of the figure. The radius of the wires is r.

replacing the conducting plane by an image charge at an equal distance below the surface of the plane to the distance of the wire above, as shown in Figure 4.10. The fields above the ground plane are the same, i.e. for either the ground plane or the image charge being present. Another way of looking at the image is that the ground plane acts like a mirror producing an 'image' of the charge.

Consider first the problem without the conducting plane, i.e. just the positively charged wire and its negatively charged image. Take a point P, which is a distance r_+ from the centre of the top wire in Figure 4.10 and a distance r_- from the image wire. Then the electric field due to the positively charged wire is

$$\overline{E}_+ = \frac{Q}{2\pi\varepsilon l r_+}\overline{r}_+, \tag{4.39}$$

where the vector representation has been used to indicate the field is radially outwards from the positively charged wire, with a length l. Using the image charge, the other electric field at P will be

$$\overline{E}_- = \frac{Q}{2\pi\varepsilon l r_-}\overline{r}_-. \tag{4.40}$$

If the point P is equidistant from the centres of both the wires, then $r_+ = r_-$ and the amplitudes of both the electric fields are equal. The horizontal components of the electric field cancel, leaving only a vertical component. The point P could be anywhere on the plane equidistant from the two wires. The normal component of the electric field is given by

$$E_\mathrm{n} = \frac{Qd}{\pi\varepsilon l r^2}\ \mathrm{Vm}^{-1} \text{ since } \cos\theta = \frac{d}{r} \text{ in Figure 4.10,} \tag{4.41}$$

where E_n is the total electric field normal to the plane. Since there is no component of the electric field tangential to this plane then there is also no change in potential

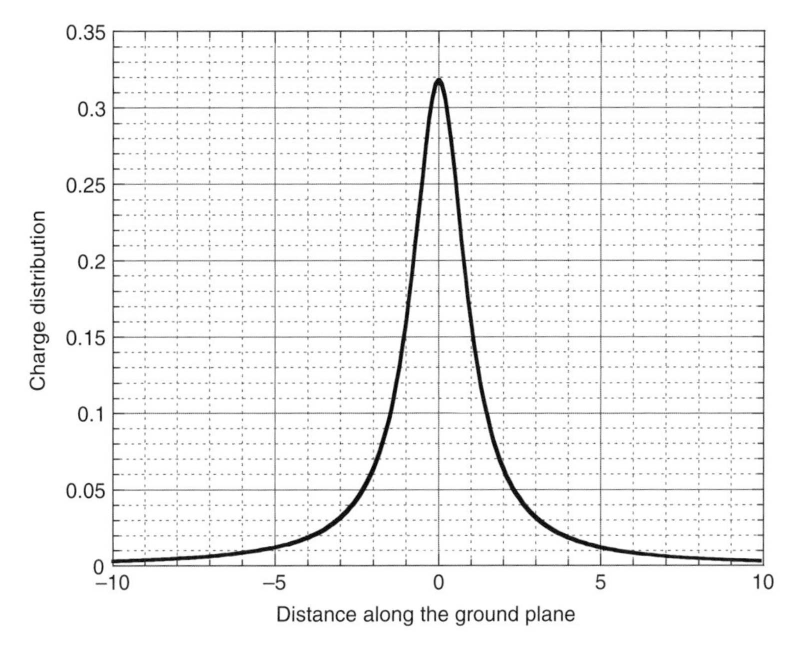

Figure 4.11 The charge distribution along a conducting ground plane underneath a charged wire. The charge has been taken as positive and the zero is the point nearest to the wire. Both Q and d have been made equal to unity. This is a graph of Equation (4.42).

along the surface of the plane. Such a plane is called an equipotential plane for this reason. By placing a conducting ground plane at the same position as this equipotential plane, the fields above the plane are undisturbed. Working backwards, the electric flux density at this plane is given by

$$D_{\mathrm{n}} = \frac{Qd}{\pi l r^2} = \frac{Qd}{\pi l\left(x^2 + d^2\right)} \ \mathrm{Cm}^{-2} \ \text{since } r^2 = d^2 + x^2. \tag{4.42}$$

If dx is an incremental length along the ground plane, transverse to the direction of the wire, i.e. in the plane of the paper nearest to the wires, then the total charge on the plane is given by the surface integral of the flux density:

$$Q_{\mathrm{T}} = -\int_{-\infty}^{\infty} \frac{Qd}{\pi l r^2}\, l \,\mathrm{d}x = -\int_{-\infty}^{\infty} \frac{Qd}{\pi\left(d^2 + x^2\right)}\,\mathrm{d}x = -Q. \tag{4.43}$$

So Equation (4.42) gives the charge distribution along the ground plane induced by a positively charged wire above it and Equation (4.43) shows that the total charge on the ground plane is equal to $-Q$. The minus sign arises because of the direction of D_{n}. This equivalence is the basis of the method of images which replaces conducting surfaces with image charges to make field calculations simpler. Figure 4.11 shows the charge distribution on the conducting ground plane using Equation (4.42).

The capacitance is somewhat easier. Using Equation (4.19) and modifying it to account for the separation of $2d$, the capacitance per metre between the wire and its image is

$$C = \frac{\pi\varepsilon}{\ln\left(\frac{2d-a}{a}\right)} \ \mathrm{Fm}^{-1}, \tag{4.44}$$

and so for the same charge and half the voltage, the capacitance per metre for the wire above a ground plane is

$$C = \frac{2\pi\varepsilon}{\ln\left(\frac{2d-a}{a}\right)} \ \mathrm{Fm}^{-1}. \tag{4.45}$$

This equation is valid for $d > 4a$, see Figure 4.4.

Example 4.8 Three parallel wires in an equilateral triangle – capacitance: In Figure 4.12 is shown the arrangement for three parallel wires where the centres of the wires form an equilateral triangle with sides of length d.

Line number 1 is charged to $+Q$ and line number 3 to $-Q$, leaving line number 2 uncharged. This arrangement is almost identical to Examples 4.3 and 4.7 except that the presence of line number 2 will have a small effect on the fields. Assuming this effect is negligible, all that is needed is to work out the voltage on this line relative to the other two. The dotted line in the diagram marks an equipotential line between the two charged lines – see Example 4.3. Since this line passes through the centre of line 2, it can be assumed that this line is at half the voltage, i.e. from Equation (4.17):

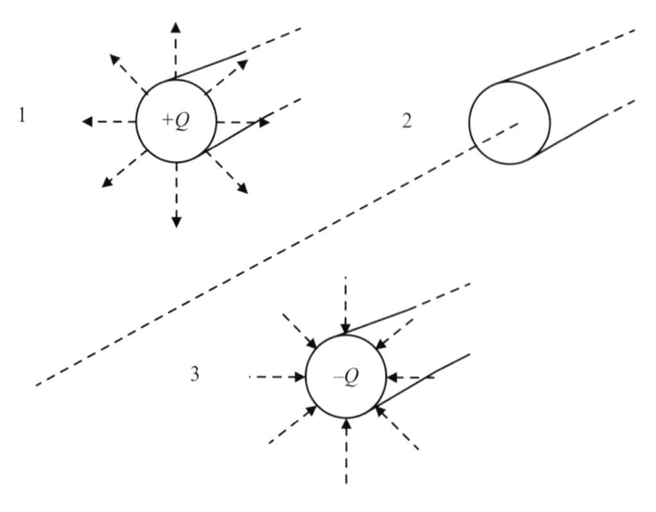

Figure 4.12 Three parallel wires with the centre of each wire at the corner of an equilateral triangle of side d, and each wire has a radius a.

$$V_2 = \frac{Q}{2\pi l\varepsilon} \ln\left(\frac{d-a}{a}\right). \tag{4.46}$$

So an elastance matrix can be constructed for these wires, taking line 3 as zero voltage, i.e. it is earthed:

$$\begin{bmatrix} V_1 \\ V_2 \end{bmatrix} = \frac{1}{\pi l\varepsilon} \ln\left(\frac{d-a}{a}\right) \begin{bmatrix} 1 & \frac{1}{2} \\ \frac{1}{2} & 1 \end{bmatrix} \begin{bmatrix} Q_1 \\ Q_2 \end{bmatrix}. \tag{4.47}$$

Inverting this matrix in Equation (4.47) gives the capacitance matrix for these wires:

$$\begin{bmatrix} Q_1 \\ Q_2 \end{bmatrix} = \frac{4\pi l\varepsilon}{3\ln\left(\dfrac{d-a}{a}\right)} \begin{bmatrix} 1 & -\frac{1}{2} \\ -\frac{1}{2} & 1 \end{bmatrix} \begin{bmatrix} V_1 \\ V_2 \end{bmatrix}. \tag{4.48}$$

To check this equation, let Q_2 be zero, then $V_2 = \frac{1}{2} V_1$ and

$$Q_1 = \frac{\pi l\varepsilon}{\ln\left(\dfrac{d-a}{a}\right)} V_1 \tag{4.49}$$

as expected. So the capacitance per metre for these three parallel lines is

$$[C] = \frac{4\pi\varepsilon}{3\ln\left(\dfrac{d-a}{a}\right)} \begin{bmatrix} 1 & -\frac{1}{2} \\ -\frac{1}{2} & 1 \end{bmatrix}. \tag{4.50}$$

Note that the capacitance C in Equation (3.1) is now different from the capacitance of just the two wires given in Equation (4.19) by a factor of 4/3. The elastance matrix in Equation (4.47) always has positive terms, so the inverse gives a negative mutual capacitance. This was used in the theory at the beginning of Chapter 3. In this case, the mutual capacitance is equal to minus half the main capacitance, C.

Example 4.9 Two parallel wires above a conducting ground plane – capacitance: This example has been added to complete the set of the common configurations of single wires. In Figure 4.13 two wires above the ground plane are shown. Since only one of them is charged, there is only one image charge below the conducting plane.

As in the previous example, the first part of the solution involves finding the voltage on line 2 arising from the charges on lines 1 and 3 shown in Figure 4.13. Using Equation (4.33) again but with $x = d$, $y = s$ and $c = -\sqrt{d^2 - r^2}$ gives

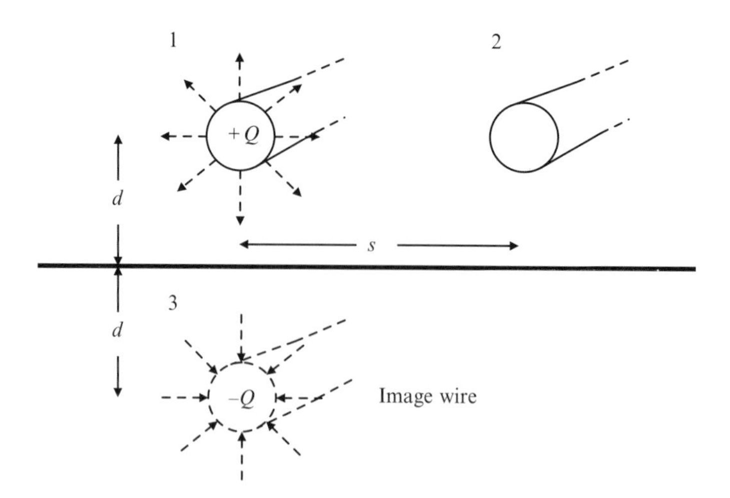

Figure 4.13 Two identical wires above a conducting ground plane, with one of them charged. The separation between the wires is s and the radius of each wire is a.

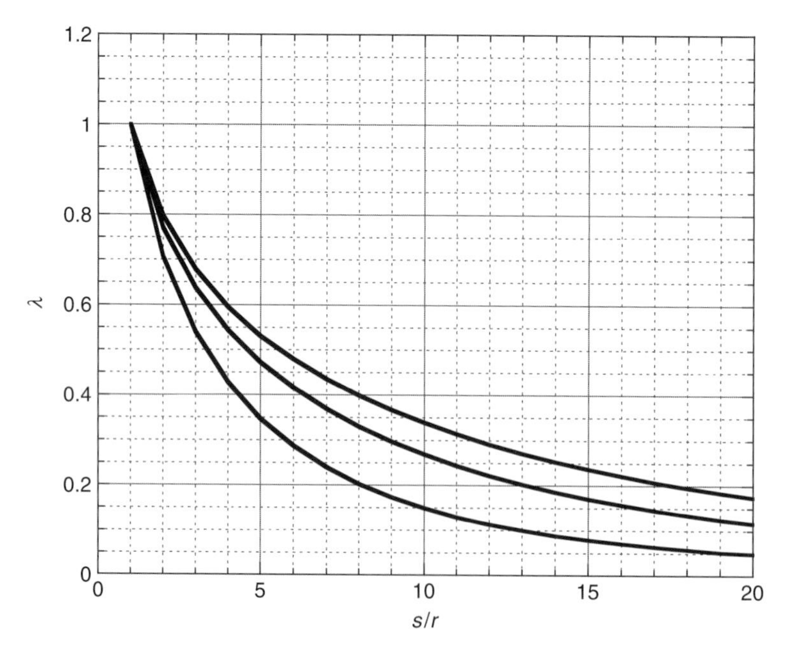

Figure 4.14 The ratio of the voltage on line 2 to the voltage on line 1(λ) for three values of d/r. The upper trace is for $d/r = 15$, the middle trace is for 10 and the lower trace is for 5.

$$V_2 = \frac{V_1}{\cosh^{-1}\left(\frac{d}{r}\right)} \ln\left(\frac{2d^2 - r^2 + s^2 + 2d\sqrt{d^2 - r^2}}{(s^2 - r^2)^2 + 4d^2s^2}\right) = \lambda V_1, \tag{4.51}$$

where the suffices refer to the wires shown in the figure. Figure 4.14 shows the variation of λ with s/r for various values of d/r.

Using Equation (4.45) with $2d$ rather than d:

$$V_1 = \frac{Q_1}{2\pi l\varepsilon}\ln\left(\frac{2d-a}{a}\right) = \kappa Q_1. \tag{4.52}$$

Now using the abbreviated forms in Equations (4.51) and (4.52), the elastance matrix is

$$\begin{bmatrix} V_1 \\ V_2 \end{bmatrix} = \kappa \begin{bmatrix} 1 & \lambda \\ \lambda & 1 \end{bmatrix} \begin{bmatrix} Q_1 \\ Q_2 \end{bmatrix} \tag{4.53}$$

and the capacitance matrix is

$$\begin{bmatrix} Q_1 \\ Q_2 \end{bmatrix} = \frac{1}{\kappa(1-\lambda^2)} \begin{bmatrix} 1 & -\lambda \\ -\lambda & 1 \end{bmatrix} \begin{bmatrix} V_1 \\ V_2 \end{bmatrix}. \tag{4.54}$$

Putting the length l equal to unity gives the distributed capacitance needed for transmission line theory:

$$[C] = \frac{1}{\kappa(1-\lambda^2)} \begin{bmatrix} 1 & -\lambda \\ -\lambda & 1 \end{bmatrix} = \frac{2\pi\varepsilon}{(1-\lambda^2)\ln\left(\frac{2d-a}{a}\right)} \begin{bmatrix} 1 & -\lambda \\ -\lambda & 1 \end{bmatrix}. \tag{4.55}$$

Example 4.10 The eccentric coaxial cable – capacitance: This final example of capacitance has been included as a further use of the conformal transformations already used. In measurements using coaxial cables it is usually assumed that the centre conductor is in the centre. However, this may not be the case, particularly with air-filled cables, and the changes in the transmission line properties may well contribute to the uncertainty of the measurement. The eccentric cable is shown in Figure 4.15.

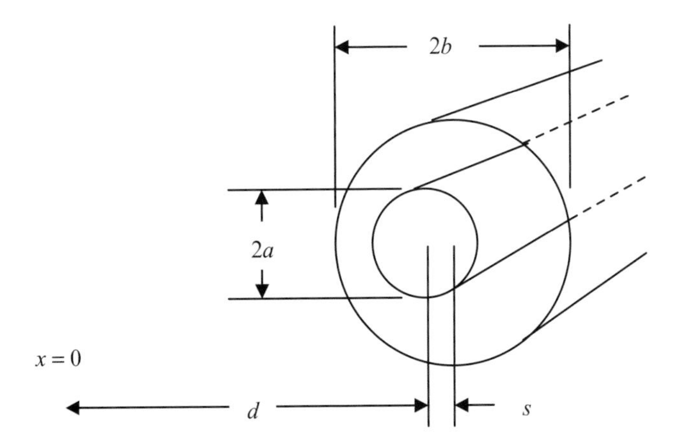

Figure 4.15 The eccentric coaxial cable. The radius of the outer conductor is b, the radius of the inner conductor is a and the shift of the central conductor from the centre is s.

Using the same technique as in Example 4.5, the transformed radius of the inner conductor is

$$R_\mathrm{i} = \frac{d}{a} + \sqrt{\left(\frac{d}{a}\right)^2 - 1}. \tag{4.56}$$

The outer conductor has a transformed radius of

$$R_\mathrm{o} = \frac{d+s}{b} + \sqrt{\left(\frac{d+s}{b}\right)^2 - 1}, \tag{4.57}$$

where the suffixes i and o refer to the inner and outer conductors. The parameter d can be eliminated from the above equations by using the second part of Equation (4.28):

$$c^2 = d^2 - a^2 = (d+s)^2 - b^2 \text{ or } d = \frac{b^2 - a^2 - s^2}{2s}, \tag{4.58}$$

so the capacitance per metre using Equation (4.31) is

$$C = \frac{2\pi\varepsilon}{\ln\left(\dfrac{\dfrac{d}{a} + \sqrt{\left(\dfrac{d}{a}\right)^2 - 1}}{\dfrac{d+s}{b} + \sqrt{\left(\dfrac{d+s}{b}\right)^2 - 1}}\right)}$$

$$= \frac{2\pi\varepsilon}{\ln\left(\dfrac{d}{a} + \sqrt{\left(\dfrac{d}{a}\right)^2 - 1}\right)\left(\dfrac{d+s}{b} - \sqrt{\left(\dfrac{d+s}{b}\right)^2 - 1}\right)},$$

which needs some careful rearranging to give

$$C = \frac{2\pi\varepsilon}{\cosh^{-1}\left(\dfrac{b^2 + a^2 - s^2}{2ab}\right)} \ \mathrm{Fm^{-1}}. \tag{4.59}$$

By combining Equations (1.7) and (1.9), eliminating L:

$$Z_0 = \frac{1}{vC}. \tag{4.60}$$

Using this relationship, it is possible to compare the characteristic impedance of the coaxial cable described in Example 4.1 with this eccentric cable. In Figure 4.16 is shown the variation of the characteristic impedance with the displacement from the centre, s, for an air-filled cable. For $s = 0$, the characteristic impedance has been taken as 50 Ω.

It can be seen in Figure 4.16 that the variation of the characteristic impedance is not very severe for small displacements. However, as the centre conductor

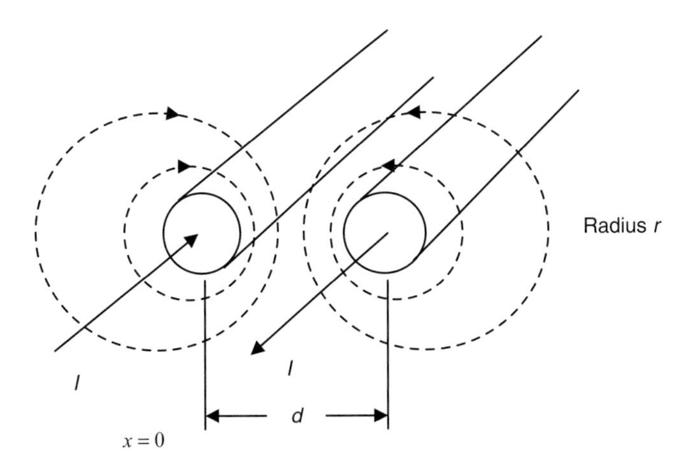

Figure 4.19 Two parallel wires carrying equal and opposite currents. The dotted lines show the magnetic field lines. The wires are separated by d and each have a radius r.

$$L = \frac{\mu}{\pi}\ln\left(\frac{d-r}{r}\right)\ \mathrm{Hm}^{-1}. \tag{4.80}$$

Once again by the same method as in the previous two examples:

$$v = \frac{1}{\sqrt{LC}} = \frac{1}{\sqrt{\frac{\mu}{\pi}\ln\left(\frac{d-r}{r}\right) \times \frac{\varepsilon\pi}{\ln\left(\frac{d-r}{r}\right)}}} = \frac{1}{\sqrt{\mu\varepsilon}} \tag{4.81}$$

and the characteristic impedance is given by

$$Z_0 = \sqrt{\frac{L}{C}} = \sqrt{\frac{\frac{\mu}{\pi}\ln\left(\frac{d-r}{r}\right)}{\frac{\varepsilon\pi}{\ln\left(\frac{d-r}{r}\right)}}} = \frac{1}{\pi}\sqrt{\frac{\mu}{\varepsilon}}\ln\left(\frac{d-r}{r}\right)\ \Omega. \tag{4.82}$$

If the exact solution discussed in Example 4.3 and Equation (4.32) is used then

$$Z_0 = \frac{1}{\pi}\sqrt{\frac{\mu}{\varepsilon}}\cosh^{-1}\left(\frac{d}{r}\right)\ \Omega. \tag{4.83}$$

Example 4.14 A single wire above a ground plane – inductance: Following Example 4.13, it is clear that only half the magnetic flux is involved in this case. So the inductance is given by

$$L = \frac{\mu}{2\pi}\ln\left(\frac{2d-a}{a}\right)\ \mathrm{Hm}^{-1} \text{ and } Z_0 = \frac{1}{2\pi}\sqrt{\frac{\mu}{\varepsilon}}\ln\left(\frac{2d-a}{a}\right)\ \Omega. \tag{4.84}$$

4.3 The link between distributed capacitance and inductance for transmission lines with a uniform dielectric

For all of the examples chosen, the dielectric between the wires has been assumed to be homogeneous with a dielectric constant ε. Most transmission lines do not have any magnetic material in them so the value of μ is the free space value of μ_0. All the examples on capacitance involved finding a voltage between the conductors by integrating along an electric field line. The voltage is independent of which path is chosen, and in all cases a linear path, r, was used. In general, the electric field is of the form

$$E = \frac{Q}{\varepsilon} f(r) \tag{4.85}$$

and so the voltage is of the form

$$V = \frac{Q}{\varepsilon} \int_a^b f(r) \mathrm{d}r. \tag{4.86}$$

Now the inductance involves a similar integral. The equipotential lines, which are orthogonal to the electric field lines, are the same as the magnetic flux lines. So the magnetic flux density is of the form

$$B = \mu I f(r) \tag{4.87}$$

and so the magnetic flux is of the form

$$\Phi_\mathrm{m} = \mu I \int_a^b f(r) \mathrm{d}r. \tag{4.88}$$

Now taking the ratio of Equations (4.86) and (4.88):

$$\frac{V}{\Phi_\mathrm{m}} = \frac{Q}{\mu \varepsilon I} \text{ or } \mu\varepsilon = \frac{\Phi_\mathrm{m}}{I} \times \frac{Q}{V} = LC, \tag{4.89}$$

which shows that the velocity for these transmission lines is universally

$$v = \frac{1}{\sqrt{\mu\varepsilon}} = \frac{1}{\sqrt{LC}}. \tag{4.90}$$

As a result of Equation (4.90),

$$Z_0 = \sqrt{\frac{L}{C}} = \frac{L}{\sqrt{\mu\varepsilon}} = \frac{\sqrt{\mu\varepsilon}}{C} = \sqrt{\frac{\mu}{\varepsilon}} \int_a^b f(r) \mathrm{d}r. \tag{4.91}$$

So all the examples on inductance so far could have been derived from the results on capacitance, without resorting to magnetostatic theory. Extending these concepts to the capacitance matrices, the voltages are related to the charges using an elastance matrix which can be expressed in the form

$$[V] = \frac{1}{\varepsilon}[s(r)]\ [Q]. \tag{4.92}$$

In a very similar way the inductance matrix can be expressed as

$$[\Phi_{\text{m}}] = \mu\ [s(r)]\ [I]. \tag{4.93}$$

An illustration of this will be given in the next example. Using Equations (4.92) and (4.93):

$$\frac{[Q]}{[V]} = [C] = \varepsilon[s(r)]^{-1} \text{ and } \frac{[\Phi_{\text{m}}]}{[I]} = [L] = \mu[s(r)].$$

$$\text{So } [L]\ [C] = \mu\varepsilon\ [1]. \tag{4.94}$$

Equation (4.94) shows that all the coupled modes have the same velocity if the dielectric surrounding them is homogeneous. The exception to this occurs when the transmission line has more than one dielectric in it, for example microstrip, when the permittivity and the permeability can have different values for each mode.

Example 4.15 Three parallel wires in an equilateral triangle – inductance: Using the results from Example 4.13, the magnetic flux between wires 1 and 3 in Figure 4.20 is

$$\Phi_{\text{m}} = \frac{\mu I}{\pi}\ln\left(\frac{d-a}{a}\right),$$

and by inspection of Figure 4.20, just half of this flux couples to the circuit formed by wires 2 and 3. Making these results into a matrix:

$$\begin{bmatrix} \Phi_{\text{m}1} \\ \Phi_{\text{m}2} \end{bmatrix} = \frac{\mu}{\pi}\ln\left(\frac{d-a}{a}\right)\begin{bmatrix} 1 & \frac{1}{2} \\ \frac{1}{2} & 1 \end{bmatrix}\begin{bmatrix} I_1 \\ I_2 \end{bmatrix}. \tag{4.95}$$

Hence the inductance matrix is

$$[L] = \frac{\mu}{\pi}\ln\left(\frac{d-a}{a}\right)\begin{bmatrix} 1 & \frac{1}{2} \\ \frac{1}{2} & 1 \end{bmatrix}\ \text{Hm}^{-1}. \tag{4.96}$$

Now from Equation (4.50), the capacitance matrix is

$$[C] = \frac{4\pi\varepsilon}{3\ln\left(\dfrac{d-a}{a}\right)}\begin{bmatrix} 1 & -\frac{1}{2} \\ -\frac{1}{2} & 1 \end{bmatrix}.$$

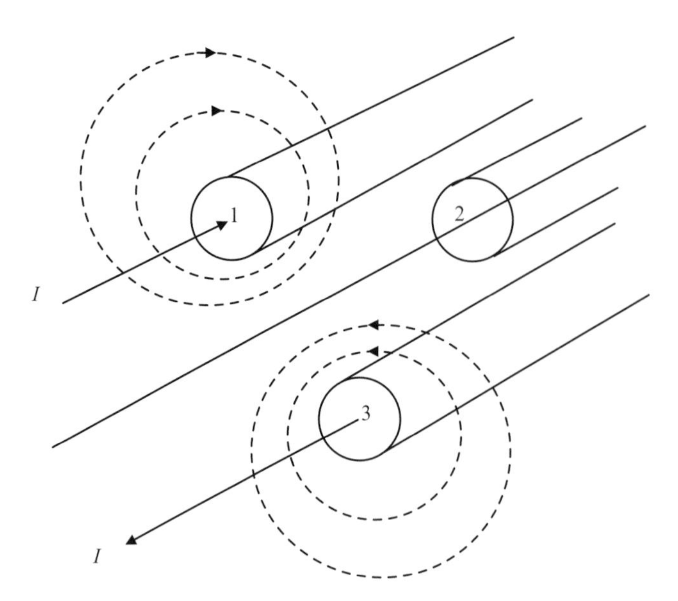

Figure 4.20 Three parallel wires in an equilateral triangle. Wire number 3 is earthed and there is no current in wire number 2. The wires are at the corners of an equilateral triangle of side d and each wire has a radius a.

In this case $[L][C] = \mu\varepsilon[1]$, which was predicted in Equation (4.94). Using Equation (3.11):

$$Z_{0\mathrm{e}} = \sqrt{\frac{L+M}{C-C_{\mathrm{M}}}} = \frac{3}{2\pi}\sqrt{\frac{\mu}{\varepsilon}}\ln\left(\frac{d-a}{a}\right)\ \Omega,$$

$$Z_{0\mathrm{o}} = \sqrt{\frac{L-M}{C+C_{\mathrm{M}}}} = \frac{1}{2\pi}\sqrt{\frac{\mu}{\varepsilon}}\ln\left(\frac{d-a}{a}\right)\ \Omega. \tag{4.97}$$

Example 4.16 Two parallel wires above a conducting ground plane – inductance: In this case, starting with Equation (4.55) for the capacitance given in Example 4.9:

$$[C] = \frac{1}{\kappa(1-\lambda^2)}\begin{bmatrix}1 & -\lambda\\ -\lambda & 1\end{bmatrix} = \frac{2\pi\varepsilon}{(1-\lambda^2)\ln\left(\dfrac{2d-a}{a}\right)}\begin{bmatrix}1 & -\lambda\\ -\lambda & 1\end{bmatrix}.$$

The inductance can be found using Equation (4.94):

$$[L] = \mu\varepsilon\begin{bmatrix}1 & 0\\ 0 & 1\end{bmatrix}[C]^{-1} = \frac{\mu}{2\pi}\ln\left(\frac{2d-a}{a}\right)\begin{bmatrix}1 & \lambda\\ \lambda & 1\end{bmatrix}\ \mathrm{Hm}^{-1}, \tag{4.98}$$

$$Z_{0\mathrm{e}} = \sqrt{\frac{L+M}{C-C_{\mathrm{M}}}} = \frac{1}{2\pi}\sqrt{\frac{\eta}{\varepsilon}}(1+\lambda)\ln\left(\frac{2d-a}{a}\right)\ \Omega,$$

$$Z_{0o} = \sqrt{\frac{L-M}{C+C_M}} = \frac{1}{2\pi}\sqrt{\frac{\eta}{\varepsilon}}(1-\lambda)\ln\left(\frac{2d-a}{a}\right)\ \Omega. \tag{4.99}$$

Example 4.17 The eccentric coaxial cable – inductance: In a similar fashion, using Equation (4.59) for the capacitance of the eccentric coaxial cable:

$$C = \frac{2\pi}{\cosh^{-1}\left(\frac{b^2+a^2-s^2}{2ab}\right)}\ \mathrm{Fm}^{-1}.$$

The inductance is given by Equation (4.94) as

$$L = \frac{\mu\varepsilon}{C} = \frac{\mu}{2\pi}\cosh^{-1}\left(\frac{b^2+a^2-s^2}{2ab}\right)\ \mathrm{Hm}^{-1} \tag{4.100}$$

and the characteristic impedance is

$$Z_0 = \frac{1}{2\pi}\sqrt{\frac{\mu}{\varepsilon}}\cosh^{-1}\left(\frac{b^2+a^2-s^2}{2ab}\right)\ \Omega. \tag{4.101}$$

Example 4.18 Stripline: Stripline is shown in Figure 4.21 and is widely used in printed circuits, particularly in multi-layer printed circuits. Since the centre conductor is entirely surrounded by a dielectric, this transmission line supports a TEM mode with the top and bottom conductors acting as earthed or ground planes.

The derivation of an empirical equation for the capacitance involves two conformal transformations. The first is the Schwarz–Christoffel transformation and a full description is given in the books listed at the end of the chapter. This solution is only valid for $t = 0$ and when the width of the dielectric substrate is very much greater than w. For other values of the thickness, only approximate equations are available.

So for $t = 0$:

$$C = 4\varepsilon_R\varepsilon_0\frac{K(k)}{K\left(\sqrt{1-k^2}\right)}, \tag{4.102}$$

where $k = \tanh\ (\pi w/4d)$ and $K(k)$ is the complete elliptic integral of the first kind. An approximate solution for the capacitance is

$$C \approx 2\varepsilon_R\varepsilon_0\left(\frac{w}{d}+\frac{4\ln(2)}{\pi}\right)\ \text{for } w > 1.12d. \tag{4.103}$$

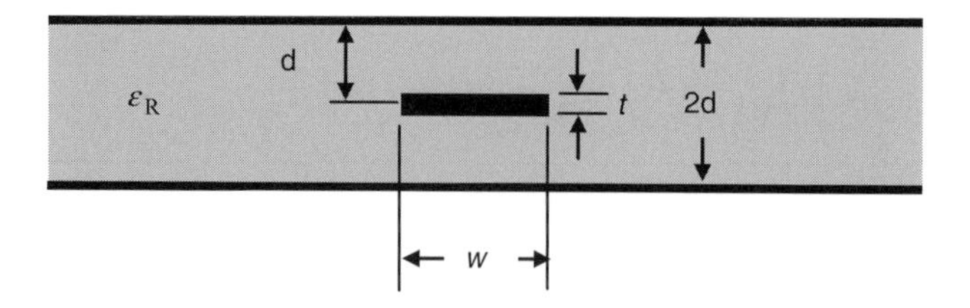

Figure 4.21 Stripline. The grey area shows the dielectric of thickness $2d$, with a relative dielectric constant ε_R. Above and below the dielectric are earthed or ground planes. The central conductor has a width w and a thickness t.

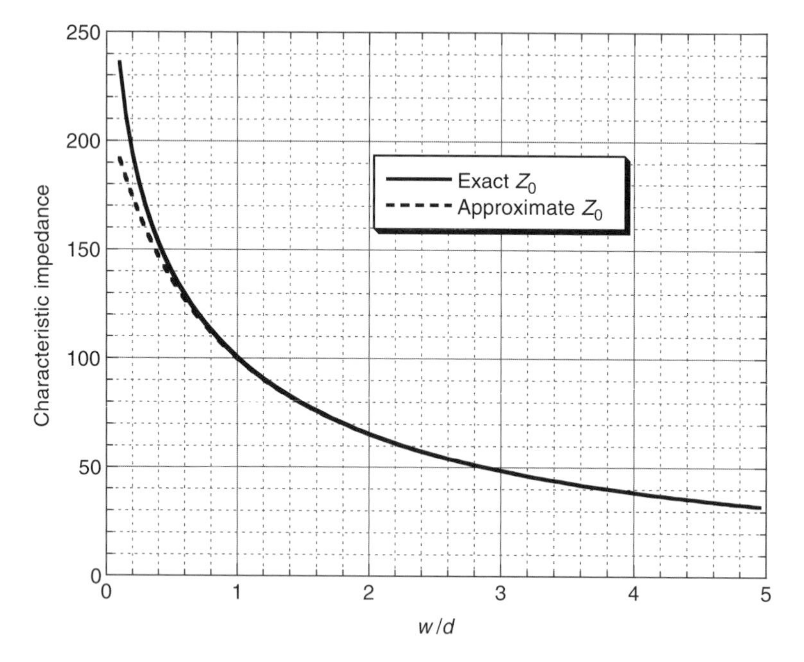

Figure 4.22 A comparison of the exact and approximate expressions for the characteristic impedance of stripline. The equations assume that the thickness of the strip is negligible, i.e. $t = 0$.

The first term of the expression in parentheses is due to the parallel plate capacitance and the second is due to the fringing fields. Using Equation (4.60), this leads to similar equations for the characteristic impedance as

$$Z_0 = \frac{\eta K\left(\sqrt{1-k^2}\right)}{4K(k)} \text{ or approximately } Z_0 \approx \frac{\eta}{2}\left(\frac{w}{d} + \frac{4\ln(2)}{\pi}\right)^{-1}. \quad (4.104)$$

The first term in the approximate equation is from Equation (4.77) and is the parallel plate characteristic impedance with a factor of two because of both ground planes. The second term is for the fringing fields.

In Figure 4.22 these two equations are plotted to show that for a wide range of w/d the approximate expression is quite adequate for values above 0.5.

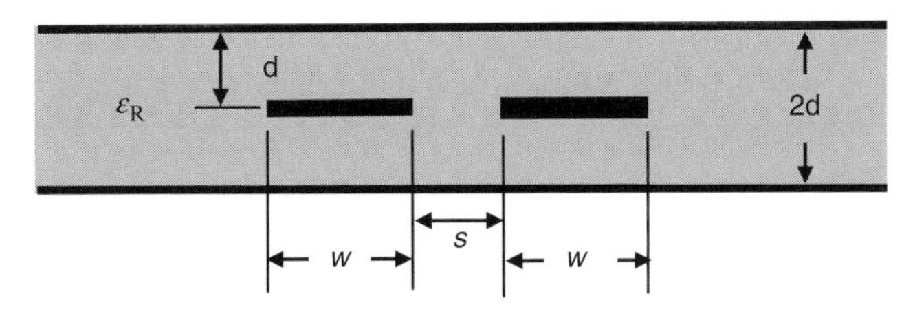

Figure 4.23 Coupled striplines.

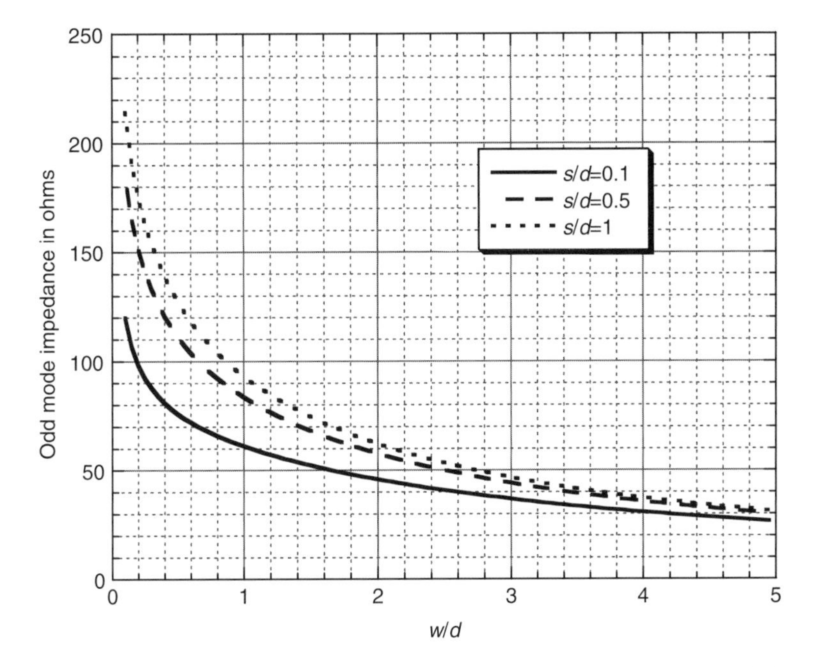

Figure 4.24 The odd mode characteristic impedance for each of two coupled striplines. The dielectric constant has been taken as unity.

In some multi-layer printed circuits, two coupled striplines are used in the odd mode. The main advantage is to avoid coupling to other circuits and higher order modes, which are discussed in Chapter 5. The coupled lines are shown in Figure 4.23 and the equation for the odd mode impedance for each line is given by

$$Z_{0o} = \frac{\eta K\left(\sqrt{1-k_0^2}\right)}{4K(k_0)}, \quad \text{where } k_0 = \tanh\left(\frac{\pi w}{4d}\right)\coth\left(\frac{\pi(w+s)}{4d}\right), \tag{4.105}$$

where these equations are exact for centre conductors with zero thickness and the dielectric region having a width very much greater than w. This equation is plotted in Figure 4.24.

If the two lines are operated in the balanced mode, then the characteristic impedance for that transmission line will be twice the odd mode impedance. So if the substrate had a dielectric constant of, say, 5 and $w/d = 1$ and $s/d = 1$ the odd mode impedance would be 42.5 Ω and the balance line impedance 85 Ω.

4.4 Transmission lines with more than one dielectric – including stripline, microstrip and coplanar waveguide

When a transmission line has more than one dielectric, the propagation is more complex than in any of the previous examples. The reason for this is that the wave has a velocity which depends on a function of both the dielectric constants. Often an effective dielectric contant is used to represent this function. However, this function is not often a constant with frequency. It is possible to find the effective dielectric constant at zero frequency using conformal transformations or computer modelling and then find approximate functions to describe the frequency variation.

Starting with the zero frequency, or a static field solution, an example using the parallel plate capacitor will be discussed. In Figure 4.25 are shown two possible arrangements of two dielectrics – one in a series configuration and the other in a parallel one. By combining the capacitances an effective dielectric constant, ε_{eff}, can be found. Since the dielectric does not affect the inductance, the above equations can still be used.

In Figure 4.25(a) the two capacitors in series are

$$C_1 = \frac{\varepsilon_1 w}{(1-\lambda)d} \text{ and } C_2 = \frac{\varepsilon_2 w}{\lambda d}$$

and the total capacitance is given by

$$C_S = \frac{\varepsilon_1 \varepsilon_2 w}{(\lambda \varepsilon_1 + (1-\lambda)\varepsilon_2)d},$$

so the effective dielectric constant is

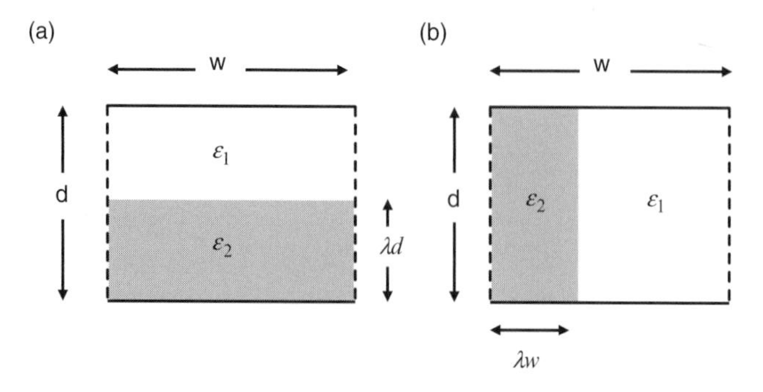

Figure 4.25 Two possible arrangements of two dielectrics between the plates of a capacitor. The series arrangement is shown in (a) and the parallel in (b). The capacitors are assumed to be one metre long.

These equations are taken from the books listed at the end of this chapter.

It is interesting to note that for very thick substrates, where $h \gg w$, coplanar waveguide is almost free from dispersion. In Figure 4.30 the parameters have been chosen to show some dispersion, but in many practical circuits it would be much smaller.

4.5 Conclusions

This chapter has attempted to derive the equations for capacitance and inductance of some of the common transmission lines. Although these have been derived for just static fields, for the transmission lines with one dielectric, these equations are valid over a wide range of frequencies. In Chapter 6 there is a discussion of the skin effect which includes the topic of self inductance. This extra inductance is usually significant at low frequencies, but for some transmission lines it cannot be neglected even at microwave frequencies. For those lines with two dielectrics, there is a change of parameters with frequency and, in particular, the effective dielectric constant has the range

$$\varepsilon_1 < \varepsilon_{\text{eff}} < \varepsilon_2. \tag{4.115}$$

As a general rule, as the frequency increases, the effective dielectric constant tends towards the value of the higher dielectric constant, i.e. $\varepsilon_{\text{eff}} \to \varepsilon_2$.

The link between the two parameters L and C and the properties of the medium between the conductors, μ and ε, has been established in Equations (4.90) and (4.94) as

$$v_{\text{p}} = \frac{1}{\sqrt{LC}} = \frac{1}{\sqrt{\mu\varepsilon}}, \quad LC = \mu\varepsilon \text{ and } [L]\ [C] = \mu\varepsilon\ [1].$$

Now the parameters L and C relate to the description of transmission lines given in the first three chapters using equivalent circuits. The parameters μ and ε will be used in the next chapters on electromagnetic waves. So this equation links the two theories. In a similar fashion, the characteristic impedance, Z_0, is given in Equation (4.91) as

$$Z_0 = \sqrt{\frac{L}{C}} = \sqrt{\frac{\mu}{\varepsilon}} \int_a^b f(r)\ \mathrm{d}r.$$

Here again is a cross-linking of parameters. The two theories are describing the same phenomenon. In the first three chapters, many of the problems discussed are solved quite satisfactorily using just an equivalent circuit. However, there are some further aspects of transmission lines which require the more complex electromagnetic theory, which follows in the next chapter.

4.6 Further reading

Capacitance and inductance

S. Ramo, J. R. Whinnery and T. Van Duzer *Fields and Waves in Communication Electronics*, Third edition, New York, Wiley, 1993. Parts of Chapters 1 and 2.

C. S. Walker *Capacitance and Inductance and Crosstalk Analysis*, Boston, Artech House, 1990.

W. Hilberg *Electrical Characteristics of Transmission Lines*, Dedham, Artech House, 1979.

F. W. Grover *Inductance Calculations*, New York, Dover Publications, reprinted by the Instrument Society of America, 1973.

W. R. Smythe *Static and Dynamic Electricity*, Third edition, New York, Hemisphere Publishing Corporation, revised 1980. Chapter 2 (elastance in section 2.15).

P. Lorrain and D. R. Corson *Electromagnetism: Principles and Applications*, San Francisco, W. H. Freeman and Company, 1978. Chapters 4 and 12.

P. C. Clemmow *An Introduction to Electromagnetic Theory*, Cambridge, Cambridge University Press, 1973.

P. Hammond *Electromagnetism for Engineers*, Oxford, Pergamon Press, 1986.

B. C. Wadell *Transmission Line Handbook*, Boston, Artech House, 1991.

D. K. Cheng *Field and Wave Electromagnetics*, New York, Addison-Wesley Publishing Company, 1989. Chapters 4 and 6.

A. K. Goel *High-Speed VLSI Interconnections*, New York, Wiley, 1994. Chapter 2.

Conformal mapping

M. R. Spiegel *Complex Variables*, Schaum's Outline Series, New York, McGraw-Hill Book Company, 1964. Chapter 8.

E. Kreyszig *Advanced Engineering Mathematics*, Seventh edition, New York, Wiley, 1993. Chapters 16 and 17.

P. V. O'Neil *Advanced Engineering Mathematics*, Second edition, Belmont, California, Wadsworth Publishing Company, 1987. Chapter 18.

K. J. Binns and P. J. Lawrenson *Electric and Magnetic Field Problems*, Oxford, Pergamon Press, 1973. Chapters 6, 7 and 8.

R. E. Collin *Foundations for Microwave Engineering*, New York, McGraw-Hill, 1992. Appendix III.

Smith chart

S. Ramo, J. R. Whinnery and T. Van Duzer *Fields and Waves in Communication Electronics*, Third edition, New York, Wiley, 1993. Chapter 5, section 5.9.

R. A. Chipman *Transmission Lines*, Schaum's Outline Series, New York, McGraw-Hill Book Company, 1968. Chapter 9.

D. K. Cheng *Fundamentals of Engineering Electromagnetics*, New York, Addison-Wesley, 1993. Chapter 8, section 8.6.

Equations for stripline, microstrip and coplanar waveguide

S. Ramo, J. R. Whinnery and T. Van Duzer *Fields and Waves in Communication Electronics*, Third edition, New York, Wiley, 1993. Chapter 8, section 8.6.

R. E. Collin *Foundations for Microwave Engineering*, New York, McGraw-Hill, 1992. Chapter 3, sections 3.11 to 3.16.

D. M. Pozar *Microwave Engineering*, New York, Wiley, 1998. Chapter 3, sections 3.7 to 3.8.

G. Matthaei, L. Young and E. M. T. Jones *Microwave Filters, Impedance-Matching Networks, and Coupling Structures*, Dedham, MA, Artech House, 1980. Chapter 5, sections 5.04 to 5.05.

B. C. Wadell *Transmission Line Handbook*, Boston, Artech House, 1991. Chapter 3, sections 3.4 to 3.6 and chapter 4, section 4.6.

5 Guided electromagnetic waves

This chapter begins with a brief description of Maxwell's equations which leads on to the theory of electromagnetic waves. More detailed treatments of this topic are given in the books listed at the end of the chapter and readers unfamiliar with this material are advised to consult them. From Maxwell's equations the electromagnetic wave equation can be derived and the solutions are then arranged into three groups. Examples are given of transmission lines or waveguides that illustrate the waves in each of these groups.

5.1 Introduction to electromagnetic waves and Maxwell's equations

Electromagnetic waves are described mathematically as solutions of the electromagnetic wave equation. This wave equation is derived from Maxwell's four electromagnetic equations and so it is appropriate to begin with these. Maxwell's equations can be written in differential or integral form and it can be useful to have both discussed to see the links. We start with Gauss' Theorem for charges and electric flux density which were mentioned in the last chapter during the discussion on capacitance:

$$Q = \int_V \rho \, \mathrm{d}V = \int_S \overline{D}.\mathrm{d}\overline{S},$$

where ρ is the charge density in a volume V. The first integral is the total charge in a volume, V, and the second integral is the sum of all the electric flux around the outside of that volume. The divergence theorem can be used as follows:

$$\int_V \nabla.\overline{D}\mathrm{d}V = \int_S \overline{D}.\mathrm{d}\overline{S} = \int_V \rho \, \mathrm{d}V,$$

where the first two terms are the divergence theorem. Comparing the two volume integrals gives

$$\nabla.\overline{D} = \rho. \tag{5.1}$$

This is the first of Maxwell's equations and it states that for every charge there is a diverging electric flux density D. It is the differential form of Gauss' Theorem.

The next equation is similar. Since there are no magnetic poles – that are the magnetic equivalent to electric charges,

$$\nabla.\overline{B} = 0. \tag{5.2}$$

In other words a magnetic field never diverges from a point. The next two equations are concerned with surfaces and the edges around the surfaces. First of all take a loop of wire through which a time-varying magnetic flux is passing. Then Faraday's Law gives

$$V = \oint \overline{E}.\mathrm{d}\overline{l} = -\frac{\partial}{\partial t}\int_S \overline{B}.\mathrm{d}\overline{S},$$

where the first integral is the electric field all around the edge of the surface and the second integral is the rate of change of the magnetic flux through the surface.

Now using Stokes' Theorem:

$$\int_S \nabla \times \overline{E}.\mathrm{d}\overline{S} = \oint \overline{E}.\mathrm{d}\overline{l} = -\frac{\partial}{\partial t}\int_S \overline{B}.\mathrm{d}\overline{S},$$

where the first two terms are Stokes' Theorem.

Comparing the two surface integrals gives the third of Maxwell's equations

$$\nabla \times \overline{E} = -\frac{\partial \overline{B}}{\partial t}. \tag{5.3}$$

Finally, taking Ampere's Law for a current passing through a surface and the magnetic field round the current:

$$I = \oint \overline{H}.\mathrm{d}\overline{l} = \int_S \overline{J}.\mathrm{d}\overline{S},$$

where J is the current density at any point in the surface. Again using Stokes' Theorem:

$$\int_S \nabla \times \overline{H}.\mathrm{d}\overline{S} = \oint \overline{H}.\mathrm{d}\overline{l} = \int_S \overline{J}.\mathrm{d}\overline{S},$$

and comparing the two surface integrals gives

$$\nabla \times \overline{H} = \overline{J}. \tag{5.4}$$

It was the genius of Maxwell that he realised that the right-hand side of Equation (5.4) need not be restricted to just ohmic currents. When a capacitor has an alternating voltage applied to it there is a capacitative current through it. Starting with the relationship between a charge and the flux density leaving the charge:

$$D = \frac{Q}{A}$$

and then differentiating with respect to time gives

$$\frac{\partial D}{\partial t} = \frac{1}{A} \times \frac{\partial Q}{\partial t} = \frac{I}{A} = J_C.$$

This capacitative current density term, J_C, can be added in to Equation (5.4), giving the final Maxwell equation:

$$\nabla \times \overline{H} = \overline{J} + \frac{\partial \overline{D}}{\partial t}. \tag{5.5}$$

Now using the vector identity:

$$\nabla \times \nabla \times \overline{A} = \nabla(\nabla . \overline{A}) - \nabla^2 \overline{A},$$

where $\overline{A}$ is any vector, and by assuming there are no charges, i.e. $\rho = 0$ in Equation (5.1), and so $\nabla . \overline{E} = 0$, and also no ohmic currents, $\overline{J} = 0$, the wave equation for electromagnetic waves becomes

$$\nabla^2 \overline{E} = -\nabla \times \nabla \times \overline{E} = \nabla \times \frac{\partial \overline{B}}{\partial t} = \frac{\partial}{\partial t}(\nabla \times \overline{B}) = \mu \frac{\partial}{\partial t}\left(\frac{\partial \overline{D}}{\partial t}\right) = \mu\varepsilon \frac{\partial^2 \overline{E}}{\partial t^2}$$

and similarly

$$\nabla^2 \overline{H} = -\nabla \times \nabla \times \overline{H} = -\nabla \times \frac{\partial \overline{D}}{\partial t} = -\frac{\partial}{\partial t}(\nabla \times \overline{D}) = \varepsilon \frac{\partial}{\partial t}\left(\frac{d\overline{B}}{dt}\right) = \varepsilon\mu \frac{\partial^2 \overline{H}}{\partial t^2}.$$

So the wave equations for electric and magnetic fields are

$$\nabla^2 \overline{E} = \mu\varepsilon \frac{\partial^2 \overline{E}}{\partial t^2} \text{ and } \nabla^2 \overline{H} = \varepsilon\mu \frac{\partial^2 \overline{H}}{\partial t^2}. \tag{5.6}$$

These equations are similar to the wave equations derived in Chapter 1, Equation (1.5). The main difference is that the electric and magnetic fields can use all three directions in space. These two extra dimensions allow for an infinite number of solutions. Each of these solutions is called a mode. For most transmission lines, it is preferable to have just one mode of propagation. In many cases there is one mode that propagates at low frequencies and the more complex modes or higher order modes come in at higher frequencies. This chapter will show how various transmission lines are designed to have a single mode of propagation. The range of frequencies over which this one mode propagates is often called the mono-mode bandwidth.

Since the discussion in this book is limited to propagation in just one direction, the coordinates are usually chosen so that this is in the z direction. The next simplification is to assume just sinusoidal waves are propagating in a homogeneous medium i.e. where μ and ε are constants. This changes the wave equations to

$$\nabla^2 \overline{E} = -\omega^2 \mu\varepsilon \overline{E} \text{ and } \nabla^2 \overline{H} = -\omega^2 \mu\varepsilon \overline{H}, \tag{5.7}$$

as $\overline{E} = \overline{E}(x, y, z) \exp(\mathrm{j}\omega t)$ and $\overline{H} = \overline{H}(x, y, z) \exp(\mathrm{j}\omega t)$. Now the left-hand side term in cartesian coordinates is

$$\nabla^2 \overline{E} = \frac{\partial^2 \overline{E}}{\partial x^2} + \frac{\partial^2 \overline{E}}{\partial y^2} + \frac{\partial^2 \overline{E}}{\partial z^2}, \tag{5.8}$$

and if the wave is propagating in the z direction this becomes

$$\nabla^2 \overline{E} = \frac{\partial^2 \overline{E}}{\partial x^2} + \frac{\partial^2 \overline{E}}{\partial y^2} - k_z^2 \overline{E} \tag{5.9}$$

as $\overline{E} = \overline{E}(x, y) \exp(-\mathrm{j}k_z z)$, where k_z is called the wavenumber in the z direction. The wavenumber and the phase constant of Chapter 2 are similar in that they are both related to the rate of change of phase with distance. The phase constant is always a scalar, but the wavenumber can also be a vector.

5.2 Three groups of electromagnetic waves: TEM, TE and TM and hybrid waves

Group 1 – TEM waves

Now there are three groups of waves which will be discussed in this chapter. The first group consists of the transverse electromagnetic waves or TEM waves which have the following constraints:

$$\nabla^2_{x,y} \overline{E} = 0, \;\; \nabla^2_{x,y} \overline{H} = 0, \qquad E_z = 0 \text{ and } H_z = 0. \tag{5.10}$$

Now in Equation (5.7), if $\omega = 0$, then

$$\nabla^2 \overline{E} = 0 \text{ and } \nabla^2 \overline{H} = 0.$$

These are the Laplace equations for electrostatic and magnetostatic fields. Equation (5.10) gives solutions to the wave Equation (5.7) which obey Laplace's equation in the transverse or x, y-plane. They have no electromagnetic fields in the direction of propagation. So the wave equation for these waves is found by substituting Equations (5.10) and (5.9) into Equation (5.7):

$$k_z^2 \overline{E} = \omega^2 \mu \varepsilon \overline{E}. \tag{5.11}$$

Now the examples discussed in the last chapter were all about static fields in the x, y-plane. So all those examples have a wave solution of Equation (5.11) as

$$\overline{E} = \overline{E}(x, y) \exp \mathrm{j}(\omega t - k_z z). \tag{5.12}$$

The phase velocity of this wave, or the velocity of a point of constant phase – say for example zero phase – is when $\omega t = k_z z$ or the velocity v_p is

$$v_\mathrm{p} = \frac{z}{t} = \frac{\omega}{k_z} = \frac{\omega}{\omega\sqrt{\mu\varepsilon}} = \frac{1}{\sqrt{\mu\varepsilon}}. \tag{5.13}$$

All these TEM waves have the same velocity given by Equation (5.13). In free space, or in a vacuum, the values of μ and ε are as follows:

$$\mu = \mu_0 = 4\pi \times 10^{-7}\ \text{Hm}^{-1} \text{ and } \varepsilon = \varepsilon_0 = 8.85419\ \text{pFm}^{-1},$$

where the suffix 0 denotes the free space value. This gives the free space velocity as

$$v_0 = c = 2.997\,924\,58 \times 10^8\ \text{ms}^{-1} \text{ or approximately } 3 \times 10^8\ \text{ms}^{-1}. \tag{5.14}$$

The effect on the velocity of any dielectric or magnetic material is to reduce this velocity as

$$\mu = \mu_R\mu_0 \text{ and } \varepsilon = \varepsilon_R\varepsilon_0 \text{ so } v = \frac{1}{\sqrt{\mu_R\mu_0\varepsilon_R\varepsilon_0}} = \frac{3 \times 10^8}{\sqrt{\mu_R\varepsilon_R}}\ \text{ms}^{-1}, \tag{5.15}$$

where the suffix R denotes the relative dielectric or magnetic constant. So here for the first time is the value of the velocity of waves on transmission lines. Equation (4.90) links this velocity to the capacitance and inductance of the lines. Thus for all transmission lines with two or more conductors and surrounded by one dielectric, there is a unique velocity given by Equation (5.15) and a TEM wave mode of propagation. This mode is unique over a frequency range from zero up to the point where the higher order modes begin. For example, in Section 5.5 the higher order modes in coaxial cable are discussed.

Group 2 – TE and TM waves

The second group of waves are ones with the following constraints:

$$\nabla^2_{x,y}\overline{E} = \nabla^2_{x,y}\overline{H} = \text{ a constant, say } -p^2\overline{E} \text{ or } -k_c^2\overline{E}. \tag{5.16}$$

If $k = \omega\sqrt{\mu\varepsilon}$ then the wave equation for these waves is

$$p^2\overline{E} + k_z^2\overline{E} = k^2\overline{E} \quad \text{or} \quad k_z^2 = k^2 - k_c^2. \tag{5.17}$$

These waves occur in well defined boundaries, for example in rectangular metallic waveguides. In general, they have usually only one electromagnetic field in the direction of propagation. So they fall into two groups. Those with no electric field in the direction of propagation, i.e. transverse electric waves or TE waves, and those without a magnetic field on the direction of propagation, i.e. transverse magnetic waves or TM waves. These will be discussed in Section 5.3.

Group 3 – Hybrid waves

Finally, the third group are the waves with no constraints which often have all six electromagnetic fields present. These waves occur, for example, in dielectric waveguides and will be discussed at the end of the chapter. In some of these modes, one of the transverse fields can be very small. In these cases, the modes can be called E^y or E^x modes, where these are the dominant fields in the solutions. In other cases, where $E_z > H_z$, the modes are called EH modes and where $E_z < H_z$, they are called HE modes. These modes occur in Example 5.8 on optical fibres.

Example 5.1 A plane TEM wave: A plane wave has just one direction for each of its electric and magnetic fields. The usual convention is to choose the direction of the electric field as the direction of polarisation. Hence the terms 'vertical' or 'horizontal' polarisation. As an example of a plane wave, let the electric field be

$$\overline{E}_x = E_0 \exp \mathrm{j}(\omega t - k_z z)\overline{x}, \tag{5.18}$$

where $\overline{x}$ is a unit vector in the x direction and E_0 is the amplitude of the wave. The next step is to find the magnetic field associated with this wave and for this we return to Maxwell's Equation (5.3):

$$\nabla \times \overline{E} = -\frac{\partial \overline{B}}{\partial t} \quad \text{or in its sinusoidal form } \nabla \times \overline{E} = -\mathrm{j}\omega\mu\overline{H}.$$

Substituting Equation (5.18) into this Maxwell equation gives

$$\nabla \times \overline{E} = \begin{vmatrix} \overline{x} & \overline{y} & \overline{z} \\ \dfrac{\partial}{\partial x} & \dfrac{\partial}{\partial y} & \dfrac{\partial}{\partial z} \\ E_x & 0 & 0 \end{vmatrix} = \frac{\partial E_x}{\partial z}\overline{y} - \frac{\partial E_x}{\partial y}\overline{z} = -\mathrm{j}k_z E_x \overline{y} = -\mathrm{j}\omega\mu H_y \overline{y}.$$

This gives $H_y = \dfrac{k_z}{\omega\mu} E_x$ and the ratio of the two fields is given by

$$\frac{E_x}{H_y} = \frac{\omega\mu}{k_z} = \frac{\omega\mu}{\omega\sqrt{\mu\varepsilon}} = \sqrt{\frac{\mu}{\varepsilon}} = \eta. \tag{5.19}$$

This ratio is called the intrinsic impedance and is given the symbol η, and in free space it has the value

$$\eta = \eta_0 = \sqrt{\frac{4\pi.10^{-7}}{8.85419.10^{-12}}} = 120\pi = 376.61 \approx 377\ \Omega. \tag{5.20}$$

So a plane wave has these two components, an electric field and a magnetic field which are at right angles in space. The ratio of the amplitudes of these fields is fixed in free space at 377 Ω. When a plane wave passes through a medium this ratio can be changed as

$$\eta = \sqrt{\frac{\mu_R \mu_0}{\varepsilon_R \varepsilon_0}} = \sqrt{\frac{\mu_R}{\varepsilon_R}} \times 377\ \Omega. \tag{5.21}$$

In Equation (4.91):

$$Z_0 = \sqrt{\frac{L}{C}} = \sqrt{\frac{\mu}{\varepsilon}} \int_a^b f(r)\mathrm{d}r = \eta_0 \sqrt{\frac{\mu_R}{\varepsilon_R}} \int_a^b f(r)\mathrm{d}r$$

it can be seen that the characteristic impedance of all the transmission lines supporting TEM waves is linked to this intrinsic impedance of the medium between the conductors. So all the equations in the previous chapter for the velocity and the characteristic impedance contain these terms linking them to electromagnetic waves. In practice, the value of Z_0 is usually less than η_0.

The power carried by a plane wave can be evaluated using Poynting's vector which will be described next.

5.3 Poynting's vector for the average power flow

A simple way to work out the amount of power in a wave is to use Poynting's vector. This vector gives both the amplitude of the power and the direction of the power flow. To establish the vector, Maxwell's Equations (5.3) and (5.5) are needed. They are

$$\nabla \times \overline{E} = -\mu \frac{\partial \overline{H}}{\partial t} \text{ and } \nabla \times \overline{H} = \overline{J} + \varepsilon \frac{\partial \overline{E}}{\partial t}. \tag{5.22}$$

Now consider the vector identity

$$\nabla.(\overline{E} \times \overline{H}) = \overline{H}.(\nabla \times \overline{E}) - \overline{E}.(\nabla \times \overline{H}). \tag{5.23}$$

Substituting Equations (5.22) into (5.23) gives

$$\nabla.(\overline{E} \times \overline{H}) = -\mu \frac{\partial}{\partial t} |\overline{H}|^2 - \overline{E}.\overline{J} - \varepsilon \frac{\partial}{\partial t} |\overline{E}|^2. \tag{5.24}$$

If the losses in the system are simple ohmic ones then $\overline{J} = \sigma \overline{E}$, where σ is the conductivity of the medium, then

$$\overline{E}.\overline{J} = \sigma |\overline{E}|^2.$$

Integrating Equation (5.24) over a volume gives

$$\int_V \nabla.(\overline{E} \times \overline{H}) \mathrm{d}V = \int_V -\sigma |\overline{E}|^2 - \frac{\partial}{\partial t}\left(\mu |\overline{H}|^2 + \varepsilon |\overline{E}|^2\right) \mathrm{d}V.$$

Finally, using the divergence theorem on the left-hand side gives

$$\int_S (\overline{E} \times \overline{H}) \mathrm{d}\overline{S} = \int_V -\sigma |\overline{E}|^2 - \frac{\partial}{\partial t}\left(\mu |\overline{H}|^2 + \varepsilon |\overline{E}|^2\right) \mathrm{d}V. \tag{5.25}$$

This is the general Poynting Theorem and the left-hand side represents the power flowing out of a volume V, the terms on the right are the ohmic losses, the change in magnetic and electric energy within the volume. Taking the left-hand side as the power flow per metre squared gives the required Poynting vector $\overline{P}$ as

$$\overline{P} = \overline{E} \times \overline{H}. \tag{5.26}$$

In order to obtain the average power flow, half the real part of this vector is taken with the complex conjugate of the magnetic field, $\overline{H}^*$, in order to average the fluctuating nature of sinusoidal power, i.e. to give the RMS value. So the Poynting's vector for sinusoidal waves is

$$\overline{P}_{\mathrm{AV}} = \frac{1}{2}\,\mathrm{Re}\left(\overline{E} \times \overline{H}^*\right)\ \mathrm{Wm}^{-2}. \tag{5.27}$$

Example 5.2 Poynting's vector and TEM waves on transmission lines: Applying this to a plane wave, where E_x and H_y are the main fields,

$$\overline{P}_{\mathrm{AV}} = \frac{1}{2}\,\mathrm{Re}\begin{vmatrix} \overline{x} & \overline{y} & \overline{z} \\ E_x & 0 & 0 \\ 0 & H_y^* & 0 \end{vmatrix} = \frac{1}{2}\,\mathrm{Re}\left(E_x H_y^*\right)\overline{z} = \frac{|E_x|^2}{2\eta}\overline{z}\ \mathrm{Wm}^{-2}. \tag{5.28}$$

Now compare this result with the parallel plate line in Example 4.2 where a plane wave exists between the plates. The power in a wave is given in Equation (2.9) as

$$\text{Power } = \frac{V^2}{2Z_0},\ \text{where } Z_0 = \sqrt{\frac{\mu}{\varepsilon}} \times \frac{d}{w}\ \text{(Equation (4.77))},$$

$$\text{Power} = \frac{V^2}{2}\sqrt{\frac{\varepsilon}{\mu}} \times \frac{w}{d} = \frac{V^2}{2\eta} \times \frac{w}{d}.$$

Rewriting this using $E = V/d$ for the parallel plate gives

$$\text{Power } = \frac{E^2}{2\eta} \times dw. \tag{5.29}$$

This is exactly the result that would have been obtained by integrating the Poynting vector given in Equation (5.28) over the area between the plates. There is a new insight here. All the energy in the wave appears to be travelling in the space between the plates. So the energy is not in the currents induced in the plates, but in the electromagnetic fields. Just to check this insight, the same can be done with a coaxial cable:

$$\text{Power} = \frac{V^2}{2Z_0},\ \text{where } Z_0 = \frac{1}{2\pi}\sqrt{\frac{\mu}{\varepsilon}}\ln\left(\frac{b}{a}\right)\ \text{(Equation (4.71))},$$

$$\text{Power} = \frac{\pi V^2}{\eta \ln\left(\dfrac{b}{a}\right)}.$$

Using Equations (4.5) and (4.6):

$$E = \frac{V}{r \ln\left(\dfrac{b}{a}\right)}.$$

Integrating the Poynting vector over the space between the conductors:

$$P = \int_0^{2\pi} \int_a^b \frac{1}{2\eta} \frac{V^2}{r^2 \ln^2\left(\frac{b}{a}\right)} r \mathrm{d}r \mathrm{d}\theta$$

gives

$$P = \frac{\pi V^2}{\eta \ln\left(\frac{b}{a}\right)},$$

which is the same result that is obtained from $\frac{V^2}{2Z_0}$.

So in both cases the energy appears to be completely between the conductors and not in the conductors themselves. This is not surprising from the integral of Equation (5.24) where the stored energy on a capacitor or an inductor can be expressed as

$$\frac{1}{2}CV^2 = \int_V \frac{1}{2}\varepsilon|\overline{E}|^2 \mathrm{d}V \text{ and } \frac{1}{2}LI^2 = \int_V \frac{1}{2}\mu|\overline{H}|^2 \mathrm{d}V. \tag{5.30}$$

In other words the energy stored on a capacitor can also be determined by the electric field between the plates, and similarly, for an inductance, the energy is a function of the magnetic field around the inductor. The problem is which is the more fundamental description and this will be discussed at the end of this chapter. It is clear from the above that the calculation of power using circuit parameters is much easier than integrating the electric or magnetic fields. For those transmission lines with a uniform medium between the conductors, all of them can be conformally transformed into a parallel plate format; so the above principles apply whatever the shape or configuration of the original conductors. For those transmission lines with two or more dielectrics, the effective dielectric constant can be used to find the power in a wave.

5.4 TE and TM waves within metallic rectangular boundaries

The most common example of waves within metallic boundaries is the rectangular waveguide, which is shown in Figure 5.1. The metallic walls cause the tangential or surface electric fields to go to zero. Clearly there is no way a static electric field can exist within this guide, and so a TEM wave is not possible. The waves are in two groups. The first assumes that there are only electric fields transverse to the direction of propagation and this group is called the TE or transverse electric group of waves. The second group is called the TM or transverse magnetic group because that has magnetic fields only in the transverse direction.

Starting with the TE waves with variation in just the x direction and referring to the coordinate axes shown in Figure 5.1, the electric field can have a solution of the form

classified by their m and n numbers as TE_{mn}. So TE_{10} means $m = 1$ and $n = 0$ and this is the first mode described in this example. This mode has the smallest value of k_c and hence the lowest cut-off frequency of all the possible TE modes.

The TM modes

Now the TM modes have as their starting point the E_z field as

$$\overline{E}_z = E_0 \sin(k_x x) \sin(k_y y) \exp \mathrm{j}(\omega t - k_z z)\overline{z}. \tag{5.45}$$

By a similar method as for the TE waves, this leads to the following field solutions:

$$E_x = \frac{k_z}{\omega\varepsilon} H_y \text{ and } E_y = -\frac{k_z}{\omega\varepsilon} H_x,$$

$$H_x = \frac{\mathrm{j}}{\left(\omega\mu - \frac{k_z^2}{\omega\varepsilon}\right)} E_0 \sin(k_x x) \cos(k_y y) \exp \mathrm{j}(\omega t - k_z z),$$

$$H_y = -\frac{\mathrm{j}}{\left(\omega\mu - \frac{k_z^2}{\omega\varepsilon}\right)} E_0 \cos(k_x x) \sin(k_y y) \exp \mathrm{j}(\omega t - k_z z).$$

When these waves are substituted in the electromagnetic wave equation the result is the same as Equations (5.43) and (5.44). The boundary conditions for this field require it to go to zero on all four walls so neither n nor m can be zero. This means that of all the modes, the TE_{10} is the one mode with the lowest cut-off frequency. In Figure 5.4(a) are shown the fields in a cross-section through the waveguide. The lines show the direction and not the amplitude of the fields. So, for instance, in the TE_{10} mode both the electric and the magnetic fields vary sinusoidally with the horizontal direction and with a peak in the centre. Except for this mode, all the other modes can be made up of the shapes seen in the figures repeated in both directions.

The cut-off frequencies of the various modes can be represented on a diagram as a ratio of the lowest cut-off frequency, as shown in Figure 5.4(b). This ratio is given by

$$\frac{f_{\mathrm{cTE}_{mn}}}{f_{\mathrm{cTE}_{10}}} = \sqrt{m^2 + 4n^2}. \tag{5.46}$$

For the next example, just the TE_{10} mode will be described as this is the most commonly used mode and the higher order modes usually are not propagating. If they are produced by some irregularities in the waveguide, then they will rapidly decay and not propagate as they will be below their cut-off frequencies. These modes can store both electric and magnetic energy. Depending on which is larger, they can be represented in an equivalent circuit as either a capacitance or an inductance. However, even though they decay away they can interact if two irregularities or discontinuities are close to each other. In this case, the equivalent circuit model can give misleading results.

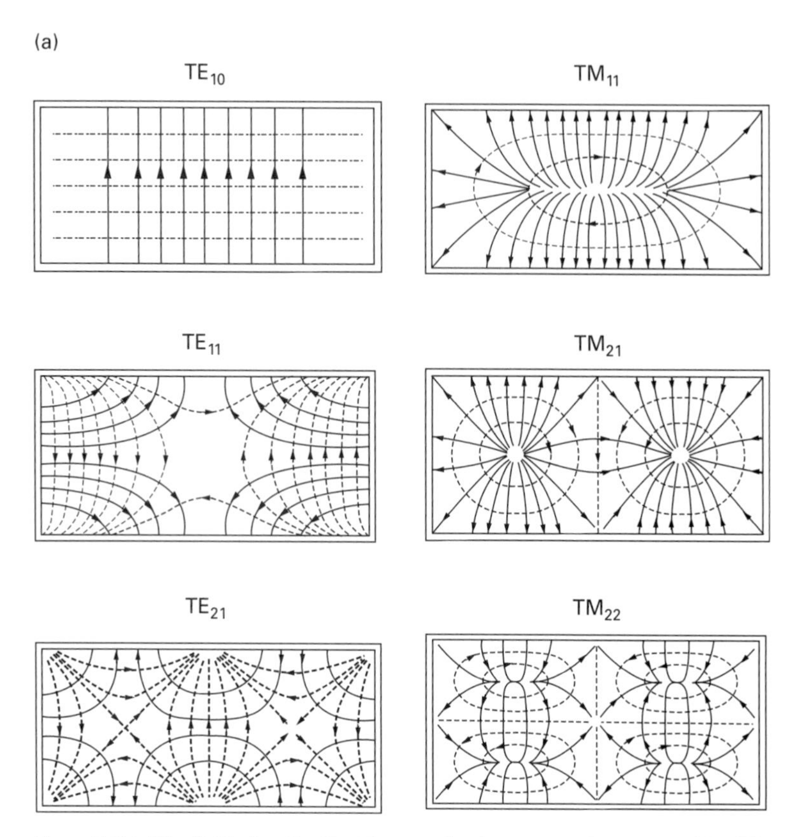

Figure 5.4(a) The fields for the first few modes in rectangular waveguide. The solid lines are the electric fields and the dotted lines the magnetic fields. These figures have been reproduced by permission of the Institution of Engineering and Technology from the *Waveguide Handbook* by N. Marcuvitz, published by the IET in 1985 as a reprint of the original publication by the McGraw-Hill Book Company in 1951.

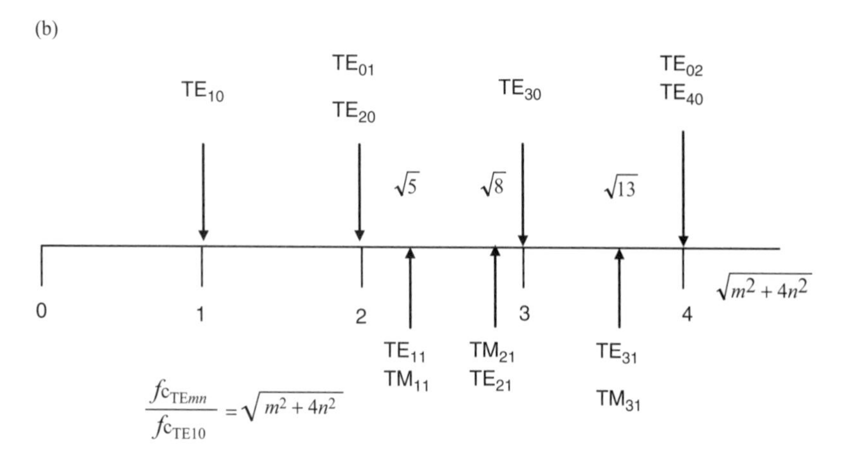

Figure 5.4(b) The cut-off frequencies of the first few modes in metallic rectangular waveguide as a ratio of the lowest cut-off frequency, i.e. that for the TE$_{10}$ mode.

Example 5.3 The TE_{10} mode in metallic rectangular waveguide: The TE_{10} mode is the lowest order mode in rectangular metallic waveguide and in this section a more complete study of this mode will be made. First of all, the associated magnetic fields that form the mode along with the electric field need to be found from Maxwell's equation (the first one in Equation (5.39) with $E_x = 0$):

$$\nabla \times \overline{E} = \begin{vmatrix} \overline{x} & \overline{y} & \overline{z} \\ \frac{\partial}{\partial x} & \frac{\partial}{\partial y} & -jk_z \\ 0 & E_y & 0 \end{vmatrix} = -j\omega\mu\overline{H}, \text{where } E_y = E_0 \sin\left(\frac{\pi x}{a}\right) \exp j(\omega t - k_z z). \tag{5.47}$$

Since there is only an electric field in the y direction, the equations for the magnetic fields are

$$jk_z E_y = -j\omega\mu H_x \text{ and } \frac{\partial E_y}{\partial x} = -j\omega\mu H_z. \tag{5.48}$$

These can be rewritten as

$$H_x = -\frac{k_z}{\omega\mu} E_0 \sin\left(\frac{\pi x}{a}\right) \exp j(\omega t - k_z z) \text{ and } H_z = \frac{j\pi}{\omega\mu a} E_0 \cos\left(\frac{\pi x}{a}\right) \exp j(\omega t - k_z z), \tag{5.49}$$

and clearly the transverse fields to the direction of propagation have the simple relationship

$$\frac{E_y}{H_x} = -\frac{\omega\mu}{k_z}. \tag{5.50}$$

These fields are shown diagrammatically in Figure 5.5.

Using Equation (5.33) to expand Equation (5.50),

$$k_z = \sqrt{k_0^2 - \left(\frac{\pi}{a}\right)^2} = k_0\sqrt{1 - \left(\frac{\lambda}{\lambda_c}\right)^2} = \omega\sqrt{\mu\varepsilon}\sqrt{1 - \left(\frac{\lambda}{\lambda_c}\right)^2} = \omega\sqrt{\mu\varepsilon}\sqrt{1 - \left(\frac{f_c}{f}\right)^2}$$

$$\text{so } \frac{E_y}{H_x} = -\sqrt{\frac{\mu}{\varepsilon}} \times \frac{1}{\sqrt{1 - \left(\frac{f_c}{f}\right)^2}} = -\frac{\eta}{\sqrt{1 - \left(\frac{f_c}{f}\right)^2}} = -Z_{TE}. \tag{5.51}$$

This ratio, Z_{TE}, is called the transverse wave impedance for metallic rectangular waveguide and it relates just the transverse fields, not the longitudinal magnetic field. It is similar to the characteristic impedance Z_0 in transmission lines except that the two fields involved are orthogonal in space. Figure 5.6 shows a typical plot for this wave impedance against frequency. In this graph the rapid variation of the wave impedance over the mono-mode bandwidth, i.e. 30–60 GHz,

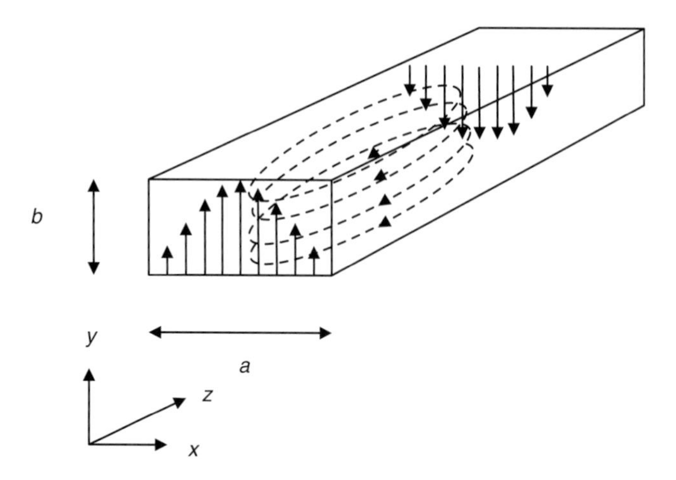

Figure 5.5 The fields inside a rectangular guide for the TE_{10} mode. The vertical lines in the y direction are the electric fields with their sine wave variation in the x direction and sinusoidal variation in the z direction. Half a wavelength in the z direction is shown. The magnetic fields form loops in the x, z-plane, and when the electric fields are at a maximum there is no magnetic field in the z direction.

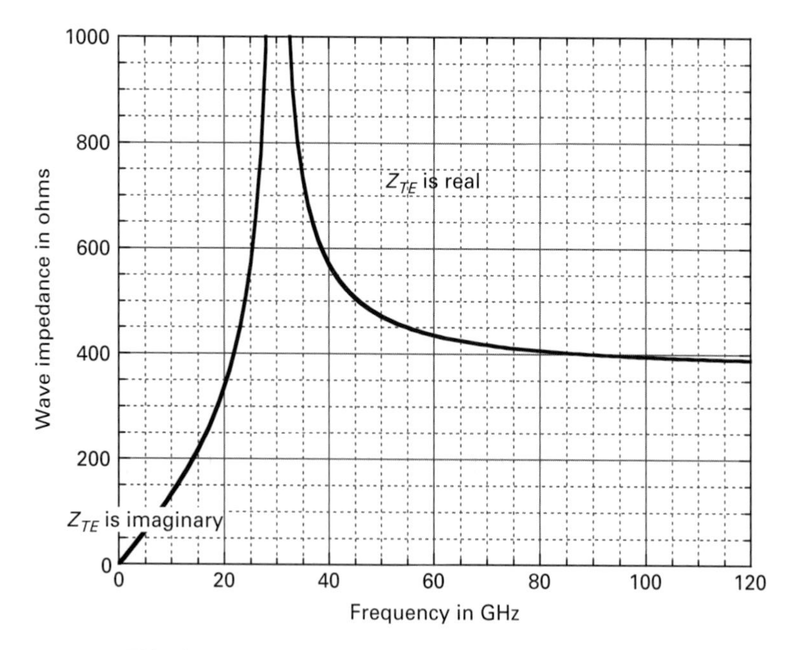

Figure 5.6 This shows the transverse wave impedance for the TE_{10} mode in metallic rectangular waveguide. The cut-off frequency is 30 GHz. Above this the wave impedance Z_{TE} is real. At the cut-off frequency it tends to infinity and above that frequency it tends to the free space value of 377 Ω. Below the cut-off frequency, there is no propagation, and the fields are evanescent, that is, they are attenuated. The wave impedance in this region is imaginary and is usually taken to be positive, that is a pure inductance. At zero frequency, this becomes a short circuit as expected.

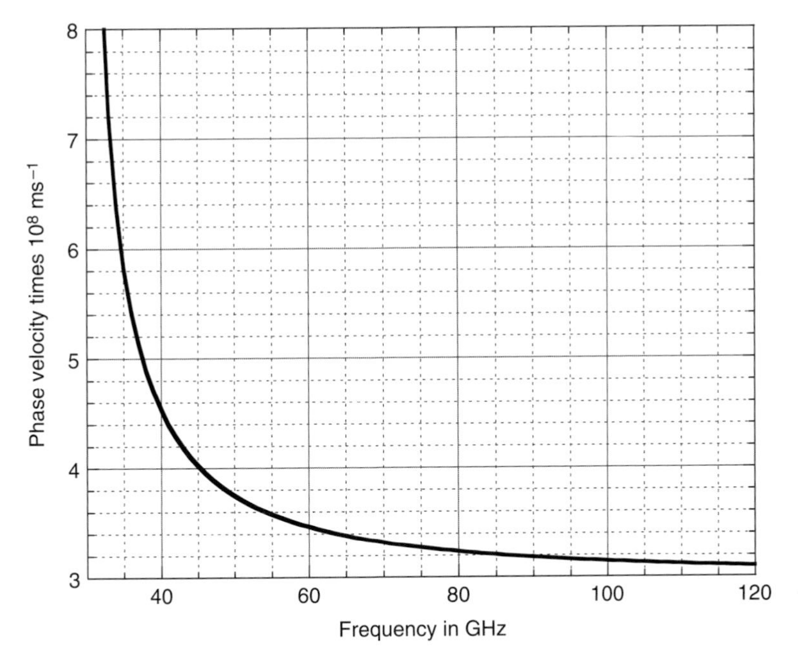

Figure 5.7 The phase velocity in metallic rectangular waveguide. The cut-off frequency is 30 GHz.

demonstrates the dispersive nature of these waveguides. In practice, the complete 30 GHz is not used and usually just the higher part, say 38.4–56.7 GHz, would be used. Over this range of frequencies, the wave impedance for an air-filled guide varies from 603 Ω to 444 Ω and is always greater than the free-space value of 377 Ω. The velocity of propagation also varies considerably over this frequency range. Figure 5.7 shows the variation for the same waveguide described in Figure 5.6. Since

$$v_z = \frac{\omega}{k_z} = \frac{\omega}{\omega\sqrt{\mu\varepsilon}\left(1 - \left(\frac{f_c}{f}\right)^2\right)^{\frac{1}{2}}} = \frac{1}{\sqrt{\mu\varepsilon}} \times \frac{1}{\left(1 - \left(\frac{f_c}{f}\right)^2\right)^{\frac{1}{2}}}, \tag{5.52}$$

the velocity in the z direction – or the direction of propagation – is always greater than the velocity of plane electromagnetic waves. This does not mean that information can be transmitted faster than the speed of light, as this is not allowed by Special Relativity. This velocity is a phase velocity and not the velocity of information. This difference will be discussed later in Chapter 6. The range of phase velocities over the same frequency range 38.4 to 56.7 GHz is from 4.8×10^8 ms^{-1} to 3.5×10^8 ms^{-1}, clearly again signifying a dispersive guide. Some typical waveguide characteristics are given in Table 5.1. The attenuation is given in dBm^{-1} which is not realistic as many of the smaller guides are only used over a few wavelengths.

Table 5.1 Rectangular waveguide data covering 8.2 to 325 GHz. The guides are air filled. The waveguides are now specified up to 3300 GHz in IEEE P1785. The IEEE guides are classified by the 'a' dimension in μm e.g. WR3 becomes WM864 in IEEE classification. The smallest guide, WM 86, is only 86 × 43 μm in cross-section and would have an attenuation of 435 dBm^{-1} or 0.047 dB/wavelength.

Band letter (US UK)	Dimensions in mm	Cut-off frequency GHz	Bandwidth GHz	Attenuation dBm^{-1}	Power kW
X WR90,WG16	22.860 × 10.160	6.570	8.20–12.4	0.110	200
J or K_u WR62,WG18	15.799 × 7.899	9.495	12.4–18.0	0.176	120
K WR42,WG20	10.668 × 4.318	14.088	18.0–26.5	0.370	43
Q or K_a WR28,WG22	7.112 × 3.556	21.184	26.5–40.0	0.583	22
F WR19,WG24	4.775 × 2.388	31.595	40.0–60.0	1.060	11
E WR12,WG26	3.099 × 1.549	48.549	60.0–90.0	2.03	4.2
N WR8,WG28	2.032 × 1.016	74.440	90.0–140.0	3.82	1.8
A or G WR5,WG30	1.295 × 0.645	116.475	140.0–220.0	7.50	0.7
R WR3, WG32	0.864 × 0.432	174.438	220.0–325.0	13.76	0.4

Finally, the power flowing in the waveguide can be found from Equation (5.27) by integrating Poynting's vector over the cross-section of the waveguide. The relevant electric and magnetic fields are given in Equations (5.47) and (5.48) so

$$\overline{P}_{\text{AV}} = \int_S \frac{1}{2}\,\text{Re}\left(\overline{E} \times \overline{H}^*\right)\text{d}S,$$

$$\text{where } \overline{E} \times \overline{H}^* = \begin{vmatrix} \overline{x} & \overline{y} & \overline{z} \\ 0 & E_y & 0 \\ H_x^* & 0 & H_z^* \end{vmatrix} = (E_y H_z^*)\overline{x} - (E_y H_x^*)\overline{z}.$$

$$\text{Now } E_y H_z^* = -\frac{\text{j}\pi}{\omega\mu a}E_0^2 \sin\left(\frac{\pi x}{a}\right)\cos\left(\frac{\pi x}{a}\right) \text{ and } E_y H_x^* = -\frac{k_z}{\omega\mu}E_0^2 \sin^2\left(\frac{\pi x}{a}\right).$$

By taking the term that is real:

$$\overline{P}_{\text{AV}} = \int_S \frac{k_z}{2\omega\mu}E_0^2 \sin^2\left(\frac{\pi x}{a}\right)\text{d}s = \frac{k_z b E_0^2}{2\omega\mu}\int_0^a \sin^2\left(\frac{\pi x}{a}\right)\text{d}x = \frac{k_z ab E_0^2}{4\omega\mu} = \frac{E_0^2}{4Z_{\text{TE}}}ab. \quad (5.53)$$

So if 1 mW is travelling down a guide where $Z_{\text{TE}} = 500\ \Omega$ and $ab = 2.323\ \text{cm}^2$, then the electric field at the centre of the guide $E_0 = 92.7\ \text{Vm}^{-1}$ or just 0.94 V at the centre.

In solving waveguide problems, it is sometimes useful to use an equivalent transmission line with a characteristic impedance given by

$$Z_0 = \frac{2bZ_{\mathrm{TE}}}{a} \quad \text{since } \overline{P}_{\mathrm{AV}} = \frac{V_0^2}{2Z_0} \text{ and } V_0 = E_0 b. \tag{5.54}$$

However, the analogy breaks down in certain cases so it needs to be used with caution. For example, when both a and b change so as to keep Z_0 constant, there will be a reflection despite the constant characteristic impedance. Needless to add, the power is travelling in the space inside the rectangular waveguide.

5.5 Waves within metallic circular boundaries

The solution for electromagnetic waves inside a circular metallic boundary is conveniently found by using circular polar coordinates rather than cartesian coordinates. As in previous examples, the propagation will be in the z direction so the appropriate wave equation is

$$\nabla_{x,y}^2 \overline{E} - k_z^2 \overline{E} = -\omega^2 \mu \varepsilon \overline{E},$$

which can be changed into polar coordinates as

$$\frac{\partial^2 \overline{E}}{\partial r^2} + \frac{1}{r}\frac{\partial \overline{E}}{\partial r} + \frac{1}{r^2}\frac{\partial^2 \overline{E}}{\partial \theta^2} - k_z^2 \overline{E} = -\omega^2 \mu \varepsilon \overline{E} = -k^2 \overline{E}. \tag{5.55}$$

The solutions are in the second group of waves, as the boundary is a fixed one, so

$$\nabla_{r,\theta}^2 \overline{E} = \left(-k^2 + k_z^2\right) \overline{E} = -k_c^2 \overline{E}. \tag{5.56}$$

So the main part of the solution is solving Equation (5.56). This is usually done by separation of the variables as follows by assuming

$$\overline{E} = R(r)\Theta(\theta). \tag{5.57}$$

Substituting this into Equation (5.56) gives

$$R''\Theta + \frac{1}{r}R'\Theta + \frac{1}{r^2}R\Theta'' = -k_c^2 R\Theta.$$

The two variables can now be separated as follows:

$$\frac{r^2}{R}R'' + \frac{r}{R}R' + r^2 k_c^2 = -\frac{\Theta''}{\Theta}. \tag{5.58}$$

Since the left-hand side of Equation (5.58) is only a function of R and the right-hand side only a function of Θ then both must be equal to a constant – say n^2. This gives two differential equations to solve:

$$R'' + \frac{1}{r}R' + \left(k_c^2 - \frac{n^2}{r^2}\right)R = 0 \text{ and } \Theta'' = -n^2\Theta. \tag{5.59}$$

The first of these equations is Bessel's equation which has solutions which are Bessel numbers, $J_n(k_c r)$. The second is the simple harmonic equation which has solutions of the form

$$\Theta = \sin(n\theta) \text{ or } \Theta = \cos(n\theta). \tag{5.60}$$

In a similar way to the rectangular waveguide, the modes are divided into two groups, those with only transverse magnetic fields, the TM modes, and those with only transverse electric fields, the TE modes.

TM modes

Starting with the TM modes, the fields in the direction of propagation can be taken as

$$\overline{E}_z = E_0 J_n(k_c r)\sin(n\theta)\overline{z}. \tag{5.61}$$

Using the Maxwell Equation (5.3) for sinusoidal fields: $\nabla \times \overline{E} = -j\omega\mu\overline{H}$, transforming the equation into cylindrical coordinates in terms of $\overline{E}_z$ and putting $\overline{H}_z = 0$ gives

$$\overline{E}_r = -\frac{jk_z}{k_c^2}\frac{\partial E_z}{\partial r}\overline{r},\quad \overline{E}_\theta = -\frac{jk_z}{rk_c^2}\frac{\partial E_z}{\partial \theta}\overline{\theta},\quad H_r = \frac{j\omega\varepsilon}{rk_c^2}\frac{\partial E_z}{\partial \theta}\overline{r} \text{ and } \overline{H}_\theta = -\frac{j\omega\varepsilon}{k_c^2}\frac{\partial E_z}{\partial r}\overline{\theta},$$

and substituting Equation (5.61) gives

$$\overline{E}_r = -\frac{jk_z}{k_c}E_0 J_n'(k_c r)\sin(n\theta)\overline{r};\qquad \overline{E}_\theta = -\frac{jnk_z}{rk_c^2}E_0 J_n(k_c r)\cos(n\theta)\overline{\theta};$$

$$\overline{H}_r = \frac{j\omega\varepsilon n}{rk_c^2}E_0 J_n(k_c r)\cos(n\theta)\overline{r};\qquad \overline{H}_\theta = -\frac{j\omega\varepsilon}{k_c}E_0 J_n'(k_c r)\sin(n\theta)\overline{\theta}. \tag{5.62}$$

At the surface of the cylindrical waveguide the tangential or surface electric fields must go to zero. If the radius of the cylinder is a, then if $J(k_c a) = 0$ this makes

$$\overline{E}_z = \overline{E}_\theta = 0. \tag{5.63}$$

Now the solutions for $J_n(k_c a) = 0$ are given by $J_n(p_{nl}) = 0$, where n is the order of the Bessel number and l is the number of the solution. In order that the solution in the θ direction is cyclic, n must have integer values including zero. So $n = 0, 1, 2, 3,\ldots$ and a few of the various solutions for $l = 1, 2, 3,\ldots$ are tabulated below:

$$\begin{array}{lll} p_{01} = 2.405; & p_{02} = 5.520; & p_{03} = 8.654, \ldots \\ p_{11} = 3.832; & p_{12} = 7.016; & p_{13} = 10.173, \ldots \\ p_{21} = 5.136; & p_{22} = 8.417; & p_{23} = 11.62, \ldots \end{array} \tag{5.64}$$

Taking $k_c a = p_{nl}$, the cut-off frequency for the various modes can be found from

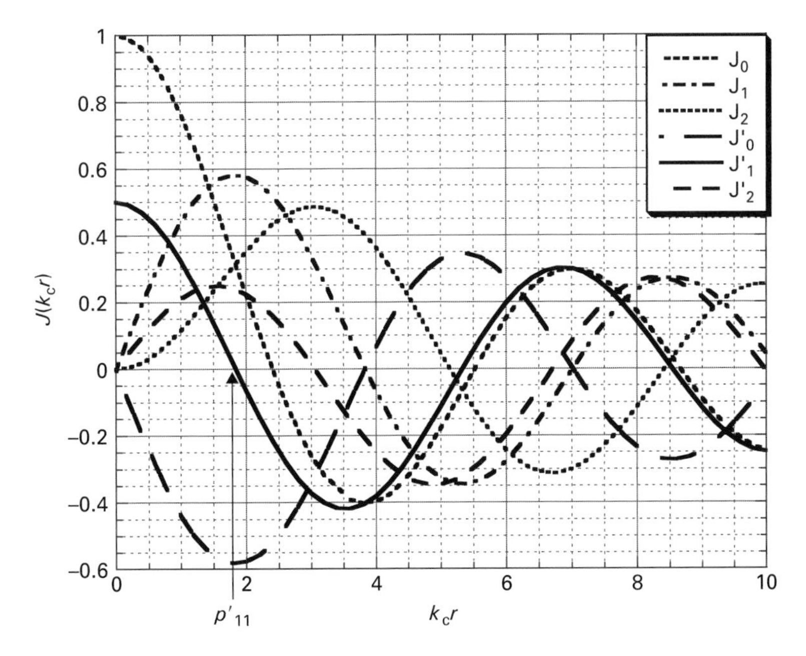

Figure 5.8 The first three Bessel functions and their derivatives. The lowest order mode, the TE_{11} mode solution, is shown where the solid line first crosses the x-axis.

$$k_c = \frac{2\pi}{\lambda_c} = \frac{2\pi f_c}{3.10^8} \text{ or } \lambda_c = \frac{2\pi a}{p_{nl}} \text{ and } f_c = \frac{3.10^8}{\lambda_c}. \tag{5.65}$$

The first three Bessel functions and their derivatives are shown in Figure 5.8.

The TE modes

The TE modes can be found by choosing a magnetic field in the z direction

$$\overline{H}_z = H_0 J_n(k_c r)\cos(n\theta)\overline{z}. \tag{5.66}$$

Rearranging the Maxwell equations in terms of $\overline{H}_z$ gives

$$\overline{E}_r = -\frac{j\omega\mu}{rk_c^2}\frac{\partial H_z}{\partial\theta}\overline{r}; \quad \overline{E}_\theta = \frac{j\omega\mu}{k_c^2}\frac{\partial H_z}{\partial r}\overline{\theta}; \quad \overline{H}_r = -\frac{jk_z}{k_c^2}\frac{\partial H_z}{\partial r}\overline{r}; \quad \overline{H}_\theta = -\frac{jk_z}{rk_c^2}\frac{\partial H_z}{\partial\theta}\overline{\theta};$$

and substituting Equation (5.66) gives

$$\overline{E}_r = \frac{jn\omega\mu}{rk_c^2}H_0 J_n(k_c r)\sin(n\theta)\overline{r}; \quad \overline{E}_\theta = \frac{j\omega\mu}{k_c}H_0 J_n'(k_c r)\cos(n\theta)\overline{\theta};$$

$$\overline{H}_r = -\frac{jk_z}{k_c}H_0 J_n'(k_c r)\cos(n\theta)\overline{r}; \quad \overline{H}_\theta = \frac{jk_z n}{rk_c^2}H_0 J_n(k_c r)\sin(n\theta)\overline{\theta}. \tag{5.67}$$

Once again the surface electrical fields must be zero at $r = a$ and in this case only the $\overline{E}_\theta$ is involved. So for these modes $J_n'(k_c r) = 0$ and the solutions are $J_n'(p_{nl}') = 0$.

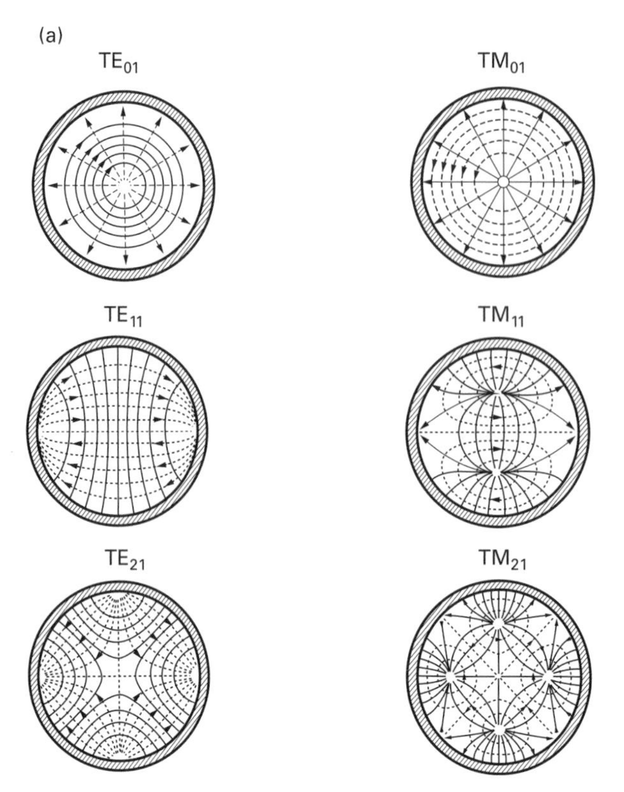

Figure 5.9(a) The fields for the first few modes in a circular waveguide. The solid lines are the electric fields and the dotted lines the magnetic fields. These figures have been reproduced by permission of the Institution of Engineering and Technology from the *Waveguide Handbook* by N. Marcuvitz, published by the IET in 1985 as a reprint of the original publication by the McGraw-Hill Book Company in 1951.

A few of the solutions are tabulated below:

$$\begin{aligned} &p'_{01} = 3.832; \quad p'_{02} = 7.01; \quad p'_{03} = 10.173; \\ &p'_{11} = 1.841; \quad p'_{12} = 5.331; \quad p'_{13} = 8.536; \\ &p'_{21} = 3.054; \quad p'_{22} = 6.706; \quad p'_{23} = 9.969. \end{aligned} \tag{5.68}$$

In Figure 5.9(a) are shown the first few circular waveguide modes. The lines show the direction of the fields and not their amplitudes. Despite the relative simplicity of the fields of the TE_{01} and TM_{01} modes they do not have the lowest cut-off frequencies. The mode with that distinction is the TE_{11} mode which has the lowest p' number and lower than any p number – see (5.64) and (5.68). Taking $k_c a = p'_{nl}$, the cut-off frequency for the various modes can be found from

$$k_c = \frac{2\pi}{\lambda_c} = \frac{2\pi f_c}{3.10^8}\ \text{ms}^{-1} \text{ or } \lambda_c = \frac{2\pi a}{p'_{nl}} \text{ and } f_c = \frac{3.10^8}{\lambda_c}\ \text{ms}^{-1}. \tag{5.69}$$

From all this data, a cut-off diagram can now be constructed, using Equations (5.65) and (5.69), as shown in Figure 5.9(b). In this case the TE_{11} mode is the one

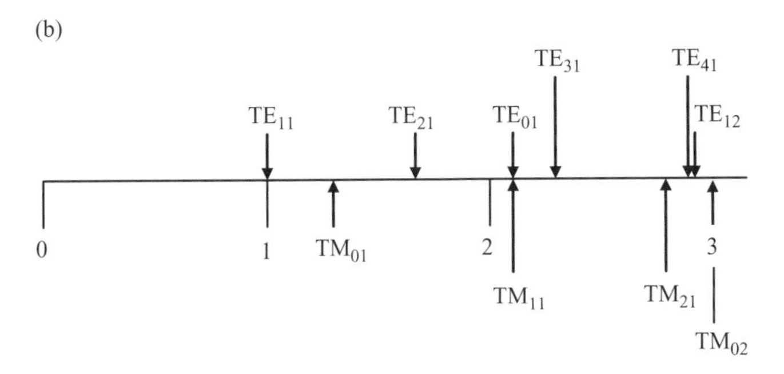

Figure 5.9(b) The cut-off frequencies of the first few modes in a circular metallic waveguide relative to the cut-off frequency of the TE_{11} mode. The small mono-mode bandwidth, just 30 % of an octave, can clearly be seen. The solutions to these modes are not harmonic like some of the rectangular ones.

with the lowest cut-off frequency, but the next higher order modes are very close and so this guide has a mono-mode bandwidth much less than an octave. However, the guide is often used over a wider bandwidth despite these other unwanted modes. The higher order modes will only propagate if there is a discontinuity in the guide which disturbs the TE_{11} mode so that energy can be coupled into them. Otherwise the orthogonality of these modes means that energy will not be coupled into them. This orthogonality is weakened when there are losses in the guide and this will be discussed in the later chapter on attenuation. The main feature of a circular guide is that the direction of the polarisation can easily be changed without any reflections. This is used to make a number of devices including isolators and attenuators. These devices are usually terminated in a rectangular waveguide via a transitional waveguide with a variable cross-section.

Example 5.4 The TE_{11} circular waveguide mode: The lowest order mode has many similarities with the lowest order mode in a rectangular waveguide, i.e. the TE_{10} mode. First of all, the transverse wave impedance is the same. Examining Equations (5.67) gives

$$\frac{E_r}{H_\theta} = \frac{\omega\mu}{k_z} \text{ and } \frac{E_\theta}{H_r} = -\frac{\omega\mu}{k_z},$$

which is the same as Equation (5.50) for the TE_{10} mode.

So the transverse wave impedance for both modes is given by Equation (5.51) as

$$Z_{\mathrm{TE}} = \frac{\eta}{\sqrt{1-(f_c/f)^2}}.$$

Then the velocity is the same. Equation (5.52) applies to both modes so in both cases

$$v_z = \frac{\omega}{k_z} = \frac{1}{\sqrt{\mu\varepsilon}} \times \frac{1}{\left(1 - (f_c/f)^2\right)^{\frac{1}{2}}}.$$

In order to make a transition between rectangular and circular waveguides, the two cut-off frequencies have to be equal and then the cross-section needs slowly changing from rectangular to circular. In practice the longer this transition is made, the less likely there will be reflections. For example, if an X-band rectangular waveguide is being changed to a circular waveguide, then, starting with the broad dimension of the X-band guide, a, as 2.286 cm, this gives the cut-off frequency for that guide as

$$f_c = \frac{3 \times 10^{10}}{\lambda_c} \text{cms}^{-1} = 6.562 \text{ GHz}, \text{ where } \quad \lambda_c = 2a.$$

Now moving to the circular waveguide:

$$\lambda_c = \frac{2\pi a}{p'_{11}} = \frac{2\pi a}{1.841},$$

so the diameter of the circular guide is 2.68 cm.

This is not much bigger than the broad dimension of the rectangular guide. So the main distinguishing characteristic of these modes is their shape of the fields as most other features are common to both. This will be further explored in a later chapter.

5.6 Higher order modes in coaxial cable

The coaxial cable is a metallic waveguide with a circular boundary and is not immune to higher order modes. These modes determine the upper limit of the mono-mode bandwidth for the cable. The theory of these modes is complex and is thoroughly covered in the texts referenced at the end of the chapter. The solutions involve not only the first solution of Bessel's equation, J, but also the second solution, N. The guides support both TE and TM waves and the equations needed to satisfy the boundary conditions at both the inner radius, a, and the outer radius, b, for TM modes are

$$AJ_n(k_c a) + BN_n(k_c a) = 0 \text{ and } AJ_n(k_c b) + BN_n(k_c b) = 0.$$

Eliminating the constants A and B gives

$$\frac{N_n(k_c a)}{J_n(k_c a)} = \frac{N_n(k_c b)}{J_n(k_c b)} \quad \text{for TM modes}$$

and for TE modes:

$$CJ'_n(k_c a) + DN'_n(k_c a) = 0 \text{ and } CJ'_n(k_c b) + DN'_n(k_c b) = 0.$$

Again, eliminating the constants C and D gives

$$\frac{N_n'(k_c a)}{J_n'(k_c a)} = \frac{N_n'(k_c b)}{J_n'(k_c b)} \quad \text{for TE modes.} \tag{5.70}$$

An approximate solution for TM modes is given by

$$\lambda_c \approx \frac{2}{p}(b - a), \; p = 1, 2, 3, \; \ldots \tag{5.71}$$

This assumes that at the cut-off frequency an integer number of half wavelengths exists between the conductors.

For TE modes the approximate solution is

$$\lambda_c = \frac{2\pi}{n}\left(\frac{a + b}{2}\right), \; n = 1, 2, 3, \; \ldots \tag{5.72}$$

This assumes that at the cut-off frequency an integer number of half wavelengths is spread round the average circumference.

Particular solutions for $b = 3a$ are shown in Figure 5.10.

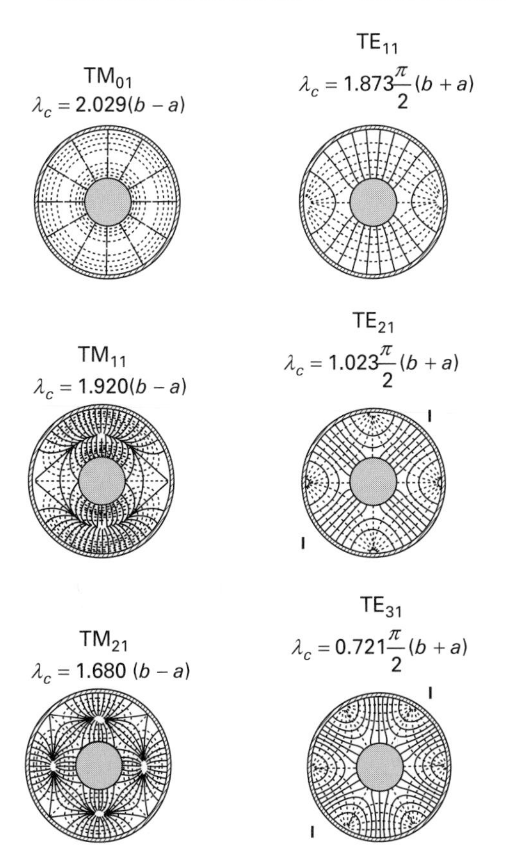

Figure 5.10 The first few higher order modes in coaxial cable. The electric fields are shown with solid lines and the magnetic fields are the dotted lines. The cut-off wavelengths are given for $b = 3a$. For other ratios of dimensions, Equation (5.70) will need to be solved for each case. These figures have been reproduced by permission of the Institution of Engineering and Technology from the *Waveguide Handbook* by N. Marcuvitz, published by the IET in 1985 as a reprint of the original publication by the McGraw-Hill Book Company in 1951.

For example, take a 50 Ω air-filled coaxial cable with an outer diameter of 7 mm. This gives the values of $b = 3.5$ mm and $a = 1.521$ mm and the cut-off frequencies of the lowest modes as

$$\mathrm{TM}_{01} = 74.7 \text{ GHz}, \quad \mathrm{TE}_{11} = 20.3 \text{ GHz}.$$

So the TE_{11} mode determines the upper frequency limit of the mono-mode bandwidth, and in order to use coaxial cables above 20 GHz, the diameter has to be reduced. For instance, if the coaxial cable was going to be used up to 100 GHz the outer diameter would need to be reduced by a factor of 5 to 1.4 mm. The effect of this reduction in size is to increase the attenuation and this will be discussed in the next chapter. These higher order modes in coaxial cable will be present in evanescent form below their cut-off frequencies and will have stored energy in them if discontinuities occur in the cable. A rough guide to the TE_{11} mode cut-off wavelength is $2\pi \times$ average radius.

5.7 Ridged waveguide

The bandwidth of rectangular metallic waveguide can be greatly increased by having a ridge down the centre of the waveguide, as shown in Figure 5.11.

An approximate theory is to assume that the ridge is a section of waveguide of reduced height d at the centre of the waveguide, then a technique of transverse resonance can be used to model this. This technique assumes that at the cut-off frequency the wave is only moving transverse to the x direction. This assumption will be discussed in a later chapter. It also assumes that this wave is effectively moving in a parallel plate guide of width one metre and height either b or d. Equation (4.77) gives the characteristic impedance of such guides as

$$\eta_0 d \ \text{ or } \eta_0 b \ \ \Omega.$$

The impedance looking sideways from the centre is shown in Figure 5.12.

Using Equations (2.21) and (2.22), the impedance at the centre of the guide looking in one direction is given by

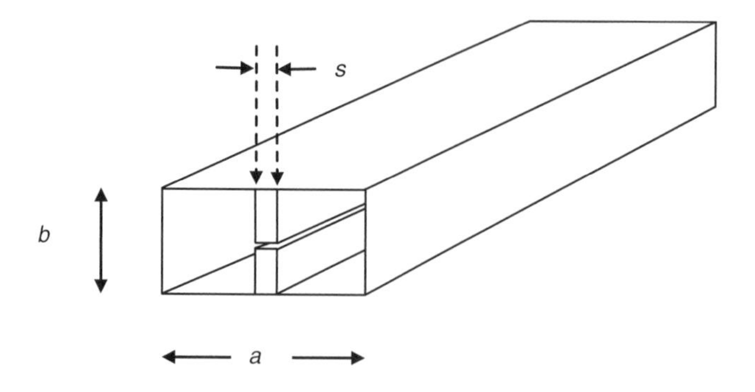

Figure 5.11 A double ridged rectangular waveguide. The gap between the ridges is d and the width of the ridges is s.

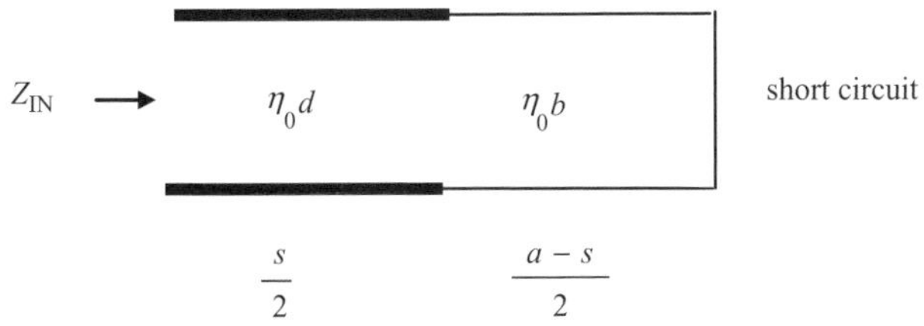

Figure 5.12 The impedance looking sideways at the cut-off frequency for a ridged waveguide.

$$Z_{\text{IN}} = \eta_0 d \left(\frac{\mathrm{j}\eta_0 b \tan\left(\frac{k_c(a-s)}{2}\right) + \mathrm{j}\eta_0 d \tan\left(\frac{k_c s}{2}\right)}{\eta_0 d - \eta_0 b \tan\left(\frac{k_c(a-s)}{2}\right) \tan\left(\frac{k_c s}{2}\right)} \right). \tag{5.73}$$

Now the TE_{10} mode has a peak of electric field at the centre of the guide. So the input impedance, Z_{IN}, has to be infinite to meet this condition. Thus the denominator of Equation (5.73) has to be equal to zero, that is

$$\frac{d}{b} = \tan\left(\frac{k_c(a-s)}{2}\right) \tan\left(\frac{k_c s}{2}\right). \tag{5.74}$$

If $k_c = 2\pi n/2a = \pi n/a$, that is n times the cut-off wavenumber without the ridge, and $s = xa$, where $0 < x < 1$, then Equation (5.74) becomes

$$\frac{d}{b} = \tan\left(\frac{n\pi\,(1-x)}{2}\right) \tan\left(\frac{n\pi x}{2}\right). \tag{5.75}$$

This equation can be solved for x and the results are plotted in Figure 5.13.

This shows that for a wide ridge with $s = a/2$ there is about a factor of 2.5 increase in the cut-off wavelength for $d/b = 0.1$ and this means that the cut-off frequency is lowered by this amount. So for an X-band waveguide, the new cut-off frequency with ridges is only 2.62 GHz. Now the next mode is the TE_{20} and this has a minimum of electric field at the centre of the guide so its cut-off frequency is not significantly changed by the presence of the ridge. Similarly, the cut-off frequency of the TE_{30} is not reduced by such a large factor. So the result is a waveguide with a mono-mode bandwidth of over an octave at 2.62–8.4 GHz. The results in Figure 5.13 are from Equation (5.75). In practice, the exact results are very similar except that the peaks occur for $d/b = 0.45$ and the curves do not converge to 2 for $s/a = 0$. The only disadvantage of a ridged waveguide is that the attenuation increases and the power rating decreases when compared with the rectangular waveguide.

5.8 Waves in dielectric waveguides

Dielectric structures as well as metallic structures can guide electromagnetic waves. The main difference is that usually the fields are totally contained within the metallic waveguide, whereas they are not in dielectric guides. This is because

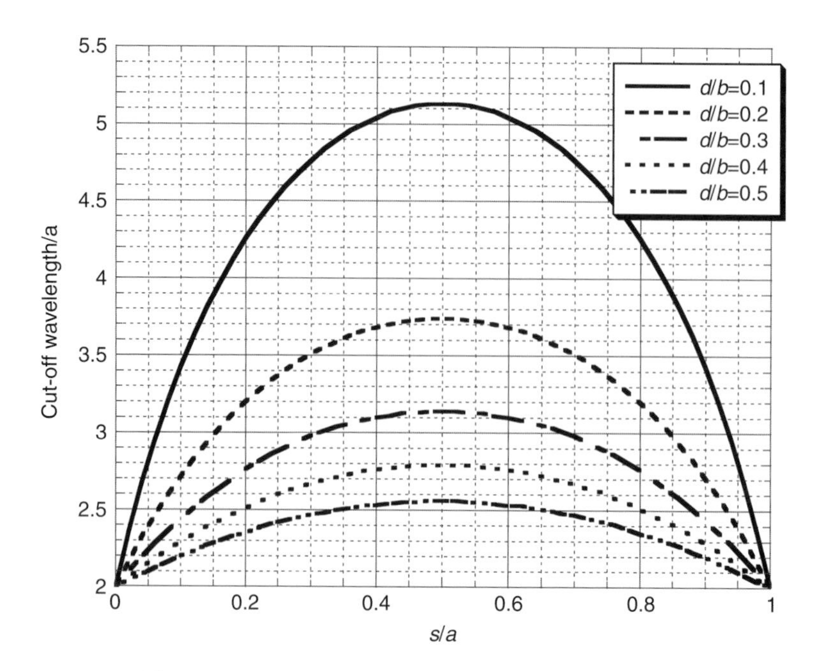

Figure 5.13 The approximate cut-off wavelength/a for a ridged waveguide for various ratios of d/b.

the boundary conditions for a metallic surface require that the surface electric fields go to zero, and this is not the case for dielectric waveguides. In the next chapter on attenuation, it will be shown that metallic structures have greater attenuation than the equivalent dielectric ones above 100 GHz. In particular, at optical frequencies, i.e. at 100 THz, the only guiding structures are all dielectric ones. The easiest structures to begin with are slab guides as they have two advantages. One is that they can be solved mathematically and the other is that they neatly illustrate the salient properties of all dielectric guides. A slab guide is any flat sheet of dielectric which is infinite in the direction of propagation, i.e. the z direction, and infinite in, say, the y direction. This may seem a little abstract but it is relatively easy to truncate the problem to a more practical size! Figure 5.14 shows such a guide with its relative dielectric constant equal to ε_1 and surrounding the guide is a medium with a relative dielectric constant ε_2. It is the main requirement for guiding a wave that $\varepsilon_1 > \varepsilon_2$. The first set of waves will be chosen to have their electric fields in the y direction, i.e. parallel with the slab.

Now the first difference between metallic guides and dielectric guides is that there are now two wave equations, one for each region, that is for the inside and outside of the slab. Let $k_1 = \sqrt{\varepsilon_1}k_0$ and $k_2 = \sqrt{\varepsilon_2}k_0$, then the two wave equations are

$$\nabla^2_{1x,y}E - k_z^2E = -k_1^2E \text{ and } \nabla^2_{2x,y}E - k_z^2E = -k_2^2E, \tag{5.76}$$

with similar equations for the magnetic fields. Now the wave must have the same velocity in the z direction in order to match the fields at the dielectric interfaces, so k_z is the same in both Equations (5.76). Now the only way this can be achieved is

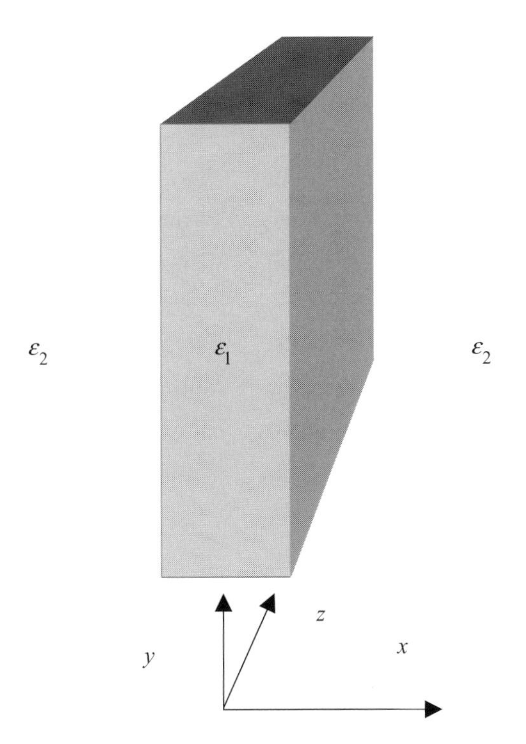

Figure 5.14 A dielectric slab guide. The thickness of the slab is usually taken as $2a$, with the y-axis at the centre of the slab.

that the part of the wave in the slab must travel faster than the plane wave velocity in the slab and the part of the wave outside must travel slower than the plane wave velocity outside. The slab part of the wave is called a 'fast' wave and the part outside is called a 'slow' wave. So if the phase velocity in the z direction is v_z then

$$\frac{c}{\sqrt{\varepsilon_1}} < v_z < \frac{c}{\sqrt{\varepsilon_2}}. \tag{5.77}$$

Now the waves inside a metallic waveguide were also fast waves as their phase velocity was greater than the velocity of light, c. In a similar fashion to a metallic rectangular guide, the modes are divided into two groups, TE and TM. The TE modes will be discussed in detail first.

Example 5.5 The TE modes in a slab guide: To start the solution for the TE modes, a cosine shape is chosen for the part inside the slab:

$$\overline{E}_y = E_0 \cos(k_x x) \exp \mathrm{j}(\omega t - k_z z)\overline{y}. \tag{5.78}$$

Substituting this into the first wave Equation (5.76) gives

$$k_x^2 + k_z^2 = k_1^2. \tag{5.79}$$

Since $k_z < k_1$ this wave will be a 'fast' wave. In order to meet the obvious boundary condition outside the slab that the waves must go to zero at infinity, an exponential solution is chosen for these fields:

$$\overline{E}_y = E_1 \exp(\mp\alpha_x x) \exp \mathrm{j}(\omega t - k_z z)\overline{y}, \tag{5.80}$$

where α_x is the decay of fields in the x direction and E_0 and E_1 are the amplitudes of the fields inside and outside the slab respectively.
Substituting this into the second wave equation of (5.76) gives

$$-\alpha_x^2 + k_z^2 = k_2^2. \tag{5.81}$$

This will be a 'slow' wave as $k_z > k_2$. The negative sign in Equation (5.80) is taken for the positive x direction. Using the sinusoidal form of Maxwell's Equation (5.3):

$$\nabla \times \overline{E} = -\mathrm{j}\omega\mu\overline{H},$$

$$\nabla \times \overline{E} = \begin{vmatrix} \overline{x} & \overline{y} & \overline{z} \\ \dfrac{\partial}{\partial x} & \dfrac{\partial}{\partial y} & \dfrac{\partial}{\partial z} \\ 0 & E_y & 0 \end{vmatrix} = -\frac{\partial E_y}{\partial z}\overline{x} + \frac{\partial E_y}{\partial x}\overline{z},$$

$$-\mathrm{j}\omega\mu H_x = -\frac{\partial E_y}{\partial z} \quad \text{and} \quad -\mathrm{j}\omega\mu H_z = \frac{\partial E_y}{\partial x}. \tag{5.82}$$

So the magnetic fields inside the slab can be found by substituting Equation (5.78) into (5.82) as follows:

$$\begin{aligned} \overline{H}_x &= -\frac{k_z}{\omega\mu} E_0 \cos(k_x x) \exp \mathrm{j}(\omega t - k_z z)\overline{x}, \\ \overline{H}_z &= -\frac{\mathrm{j}k_x}{\omega\mu} E_0 \sin(k_x x) \exp \mathrm{j}(\omega t - k_z z)\overline{z}. \end{aligned} \tag{5.83}$$

Similarly, the magnetic fields outside the slab are

$$\begin{aligned} \overline{H}_x &= -\frac{k_z}{\omega\mu} E_1 \exp(\mp\alpha_x x) \exp \mathrm{j}(\omega t - k_z z)\overline{x}, \\ \overline{H}_z &= \mp\frac{\mathrm{j}\alpha_x}{\omega\mu} E_1 \exp(\mp\alpha_x x) \exp \mathrm{j}(\omega t - k_z z)\overline{z}. \end{aligned} \tag{5.84}$$

Now all that remains is to equate the fields at the two interfaces at $x = \pm a$. Both the magnetic and the electric fields will be continuous at the interface as they are all tangential to the surfaces. Omitting the common factors, equating E_y gives

$$E_0 \cos(k_x a) = E_1 \exp(-\alpha_x a). \tag{5.85}$$

The positive sign relates to a negative x, so both interfaces give the same equation. Since the magnetic field H_x is proportional to the electric field on both sides of the boundary, equating these fields gives the same equation as (5.85). Finally, equating the magnetic field H_z gives

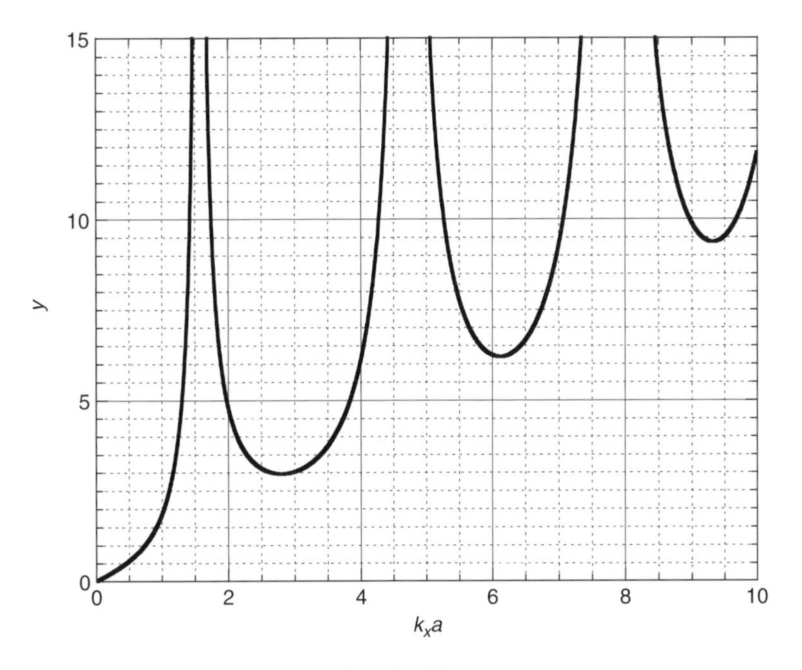

Figure 5.15 The graph of $y = |k_xa/\cos(k_xa)|$.

$$k_xE_0 \sin(k_xa) = \alpha_xE_1 \exp(-\alpha_xa). \tag{5.86}$$

By dividing Equation (5.86) by Equation (5.85), an equation in just k_x and α_x results:

$$k_x \tan(k_xa) = \alpha_x. \tag{5.87}$$

Now α_x can be eliminated from Equation (5.87) by using Equations (5.79) and (5.81) as follows:

$$\alpha_x^2 = k_z^2 - k_2^2 = k_1^2 - k_x^2 - k_2^2 = (\varepsilon_1 - \varepsilon_2)k_0^2 - k_x^2,$$

$$\alpha_x^2 = k_x^2 \tan^2(k_xa) = (\varepsilon_1 - \varepsilon_2)k_0^2 - k_x^2 \text{ so } k_x^2(1 + \tan^2(k_xa)) = (\varepsilon_1 - \varepsilon_2)k_0^2,$$

and finally

$$\frac{k_xa}{\cos(k_xa)} = \pm\sqrt{(\varepsilon_1 - \varepsilon_2)}k_0a. \tag{5.88}$$

This is called the characteristic equation for TE modes in a slab guide. The right-hand side of this equation is proportional to frequency, so it can be solved in terms of frequency. From the value of k_x all the other parameters can be found. Figure 5.15 shows a graph of the modulus of the left-hand side of Equation (5.88).

Since the right-hand side of Equation (5.88) can be taken as positive, taking the modulus of the left-hand side of the equation is appropriate. The first, or lowest order, mode is in the region $0 < k_xa < \pi/2$. The next mode is in the region $\pi < k_xa < 3\pi/2$ and so on. These regions give rise to proper modes where α_x is positive,

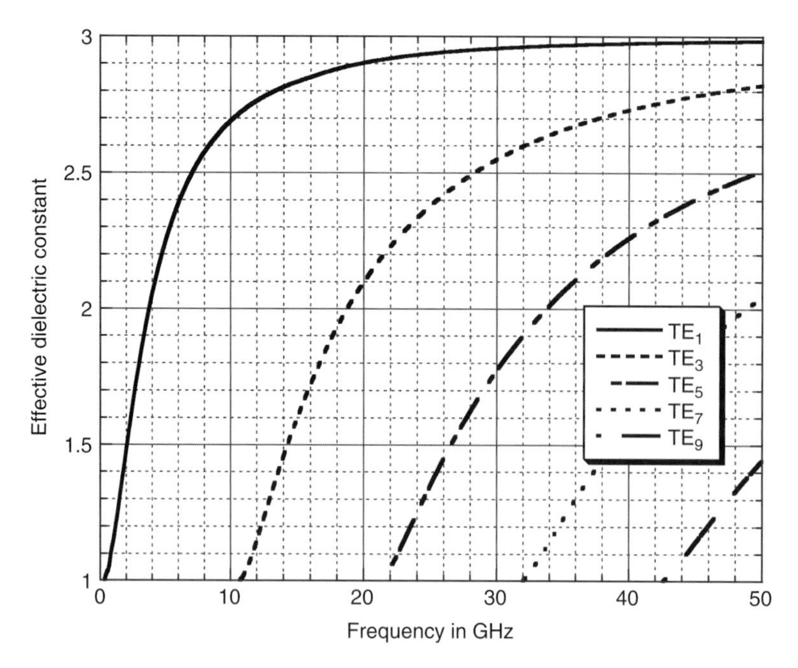

Figure 5.16 The effective dielectric constant of the TE_n modes on slab guides for $n = 1, 3, 5, \ldots$ The slab is 2 cm thick and has a relative dielectric constant of 3. It is surrounded by a medium with a relative dielectric constant of 1.

where a proper mode has the external fields decaying to zero at infinity. In between, the modes are called improper as α_x is negative and this gives rise to infinite fields at infinity. These improper modes may occur between certain boundaries, but this text will only consider the proper modes. If the effective dielectric constant is $\varepsilon_{\rm eff}$ then

$$k_z^2 = \varepsilon_{\rm eff} k_0^2 = \varepsilon_1 k_0^2 - k_x^2 \quad \text{so } \varepsilon_{\rm eff} = \varepsilon_1 - \frac{k_x^2}{k_0^2}. \tag{5.89}$$

A graph of $\varepsilon_{\rm eff}$ against frequency is shown in Figure 5.16.

One of the distinguishing characteristics of dielectric guides is that the effective dielectric constant changes from the lower dielectric constant, ε_2, to the higher, ε_1, as the frequency increases. In Figure 5.17, the amplitude of the E_y field for the TE_1 mode is plotted for various values of $k_x a$. Now $k_x a = 0$ corresponds to zero frequency and the fields are infinite in extent with a value of $\alpha_x = 0$. Not surprisingly, since the vast majority of the field is in the dielectric outside, $\varepsilon_{\rm eff} = \varepsilon_2$. As the frequency increases, the value of α_x increases and so the fields become more located in the slab. At a very high frequency, there will be a negligible field outside the slab and so $\varepsilon_{\rm eff} = \varepsilon_1$. The fields inside the slab are very similar to those in the TE_{10} mode in a rectangular waveguide; the difference being that only the central fraction of the field's pattern is present instead of the complete cosine wave. As the frequency increases, more of the cosine wave is inside the slab and

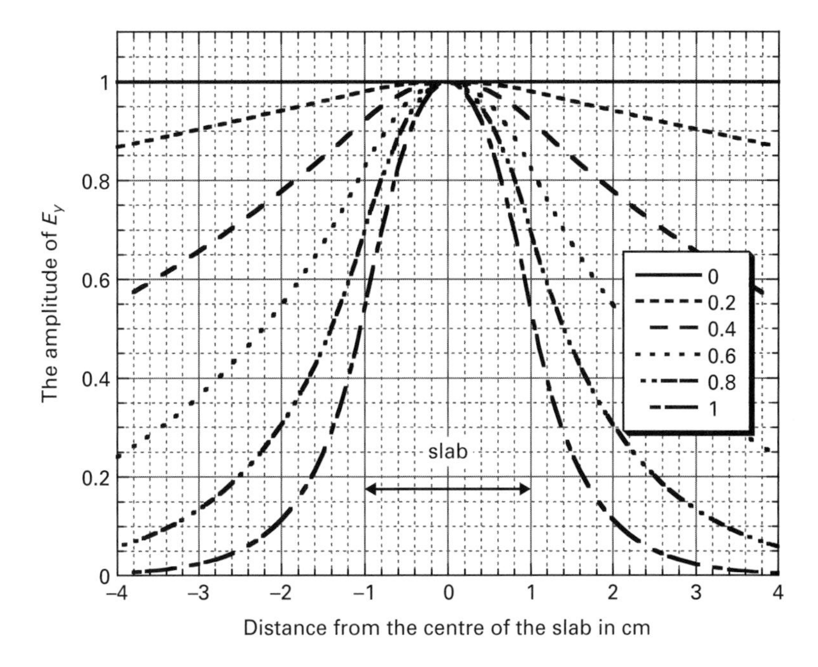

Figure 5.17 The E_y field for the lowest order TE_1 mode in a slab guide for various values of $k_x a$. The slab has a thickness of 2 cm and a relative dielectric constant of 3. It is surrounded by a medium with a relative dielectric constant of unity.

finally, at very high frequency, all of the cosine wave is inside. One way of visualising this is to imagine a metallic waveguide filled with a dielectric identical to the slab's dielectric. Now if the waveguide's broad dimension is allowed to vary with frequency, it can be made equivalent to the dielectric guide. Let the equivalent waveguide have a broad dimension of $2an$, where $2a$ is the width of the slab. Then, to be equivalent, $k_x na = \pi/2$, and the cut-off wavelength of this equivalent guide is $4na$. Using Equations (5.88) and (5.89), it can be shown that

$$\varepsilon_{\text{eff}} = \varepsilon_1 - (\varepsilon_1 - \varepsilon_2)\cos^2(k_x a). \tag{5.90}$$

For the equivalent waveguide with the same velocity:

$$v_{\text{p}} = \frac{\dfrac{3 \times 10^{10}}{\sqrt{\varepsilon_1}}}{\left(1 - \left(\dfrac{\lambda}{4na}\right)^2\right)^{\frac{1}{2}}} = \frac{3 \times 10^{10}}{\sqrt{\varepsilon_{\text{eff}}}}\ \text{cms}^{-1}. \tag{5.91}$$

To complete this account of the TE modes in a slab guide, the other possible solution to the boundary conditions is similar to Equation (5.78), but with a sine rather than a cosine, i.e.

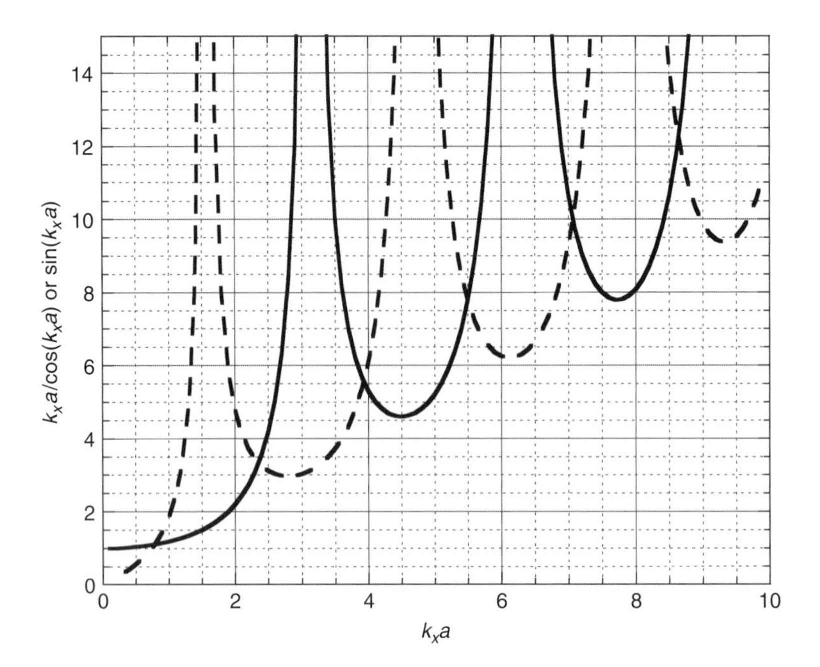

Figure 5.18 Equations (5.88) and (5.93). The dotted lines are Figure 5.15 repeated for TE_n modes with $n = 1, 2, 3,...$ The solid lines are TE_n with $n = 2, 4, 6,....$

$$\overline{E}_y = E_0 \sin(k_x x) \exp j(\omega t - k_z z). \tag{5.92}$$

This supplies the TE_n modes with $n = 2, 4, 6$. Figure 5.18 shows the solutions to the equivalent equation to (5.88), which is

$$\frac{k_x a}{\sin(k_x a)} = \pm\sqrt{(\varepsilon_1 - \varepsilon_2)} k_0 a. \tag{5.93}$$

Just like the previous set of modes, these have regions where α_x is positive and this gives rise to improper modes. The proper modes start at $k_x a = \pi/2$, for lower values of $k_x a$ it gives rise to an improper mode. So these proper modes are 'interleaved' between the other solutions. Figure 5.16 can now be repeated with these extra modes added to form the complete set. This is shown in Figure 5.19.

Finally, to complete the set of figures, the amplitude of the E_y of the TE_{20} mode for various values of $k_x a$ is shown in Figure 5.20. The change from a half sine variation within the slab for $k_x = \pi/2$ to a full sine as $k_x = \pi$ can be seen in this figure. In addition the value α_x changes from 0 to approaching infinity over this region. Comparing Figures 5.17 and 5.20 shows that the two modes are orthogonal and energy in the lowest order mode is unlikely to couple into this mode unless there is an asymmetrical discontinuity.

Using Equations (5.78) and (5.83) for the inside of the slab:

$$\overline{P}_{\mathrm{AV}} = \frac{1}{2}\mathrm{Re}\left(\frac{\mathrm{j}k_x E_0^2}{2\omega\mu}\sin(2k_x x)\overline{x} + \frac{k_z E_0^2}{\omega\mu}\cos^2(k_x x)\overline{z}\right).$$

So integrating over half the width of the slab and one metre in the y direction, the power travelling inside the slab per metre, P_{IN}, is

$$P_{\mathrm{IN}} = \frac{1}{2}\int_0^a \frac{k_z E_0^2}{\omega\mu}\cos^2(k_x x)\mathrm{d}x = \frac{k_z E_0^2}{4\omega\mu}\left(a + \frac{\sin(2k_x a)}{2k_x}\right). \qquad (5.106)$$

Then, using Equations (5.80) and (5.84) for outside the slab

$$\overline{P}_{\mathrm{AV}} = \frac{1}{2}\mathrm{Re}\left(\pm\frac{\mathrm{j}\alpha_x E_1^2}{\omega\mu}\exp(\mp 2\alpha_x x)\overline{x} + \frac{k_z E_1^2}{\omega\mu}\exp(\mp 2\alpha_x x)\overline{z}\right).$$

Again integrating this over just one side of the slab, the total power outside the slab, P_{OUT}, is given by

$$P_{\mathrm{OUT}} = \frac{1}{2}\int_a^\infty \frac{k_z E_1^2}{\omega\mu}\exp(-2\alpha_x x)\mathrm{d}x = \frac{k_z E_1^2}{4\omega\mu\alpha_x}\exp(-2\alpha_x a). \qquad (5.107)$$

Using Equations (5.85), (5.87), (5.106) and (5.107), the ratio of the power outside the slab to the total power for the TE_1 mode is given by

$$\frac{\cos^2(k_x a)}{1 + k_x a\tan(k_x a)}. \qquad (5.108)$$

This is plotted in Figure 5.23. For just half the total power to be travelling external to the slab guide, that is a quarter on either side, $k_x a = 0.585$. For the slab guide discussed so far, that is with a width of 2 cm and a relative dielectric constant of 3, this value of $k_x a$ corresponds to a frequency of 2.38 GHz and an effective dielectric constant of 1.61. Now the next higher order mode is the TE_2 mode which is unlikely to propagate when $k_x a = \pi/2$ as its fields will be infinite in extent. So the same power ratio curve is plotted for the TE_2 mode as well in Figure 5.23. The half power point is at $k_x a = 1.940$ (the figure is shifted by $\pi/2$) if the same criterion is used for this mode. The upper frequency for the TE_1 mode will be when just half the power is outside the guide for the TE_2 mode, that is at 7.022 GHz; in other words well over an octave bandwidth.

Finally, the dispersion of the slab guide is the change in the phase velocity over the bandwidth. At the lower frequency of 2.38 GHz, the phase velocity is $2.36.10^8\,\mathrm{ms}^{-1}$ and at the upper frequency of 7.022 GHz, the phase velocity is $1.9.10^8\,\mathrm{ms}^{-1}$. This is considerably less dispersive than rectangular waveguide, but not as good as coaxial cable. Over the same frequency range, the TE_1 wave impedance varies from 297 Ω to 238 Ω, which again is better than metallic waveguide.

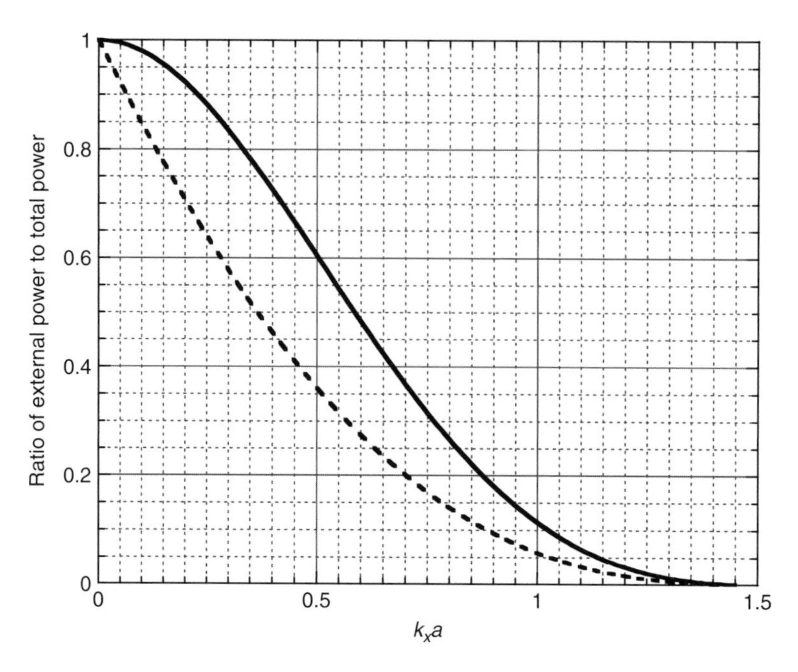

Figure 5.23 The ratio of the power outside the slab guide to the total power for the TE_1 mode against k_xa (solid line). The dotted line is the same ratio for the TE_2 but shifted by $\pi/2$ in order to compare the results.

Example 5.7 Higher order modes in microstrip, coplanar waveguide and stripline microstrip: All transmission lines have higher order modes, which restrict their mono-mode bandwidth. For microstrip, described in Example 4.20, the first higher order mode that appears is the HE_1 mode which has a cut-off frequency given by

$$f_C(HE_1) = \frac{3.10^8 Z_0}{2\eta_0 h} \approx \frac{3.10^8}{\sqrt{\varepsilon_R} \times 2w} \text{ ms}^{-1}. \qquad (5.109)$$

This mode is a variation of approximately a half wavelength along the width of the microstrip top conductor, i.e. a type of waveguide mode like the TE_{10} but with a maximum of electric field at the edges, rather than a minimum. For the microstrip given as an example in Figure 4.28, this frequency, $f_C(HE_1)$, is approximately 50 GHz. So the mono-mode bandwidth of this particular microstrip circuit is about 50 GHz. However, this is not the whole picture. Since microstrip is mounted on a dielectric substrate, there are other possible modes which can occur. One possibility are the slab guide modes discussed in Examples 5.5 and 5.6. They will not be quite the modes described in these examples as the substrate will be finite in extent. However, the effect of the substrate's finite size can be neglected if the width of the substrate is much greater than the thickness. Clearly, if energy is coupled from a microstrip mode to a slab mode, or as they are often called, substrate mode, the latter is likely to be a resonant mode due to reflections from the edges of the substrate or, if the circuit is in a metal box, the walls of the box. Now the two principles of mode

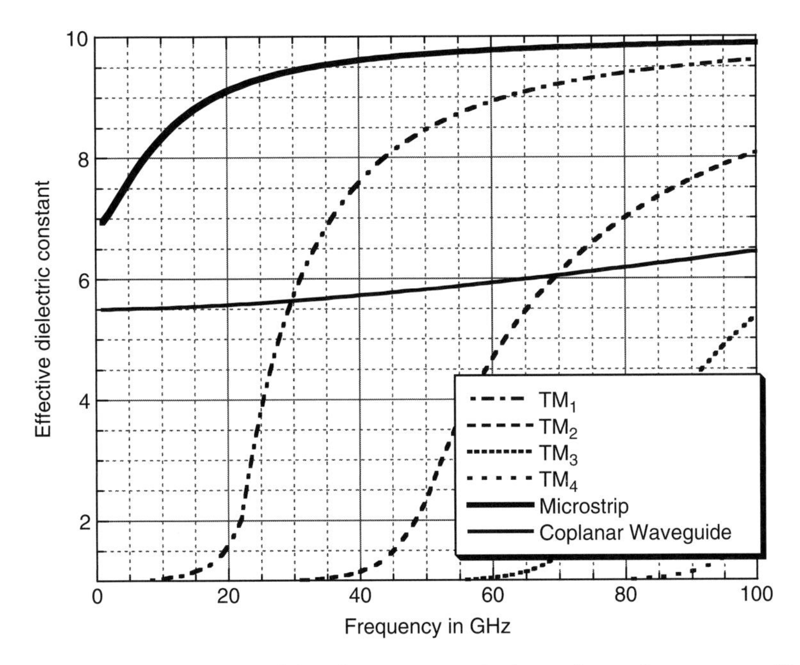

Figure 5.24 The effective dielectric constants of microstrip, coplanar waveguide and slab or substrate modes against frequency. The microstrip graph has been repeated from Figure 4.28 and the graph for coplanar waveguide is from Figure 4.30. The slab modes are derived from equations in Section 5.6.

coupling are that a slower mode cannot couple to a faster mode, and the fields of the two modes must not be orthogonal. In Figure 5.24 the frequency variation of the effective dielectric constant has been repeated from Figure 4.28. In addition some substrate modes, i.e. the slab guide modes assuming an infinite substrate, have been added. The ground plane of microstrip will prevent any modes which have a tangential electric field. So only the TM modes have been shown. The ground plane acts as a mirror, so the slab or substrate appears to be 2 mm thick electrically, rather than 1 mm. It can be seen in Figure 5.24 that none of the TM modes intersect the microstrip graph. This means that the waveguide type mode, discussed at the beginning of this section, determines the bandwidth of this transmission line. The only way to increase the bandwidth above the 50 GHz in the example is to reduce the values of both the width of the top conductor and the substrate thickness. In practice this increases the attenuation, but for small circuits this is usually not a major problem.

Coplanar waveguide

The picture for coplanar waveguide, described in Example 4.21, is different to that of microstrip. Figure 5.24 clearly shows that the TM_1 slab mode intersects with the coplanar waveguide graph at 30 GHz. This means that, from this frequency and above, the coplanar waveguide can lose energy to substrate modes. Above this frequency, the energy will be released at an angle so that the substrate modes and the resolved component of the coplanar waveguide mode have the same velocity. About the only way to increase the bandwidth in this case is to reduce the thickness of the substrate. The addition of a ground plane can prove a further problem as it both doubles the electrical thickness of the substrate and introduces a further possibility of a microstrip mode. The latter is really undesirable as the microstrip mode has the highest effective dielectric constant and hence is the slowest mode, which means that all faster modes may couple some energy into it, given the correct conditions.

The amount of coupling depends on the orthogonality of the two modes concerned. In general, the solution to a single transmission line should produce only orthogonal modes. This is clearly seen in rectangular waveguide where the solutions are sines and cosines, and integrating a product of the modes over the area inside the guide gives a zero result. In other words, there will be no coupling between the modes. However, this is not the case for different transmission lines, and so there will be some coupling depending on the degree of non-orthogonality. It should be noticed that, even in rectangular waveguide, the attenuation in the walls of the guide can remove some of the orthogonality of modes and this is discussed in the books listed at the end of this chapter.

Finally, it is often the case that microstrip and coplanar waveguide circuits are mounted in metallic boxes. Since both transmission lines are open structures, they are capable of radiating at both bends and discontinuities. This radiation is then trapped inside the box, which will become a resonant cavity at certain frequencies. These resonances are often called box modes. The result of these resonances can be that the performance of the circuit is reduced to zero at certain discrete frequencies. In many cases, absorbant material is included in the lid of these boxes to remove these effects.

Stripline

Stripline, described in Example 4.18, has three conductors and therefore two TEM modes of propagation. The unwanted mode is between the parallel plate conductors. Fortunately, the electromagnetic fields in a stripline mode are equal and opposite above and below the centre conductor. This means that these fields are orthogonal to those of the unwanted mode, assuming the centre conductor is really central. The first higher order mode, which determines the mono-mode bandwidth for stripline is the parallel plate waveguide mode. The cut-off wavelengths and the associated frequencies are given by

$$\lambda_C = \frac{4d}{m}, \quad f_C = \frac{m}{4d\sqrt{\varepsilon_R}} \times 3.10^8 \text{ ms}^{-1}, \quad \text{where } m = 1, 2, 3, \ldots \tag{5.110}$$

If the substrate has a dielectric constant of, say, 5 and $d = 0.2$ mm, then the first mode starts at 55 GHz. The electric fields of the TE_1 mode are almost orthogonal

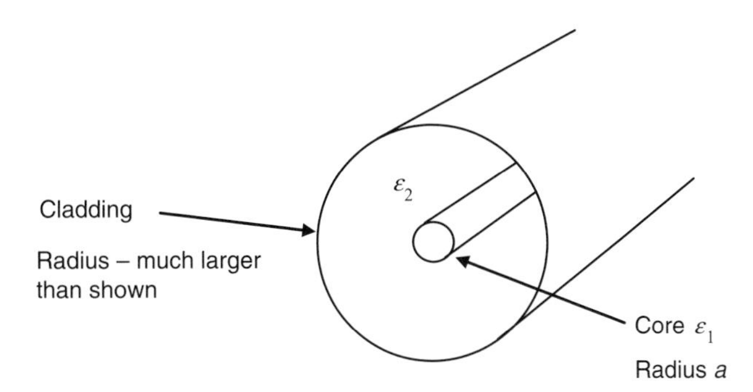

Figure 5.25 A step index optical fibre. A typical value for the diameter of the core is 6 μm and for the cladding 150 μm. In this figure, the cladding should be over four times larger in diameter.

to the stripline mode and so the TM_1 mode is the most likely to propagate. The use of coupled striplines in a balanced mode does not change the above arguments. The main reason for the use of a balanced mode is to prevent coupling to adjacent stripline circuits between the same conducting planes.

Example 5.8 Optical fibre: A very large amount has been written about optical fibre and so only a few comments are included here to complete this survey of guided waves in dielectric structures. Optical fibre is the prime example of a purely dielectric structure for guiding electromagnetic waves and is now used extensively over huge distances and is shown in Figure 5.25.

The main design problem with large distances is to avoid both dispersion and attenuation. Most fibres are now designed to support just one mode, which eliminates the dispersion caused by the different velocities of the various modes. The wavelength at which they are operated is also chosen to reduce both the attenuation and the dispersion caused by changes in the dielectric constants with frequency. At optical frequencies a large bandwidth is not so important as it is at lower frequencies. The material chosen for the core of the fibre and its cladding is usually very pure and has an exceptionally low attenuation. The relative dielectric constant of the cladding is usually just less than that of the core. This gives a reasonably large diameter for the core, typically around 6 μm, but has the disadvantage that the fields are quite extensive in the cladding. The diameter of the cladding is made sufficiently large to ensure that all the fields are negligible at the outer surface. A typical value of the diameter might be 150 μm. When the two dielectric constants are close this is often called a 'weakly guided mode'. Since sharp corners are avoided in optical fibre, as the cladding is so much larger than the core, this 'weak guiding' is usually not a problem. From the author's experience, when an optical fibre was wound round his finger just once, there was a loss of about half of the optical signal.

Now the analysis of the modes in an optical fibre is made more complex as there is no simple division into just TE and TM modes. When all six fields are present there are other modes called hybrid modes with both $\overline{E}_z$ and $\overline{H}_z$. If $\overline{H}_z > \overline{E}_z$ then this hybrid mode is usually called an HE mode. For the other condition, $\overline{E}_z > \overline{H}_z$, the mode is called an EH mode. So to begin, the fields inside the core will have the same form as equations in Section 5.5, but combining both the TE and TM solutions.

From Equations (5.61) and (5.67):

$$\overline{E}_z = E_0 J_n(k_r r)\sin(n\theta)\overline{z} \text{ and } \overline{H}_z = H_0 J_n(k_r r)\cos(n\theta)\overline{z},$$

where k_c (a fixed quantity) has been replaced by k_r (a variable). The modes are classified as HE_{nm} or EH_{nm} where n is the order of the Bessel function and m is the number of the solution of the Bessel equation, i.e. $m = 1, 2, 3, \ldots$

From Equation (5.62) and the equation preceding Equation (5.67):

$$\overline{E}_r = \left(-\frac{\mathrm{j}k_z}{k_r} E_0 J_n'(k_r r)\sin(n\theta) + \frac{\mathrm{j}n\omega\mu}{rk_r^2} H_0 J_n(k_r r)\sin(n\theta)\right)\overline{r},$$

$$\overline{E}_\theta = \left(-\frac{\mathrm{j}nk_z}{rk_r^2} E_0 J_n(k_r r)\cos(n\theta) + \frac{\mathrm{j}\omega\mu}{k_r} H_0 J_n'(k_r r)\cos(n\theta)\right)\overline{\theta},$$

$$\overline{H}_r = \left(\frac{\mathrm{j}\omega\varepsilon n}{rk_r^2} E_0 J_n(k_r r)\cos(n\theta) - \frac{\mathrm{j}k_z}{k_r} H_0 J_n'(k_r r)\cos(n\theta)\right)\overline{r},$$

$$\overline{H}_\theta = \left(-\frac{\mathrm{j}\omega\varepsilon}{k_r} E_0 J_n'(k_r r)\sin(n\theta) + \frac{\mathrm{j}k_z n}{rk_r^2} H_0 J_n(k_r r)\sin(n\theta)\right)\overline{\theta}. \tag{5.111}$$

In these equations, $k_r^2 + k_z^2 = \varepsilon_1 k_0^2$ and k_r is a radial wavenumber.

The fields outside the slab must uniformly decay to zero as $r \to \infty$ and so the Bessel functions used above will not satisfy these conditions. Instead a modified Bessel function, called a K function, is used, which decays to zero for large arguments as

$$K_n(\alpha_r r) \to \sqrt{\frac{\pi}{2\alpha_r r}}\exp(-k_c r) \text{ as } \alpha_r r \to \infty. \tag{5.112}$$

The function is infinite when the argument is zero, but since it will be used only outside the core where $r > a$, this is not a limitation. The first two K functions are shown in Figure 5.26. In Equation (5.112), α_r is a radial wavenumber for the region outside the central core of the fibre. So

$$-\alpha_r^2 + k_z^2 = \varepsilon_2 k_0^2.$$

The fields outside the core are

$$\overline{E}_z = E_1 K_n(\alpha_r r)\sin(n\theta)\overline{z},\ \overline{H}_z = H_1 K_n(\alpha_r r)\cos(n\theta)\overline{z},$$

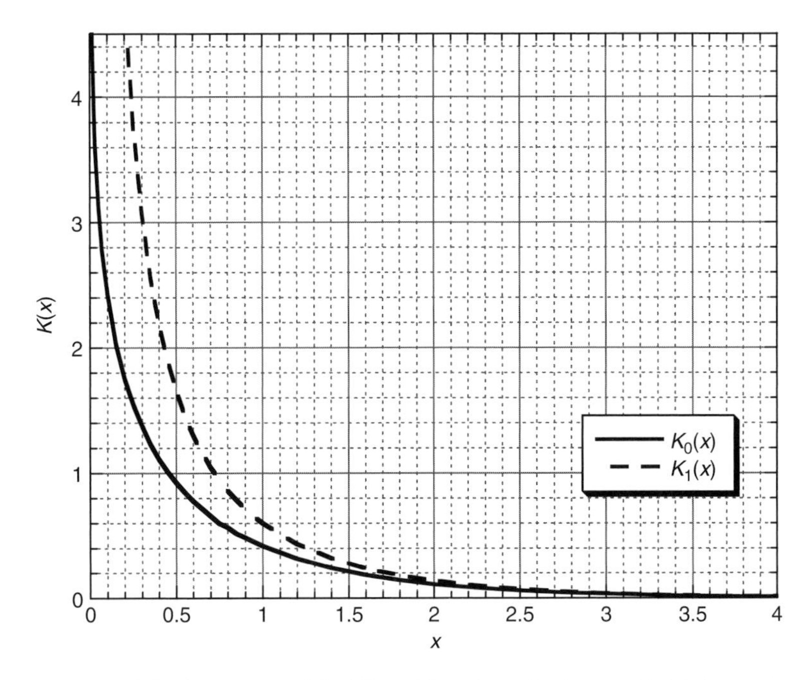

Figure 5.26 The first two modified Bessel functions $K_0(x)$ and $K_1(x)$. They are both infinite for $x = 0$ and rapidly decay to zero for $x > 5$.

$$\overline{E}_r = \left(\frac{\mathrm{j}k_z}{\alpha_r}E_1K_n'(\alpha_r r)\,\sin(n\theta) - \frac{\mathrm{j}n\omega\mu}{r\alpha_r^2}H_1K_n(\alpha_r r)\,\sin(n\theta)\right)\overline{r},$$

$$\overline{E}_\theta = \left(\frac{\mathrm{j}nk_z}{r\alpha_r^2}E_1K_n(\alpha_r r)\,\cos(n\theta) - \frac{\mathrm{j}\omega\mu}{\alpha_r}H_1K_n'(\alpha_r r)\,\cos(n\theta)\right)\overline{\theta},$$

$$\overline{H}_r = \left(\frac{\mathrm{j}\omega\varepsilon n}{r\alpha_r^2}E_1K_n(\alpha_r r)\,\cos(n\theta) + \frac{\mathrm{j}k_z}{\alpha_r}H_1K_n'(\alpha_r r)\,\cos(n\theta)\right)\overline{r},$$

$$\overline{H}_\theta = \left(\frac{\mathrm{j}\omega\varepsilon}{\alpha_r}E_1K_n'(\alpha_r r)\,\sin(n\theta) - \frac{\mathrm{j}k_z n}{r\alpha_r^2}H_1K_n(\alpha_r r)\,\sin(n\theta)\right)\overline{\theta}. \tag{5.113}$$

Now equating the fields inside the core to those outside at the boundary $r = a$ gives six equations. These can be rearranged, with some effort, to eliminate the four amplitudes E_0, E_1, H_0 and H_1 to give a characteristic equation for this optical fibre:

$$k_0^2\left(\frac{J_n'(k_r a)}{k_r J_n(k_r a)} + \frac{K_n'(\alpha_r a)}{\alpha_r K_n(\alpha_r a)}\right)\left(\frac{\varepsilon_1 J_n'(k_r a)}{k_r J_n(k_r a)} + \frac{\varepsilon_2 K_n'(\alpha_r a)}{\alpha_r K_n(\alpha_r a)}\right) = \frac{n^2k_z^2}{a^2}\left(\frac{1}{k_r^2} + \frac{1}{\alpha_r^2}\right)^2. \tag{5.114}$$

Equation (5.114) reveals the complex nature of these modes. The lowest order mode, which propagates right down to zero frequency, is the HE_{11} mode. This is the main mode in optical fibre and is almost mono-mode. Unfortunately, since it can have any orientation in the fibre, due to its circular symmetry, it is not really a mono-mode like the equivalent in a rectangular waveguide. When the fields

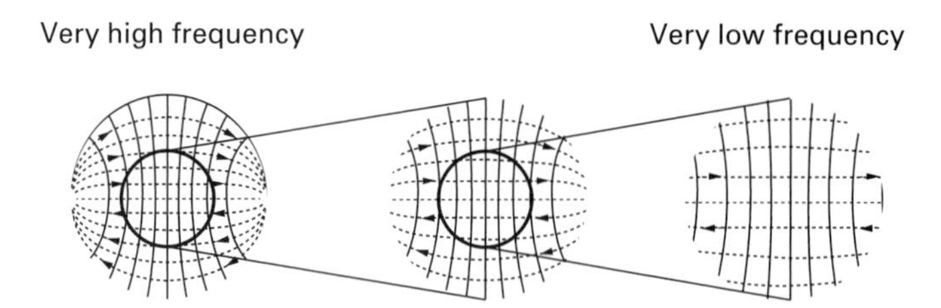

Figure 5.27 The fields inside the core of an optical fibre, for the dominant HE_{11} mode. The electric fields are the solid lines and are mainly vertical. The dotted lines are the magnetic field lines.

are nearly confined inside the core of the fibre, at very high frequencies, the fields are identical to the TE_{11} modes of metallic circular waveguide. Figure 5.27 repeats the relevant part of Figure 5.9(a) for the TE_{11} mode.

The first circle on the left shows the fields for very high frequencies, when all of the energy is in the core and the shape is the same as the TE_{11} mode in metallic circular waveguide. At lower frequencies, only the middle section of the mode is in the core. This circle in the middle shows just this part of the first circle enlarged. As the frequency goes even lower, only the central part of the mode is in the core. The right-hand circle shows this part enlarged even more. At zero frequency, the electric fields would be entirely in one linear direction; in the diagram this is the vertical direction. This phenomenon is similar to the first order mode in the slab guide shown in Figure 5.17. In that figure, at very high frequencies all the energy is in the slab with a half sine shape. At zero frequency, the fields are of infinite extent outside the core.

In Equation (5.114), if $n = 0$, then the right-hand side goes to zero and the fibre supports only TE or TM modes. For the TE_{0m} modes, the first bracket on the left-hand side of Equation (5.114) goes to zero, i.e.

$$\left(\frac{J_0'(k_r a)}{k_r J_0(k_r a)} + \frac{K_0'(\alpha_r a)}{\alpha_r K_0(\alpha_r a)}\right) = 0. \tag{5.115}$$

Similarly, for TM_{0m} modes, the second bracket of Equation (5.114) goes to zero, i.e.

$$\left(\frac{\varepsilon_1 J_0'(k_r a)}{k_r J_0(k_r a)} + \frac{\varepsilon_2 K_0'(\alpha_r a)}{\alpha_r K_0(\alpha_r a)}\right) = 0. \tag{5.116}$$

Just as in the slab guide, at the mode cut-off, the fields outside the core are infinite and $\alpha_r = 0$. Equations (5.115) and (5.116) can only be satisfied at the cut-off frequency if

$$J_0(k_r a) = 0 \text{ or } k_r a = 2.405,\ 5.520,\ 8.654,\ \ldots, \tag{5.117}$$

where these solutions are given in Equation (5.64). These cut-off conditions give the frequencies of the higher order modes in the fibre. Now both k_r and α_r have to be positive for a proper mode. Since $K_0(\alpha_r a)$ is always positive and $K_0'(\alpha_r a)$ is always negative, the solutions begin at $k_r a = 2.405$ where $J_0(k_r a)$ goes negative while $J_0'(k_r a)$ is still negative, see Figure 5.8. This is similar to a slab guide, where there is a region for a proper mode solution.

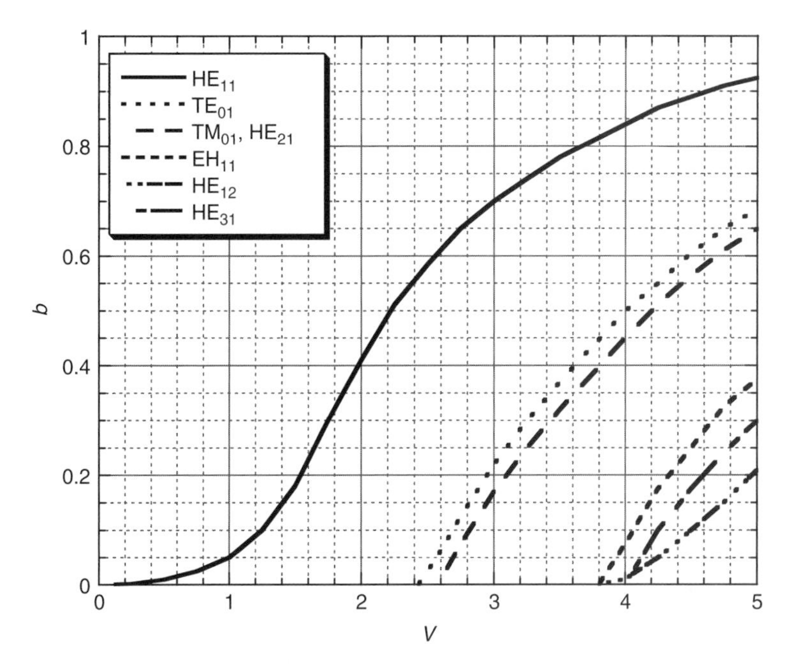

Figure 5.28 The normalised effective dielectric constant, b, against the V number, for the first few modes in an optical fibre.

Now $\varepsilon_1 k_0^2 = k_r^2 + k_z^2$ and $\varepsilon_2 k_0^2 = -\alpha_r^2 + k_z^2$ so $(\varepsilon_1 - \varepsilon_2)k_0^2 = k_r^2 + \alpha_r^2$, so at the first TE or TM cut-off frequency where $\alpha_r = 0$:

$$(\varepsilon_1 - \varepsilon_2)^{\frac{1}{2}} k_c a = k_r a = 2.405. \tag{5.118}$$

In many optical texts, a normalised frequency – called the V number – and a normalised effective dielectric constant – given the symbol b – are often used. These are defined below as well as the equivalent definition using the refractive indices, n:

$$V = (\varepsilon_1 - \varepsilon_2)^{\frac{1}{2}} k_0 a, \qquad b = \frac{\sqrt{\varepsilon_{\mathrm{EFF}}} - \sqrt{\varepsilon_2}}{\sqrt{\varepsilon_1} - \sqrt{\varepsilon_2}} = \frac{n_{\mathrm{EFF}} - n_2}{n_1 - n_2}. \tag{5.119}$$

In Figure 5.28 the solution for this normalised effective dielectric constant for the first few modes of an optical fibre against the V number are shown. It can be seen that the TE and TM modes start at $V = 2.405$ and a hybrid mode starts there as well.

Weakly guided or linearly polarised modes

For optical fibres where the difference between the two dielectric constants is very small, many of these modes become degenerate, i.e. they have almost the same velocity of propagation and effectively combine to form a composite mode. These optical fibres are often described as 'weakly guiding' as the fields in the cladding are more extensive. Many of these combinations turn out to also be linearly polarised

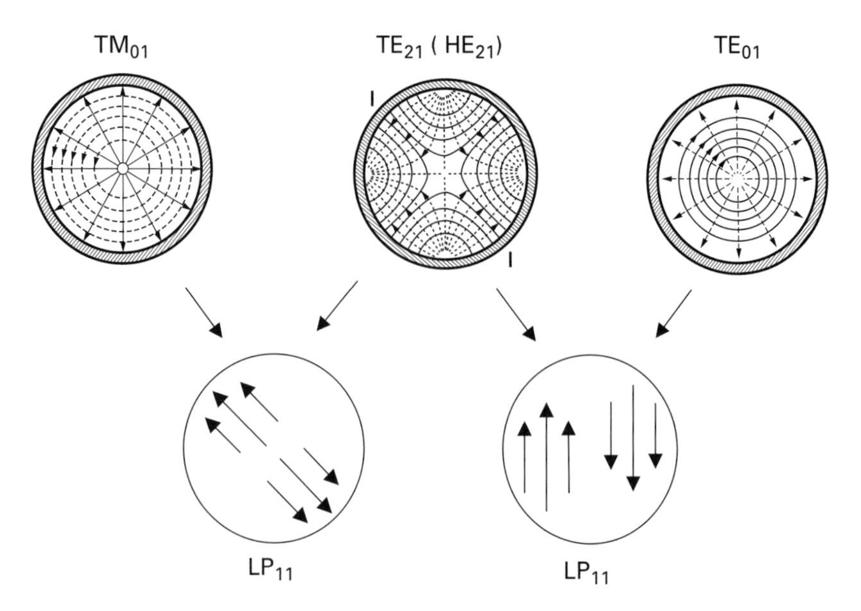

Figure 5.29 The formation of linearly polarised modes in optical fibre. Both modes are classified as LP_{11} modes and they are the first higher order modes after the dominant HE_{11} mode. These figures have been reproduced by permission of the Institution of Engineering and Technology from the *Waveguide Handbook* by N. Marcuvitz, published by the IET in 1985 as a reprint of the original publication by the McGraw-Hill Book Company in 1951.

and have been classified as LP modes. The HE_{11} mode remains the dominant mode and is called the LP_{01} mode in this limiting case where $\varepsilon_1 \to \varepsilon_2$. Figure 5.9(a) shows that TE_{11}, which is the high frequency limit of the HE_{11}, is almost linearly polarised in the vertical direction. It could be in any transverse direction. The next cluster of three modes can combine, as shown in Figure 5.29; they are the TE_{01}, TM_{01} and HE_{21} modes. Just like the HE_{11} mode in Figure 5.28, these modes also become like their equivalent modes in circular metallic waveguide mode at high frequencies. In Figure 5.29, these circular metallic waveguide modes have been repeated from Figure 5.9(a) to illustrate the addition of modes.

The HE_{11} mode in optical fibre

Figure 5.28 shows that the HE_{11} mode propagates as a single mode between $V = 0$ and $V = 2.405$. At small V numbers, i.e. at very low frequencies, the fields outside the core are very extensive. So a typical value of V might be 2 to make sure a reasonable amount of the energy is actually in the core and to maintain the mono-mode conditions. If the free space wavelength is taken as 1.1 μm and $\varepsilon_1 = 2.340$ and $\varepsilon_2 = 2.325$, then from Equation (5.119):

$$V = (\varepsilon_1 - \varepsilon_2)^{\frac{1}{2}} k_0 a, \text{ which gives}$$

$$a = \frac{2 \times 1.1}{2\pi (2.340 - 2.325)^{\frac{1}{2}}} = 2.86 \ \mu\text{m}.$$

This gives a diameter of the core at about 6 μm, which is a typical value. Also from Figure 5.28 the b value is 0.4. Again using the equations in (5.119):

$$b = \frac{\sqrt{\varepsilon_{EFF}} - \sqrt{\varepsilon_2}}{\sqrt{\varepsilon_1} - \sqrt{\varepsilon_2}}$$

or $\sqrt{\varepsilon_{EFF}} = b\sqrt{\varepsilon_1} + (1 - b)\sqrt{\varepsilon_2}$, which for this case gives $\varepsilon_{EFF} = 2.331$.

Now using $-\alpha_r^2 + k_z^2 = \varepsilon_2 k_0^2$, a value for α_r can be found:

$$\alpha_r = \sqrt{\varepsilon_{EFF} - \varepsilon_2} \times \frac{2\pi}{\lambda} \times 8.686 = 0.43 \text{ dB } \mu\text{m}^{-1},$$

which for a cladding of radius 75 μm and assuming an exponential decay gives 32 dB. However, the K functions would give a value higher than just an exponential decay, so the fields at the outer edge of the cladding are negligible.

Example 5.9 Rectangular dielectric waveguide: Rectangular dielectric waveguide is used for transmission lines above 100 GHz and in optical circuits. Unfortunately it does not have such a complete solution as the circular optical fibre just described. The problem is that, inside the dielectric, the obvious coordinates would be the normal cartesian ones where the solutions would be in terms of the rectangular harmonics, namely sine and cosine functions. However, outside the dielectric, cylindrical polar coordinates are needed to describe the distant fields. These two systems of coordinates do not meet easily at the boundaries. Figure 5.30 shows a typical rectangular dielectric waveguide surrounded by a single dielectric – often air in the microwave applications. In optical circuits the surrounding dielectric may be more complicated.

There have been several methods used to solve this guide. One of the most common is the effective dielectric constant method, which will be described in this section. Almost all of the others involve quite complex computing. For example, an early solution by Goell used an infinite sum of Bessel's J functions to give the fields inside the guide and these were matched at the boundary to an infinite sum of Bessel's K functions outside the guide. A more modern technique is to use numerical techniques to solve directly the electromagnetic wave equation – with the distinct problem that since there are infinite solutions, some prior knowledge of the particular solution is required. The effective dielectric constant method is an approximation but it does give a sense of the mode shape and is reasonably accurate at higher frequencies and can be easily calculated. As such, it proves useful in checking the more sophisticated solutions that result from electromagnetic wave computations.

The effective dielectric constant method relies heavily on the slab guide solutions which have been fully discussed in Section 5.8, Examples 5.5 and 5.6. The technique finds two effective dielectric constants and then in the simplest approach takes their average. So, to begin with, if we take a slab guide of width $2a$

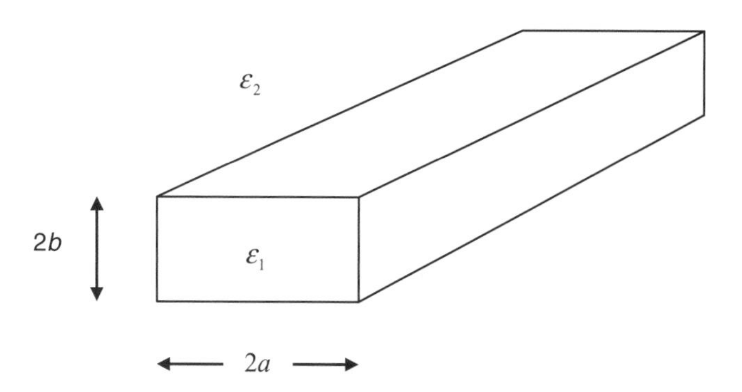

Figure 5.30 A rectangular dielectric waveguide. The guide dimensions are $2a \times 2b$ with the cartesian coordinates having their origin at the centre of the guide. The dimension in the x direction is $2a$ and in the y direction $2b$.

and infinite in the y direction, it will be relatively easy to solve for the TE_n modes using Example 5.5. As a result of this exercise, an effective dielectric constant ε_{TE} can be found. Now take a slab of thickness $2b$ and infinite in the x direction, but with a dielectric constant not ε_1 as before but ε_{TE}. Solving this for TM_m modes using Example 5.6 gives a further effective dielectric constant, $\varepsilon_{\text{TETM}}$, where

$$\varepsilon_2 < \varepsilon_{\text{TETM}} < \varepsilon_{\text{TE}} < \varepsilon_1. \tag{5.120}$$

The whole process is then repeated by starting with a slab of thickness $2b$ and infinite in the x direction. Solve this for TM_n modes using Example 5.6. The result for each mode will be some effective dielectric constant, ε_{TM}. Now taking a slab which is infinite in the y direction, and with a dielectric constant of ε_{TM} and solving this for TE_m modes gives yet another effective dielectric constant, $\varepsilon_{\text{TMTE}}$, where

$$\varepsilon_2 < \varepsilon_{\text{TMTE}} < \varepsilon_{\text{TM}} < \varepsilon_1. \tag{5.121}$$

The approximate effective dielectric constant is now found from

$$\varepsilon_{\text{EFF}} = \frac{1}{2}(\varepsilon_{\text{TETM}} + \varepsilon_{\text{TMTE}}). \tag{5.122}$$

There are more complex methods for combining these effective dielectric constants, but the simple average has been used extensively and is reasonably accurate, particularly at the frequencies well above the cut-off frequency for the mode. In Figure 5.31, the first few modes are shown for the rectangular dielectric waveguide. The mode classification is for hybrid modes, but rather different from the classification used for circular optical fibres. Here the dominant transverse electric field is taken and the direction is given as a superscript. The subscripts give the variations in the x and y directions respectively. These variations are not complete sine or cosine waves, except at very high frequencies, where the solutions become identical to those for metallic rectangular waveguide. The horizontal axis is a normalised frequency parameter, u, used by Goell, and is similar to the V parameter used in optical fibres. The equation for u is

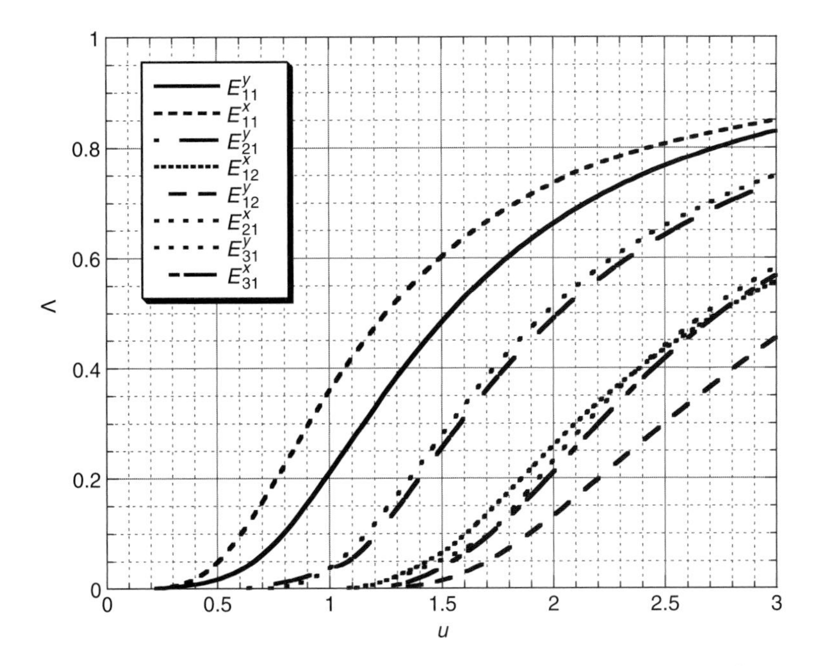

Figure 5.31 Rectangular dielectric guide modes. The ratio of the sides is $\frac{2a}{2b} = 2$, the dielectric constant of the guide is ε_1 and the surrounding medium has a dielectric constant of ε_2. The graph is plotted using Goell's parameters, so that a comparison can be made.

$$u = \frac{2b_1}{\lambda_0}(\varepsilon_1 - \varepsilon_2)^{\frac{1}{2}},$$

where the b_1 in Goell's work is equal to the shorter dimension on the waveguide. So from Figure 5.28, $2b = b_1$ and in the calculations for Figure 5.27,

$$u = \frac{4b}{\lambda_0}(\varepsilon_1 - \varepsilon_2)^{\frac{1}{2}} \text{ since } \frac{1}{\lambda_0} = \frac{k_0}{2\pi},$$

and in the example chosen $a = 2b$, then

$$u = \frac{ak_0}{\pi}(\varepsilon_1 - \varepsilon_2)^{\frac{1}{2}} = \frac{V}{\pi}. \tag{5.123}$$

Finally, the vertical axis in Goell's work is a normalised effective dielectric constant, Λ, defined as

$$\Lambda = \frac{\varepsilon_{\text{EFF}} - \varepsilon_2}{\varepsilon_1 - \varepsilon_2}. \tag{5.124}$$

This is similar to the b parameter in Figure 5.28, which is a normalised refractive index. There is not a simple relationship between Λ and b.

It can be seen in Figure 5.31 that the lowest order mode is the E^x_{11} mode and the next mode is the E^y_{11} mode. If the dielectric waveguide is connected to a metallic waveguide of similar dimensions, then this second mode is the most likely one to propagate. So a genuine mono-mode propagation is not possible for this waveguide. However, since the main electric and magnetic fields are orthogonal, coupling between these modes is unlikely to occur unless a discontinuity is introduced which changes the polarisation direction. For example, a metal plate at 45° to the x-axis would cause coupling. In a similar argument to that for slab guide, the E_{11}^y mode is not properly guided until over half of the power is inside the guide. The modes above this mode, the E_{21}^x mode and the E_{21}^y mode, are a possible limitation on the bandwidth. However, the E_{21}^x mode may not propagate for the same reasons given for the E_{11}^x mode above. The E_{21}^y mode has a zero of electric field in the centre and is clearly orthogonal to the E_{11}^y mode and in practice is not easily propagated, with the exception of special discontinuities. So, curiously, the first higher order mode which is likely to propagate is the E_{31}^y mode, and this means that the bandwidth of rectangular dielectric waveguide can be much larger than the equivalent metallic waveguide, assuming the circuits do not contain mode coupling elements.

Example 5.10 Beyond cut-off attenuators: The beyond cut-off attenuator is an interesting example of the use of a metallic waveguide at a frequency well below its cut-off frequency. At this frequency, a waveguide mode is not propagating but is evanescent, i.e. its fields decay exponentially along the guide. Also the wave impedance is purely imaginary, as shown in the left-hand part of Figure 5.6. In order to analyse this device, it is easiest to return to an equivalent circuit using wave impedances rather than characteristic impedances. Figure 5.32 shows the basic scheme of this attenuator.

Using a similar technique to the method given in Chapter 3 following Figure 3.11 in Example 3.7, the analysis starts at B. When a wave, E_+, arrives at B it is reflected with a reflection coefficient ρ_B, which is

$$\rho_B = \frac{Z_{TE} - jZ_{BC}}{Z_{TE} + jZ_{BC}}. \tag{5.125}$$

The incident wave at B is reflected as $\rho_B E_+$ and then its fields decay exponentially as it 'moves' back to A. So the electric field at A, E_A, is given by

$$E_A = E_+ \exp(\alpha D) + \rho_B E_+ \exp(-\alpha D), \tag{5.126}$$

where α is the attenuation constant for the decaying waves in the centre section of the guide. Using the usual conventions, the magnetic field at A, H_A, is given by

$$H_A = \frac{E_+}{jZ_{BC}} \exp(\alpha D) - \frac{\rho_B E_+}{jZ_{BC}} \exp(-\alpha D), \tag{5.127}$$

and hence the input impedance at A is given by

and taking the square root:

$$\alpha = \frac{R}{2}\sqrt{\frac{C}{L}} + \frac{G}{2}\sqrt{\frac{L}{C}} + \cdots = \frac{R}{2Z_0} + \frac{GZ_0}{2} + \cdots \text{ nepers m}^{-1} \text{ or}$$
$$\alpha = 8.686\left(\frac{R}{2Z_0} + \frac{GZ_0}{2} + \cdots\right) \text{ dBm}^{-1}. \tag{6.10}$$

Now for β, the procedure is similar: neglecting the RG term Equation (6.9) gives

$$\beta^2 = \frac{1}{2}\left(\omega^2 LC + \left(\left(-\omega^2 LC\right)^2 + \omega^2(RC + LG)^2\right)^{\frac{1}{2}}\right),$$
$$\beta^2 = \frac{1}{2}\left(\omega^2 LC + \omega^2 LC\left(1 + \frac{(RC + LG)^2}{(\omega LC)^2}\right)^{\frac{1}{2}}\right) \tag{6.11}$$
and $\beta = \omega\sqrt{LC} + \cdots$

Now for very low frequencies when $\omega^2 LC < \omega(RC + LG) < RG$ the term $\omega^2 LC$ can be neglected, so

$$\alpha^2 = \frac{1}{2}\left(RG + \left((RG)^2 + \omega^2(RC + LG)^2\right)^{\frac{1}{2}}\right) \text{ and}$$

$$\beta^2 = \frac{1}{2}\left(-RG + \left((RG)^2 + \omega^2(RC + LG)^2\right)^{\frac{1}{2}}\right)$$

and using the binomial theorem again gives

$$\alpha = \sqrt{RG} + \cdots \text{ nepers m}^{-1} \text{ or } \alpha = 8.686\sqrt{RG} + \cdots \text{ dBm}^{-1}, \tag{6.12}$$

$$\beta^2 = \frac{1}{2}\left(-RG + RG\left(1 + \frac{(\omega(RC + LG))^2}{(RG)^2}\right)^{\frac{1}{2}}\right) = \frac{1}{2}\left(\frac{\left(\omega^2(RC + LG)^2\right)}{2RG}\right),$$

$$\beta = \frac{\omega(RC + LG)}{2\sqrt{RG}} \text{ radians m}^{-1} \text{ and } v_p = \frac{\omega}{\beta} = \frac{2\sqrt{RG}}{(RC + LG)} \text{ ms}^{-1}. \tag{6.13}$$

Example 6.1 A transmission line has the following distributed parameters:

$R = 5.10^{-2}\ \Omega\text{m}^{-1}$, $C = 40\ \text{pFm}^{-1}$, $G = 10^{-9}\ \text{Sm}^{-1}$ and $L = 0.5\ \mu\text{Hm}^{-1}$.

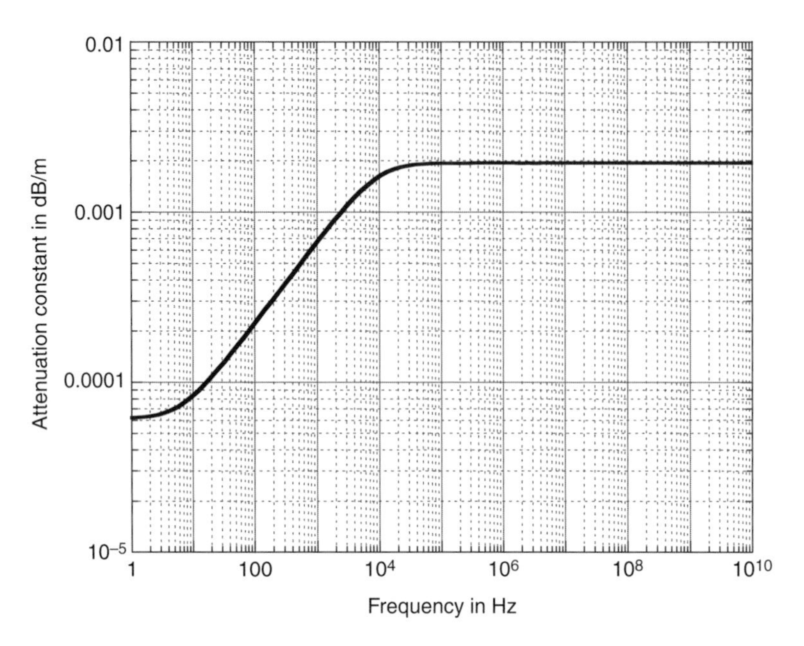

Figure 6.2 The graph of the attenuation constant against frequency for the parameters given in Example 6.1.

Calculate the lower frequency limit for the 'high frequency condition'. At a frequency of 100 MHz, calculate Z_0, α, β and v_p. Using the conditions given in (6.7):

$$f_{c1} = \frac{R}{2\pi L} = 15.9 \text{ kHz and } f_{c2} = \frac{G}{2\pi C} = 3.98 \text{ Hz}, \qquad Z_0 = \sqrt{\frac{L}{C}} = 111.85\ \Omega,$$

$$\alpha = 8.686\left(\frac{R}{2Z_0} + \frac{GZ_0}{2}\right) = 8.686\left(2.24 \times 10^{-4} + 5.59 \times 10^{-8}\right) = 0.00194 \text{ dBm}^{-1},$$

$$\beta = \omega\sqrt{LC} = 2.81 \text{ radians m}^{-1} \text{ and } v_p = \frac{1}{\sqrt{LC}} = 2.23.10^8 \text{ ms}^{-1}.$$

Note that the losses due to dielectric are often much smaller than the resistive losses in the conductor. It is the latter than causes the lower limit to the high frequency condition and is the main contributor to the attenuation. In Figures 6.2 and 6.3 are shown the graphs of both α and β using the values of the parameters given at the beginning of this example. The low frequency values are given in Equations (6.2) and (6.3) as

$$\alpha = 8.686\sqrt{RG} = 6.14 \times 10^{-5} \text{ dBm}^{-1} \text{ and } v_p = \frac{2\sqrt{RG}}{(RC + LG)} = 6.90 \times 10^6 \text{ ms}^{-1}.$$

Looking at Figure 6.2, the two 'turning points' are at the frequencies calculated above, namely 16 kHz and 4 Hz. The region above 16 kHz has a fixed value of

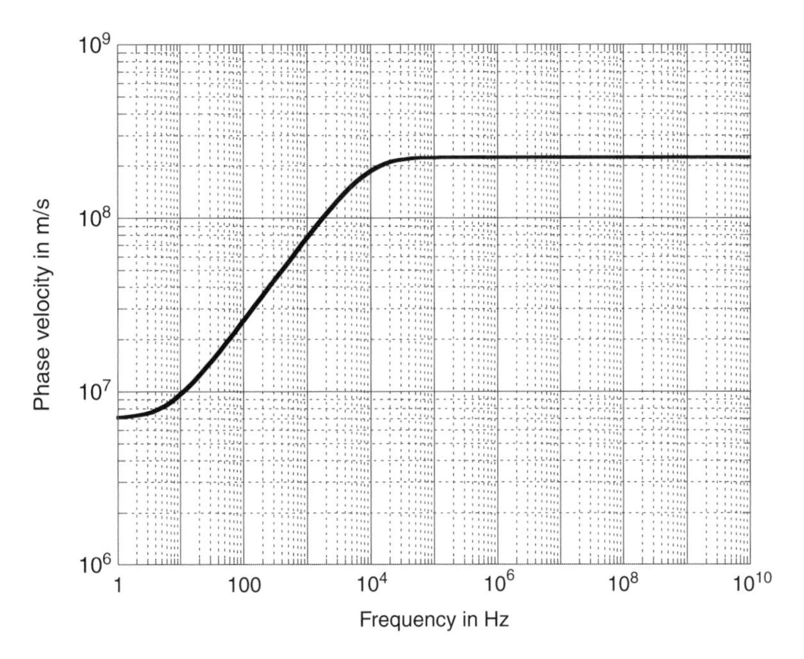

Figure 6.3 The phase velocity against frequency for parameters given in Example 6.1.

attenuation of 0.002 dBm^{-1} and, rather surprisingly, the attenuation falls below 16 kHz to the low frequency value of 0.00006 dBm^{-1}. Figure 6.3 has similar 'turning points' and the velocity above 16 kHz is $2.23.10^8\,ms^{-1}$, whereas the velocity is reduced at the low frequencies to only $7.10^6\,ms^{-1}$. Now in the region above 16 kHz, the velocity and the attenuation are fixed and so this gives no dispersion or distortion of signals. In practice both R and G increase with frequency and as a result there is an upper limit to the frequency range; this will be discussed later in this chapter. The early uses of transmission lines, telegraphy and telephony, all took place at the frequencies below 16 kHz and as a result this distortion of speech signals was a major problem. Various ingenious methods were used to overcome this, including loading coils placed along the transmission line. These were to increase L so that $RC = LG$, which condition simplifies the equations. However, these loading coils proved counter-productive above these frequencies and are now never used. The technique of carrier frequencies used in today's telephony means that the practical frequencies never go below this lower frequency limit. It is instructive to note that Wheatstone in 1833 measured the velocity of electricity and found a value of 288 000 miles a second or $4.64.10^8\,ms^{-1}$, which he claimed at the time made his new telegraphic apparatus work faster than light! The spark pulses he used in his measurement may well have travelled near the speed of light, but the telegraphic signals will have been a little more 'sluggish'.

6.2 The characteristic impedance of transmission lines with losses

Now the characteristic impedance can be defined for sinusoidal waves and with this limitation, the Telegraphists' Equations (6.3) become

$$\frac{\partial V}{\partial x} = -(R + \mathrm{j}\omega L)I \text{ and } \frac{\partial I}{\partial x} = -(G + \mathrm{j}\omega C)V.$$

From the solutions of Equation (6.4) given in (6.5):

$$\gamma V = (R + \mathrm{j}\omega L)I \text{ and } \gamma I = (G + \mathrm{j}\omega C)V. \tag{6.14}$$

Now $\gamma = \sqrt{-\omega^2 LC + \mathrm{j}\omega(RC + LG) + RG} = \sqrt{(R + \mathrm{j}\omega L)(G + \mathrm{j}\omega C)}$, which gives the characteristic impedance from either equation in (6.14) as

$$Z_0 = \sqrt{\frac{(R + \mathrm{j}\omega L)}{(G + \mathrm{j}\omega C)}}. \tag{6.15}$$

Now if the high frequency conditions apply to $R < \omega L$ and $G < \omega C$, the characteristic impedance is the normal equation $Z_0 = \sqrt{L/C}$, which was used in Example 6.1. However, it is useful to note what happens to the characteristic impedance at low frequencies. These are plotted in Figure 6.4 using the same parameters as in Example 6.1. The low frequency value of the characteristic impedance is given by

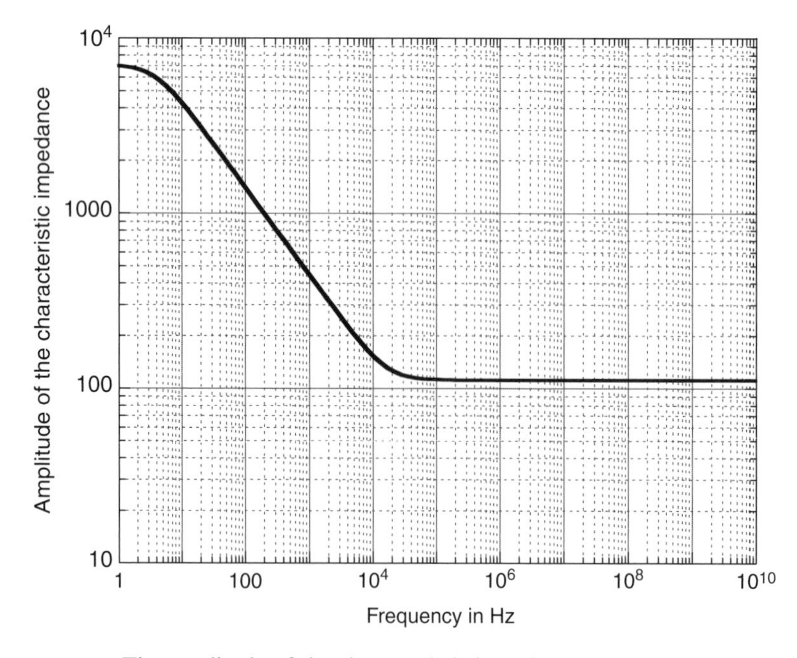

Figure 6.4 The amplitude of the characteristic impedance against frequency using the same parameters as in Example 6.1.

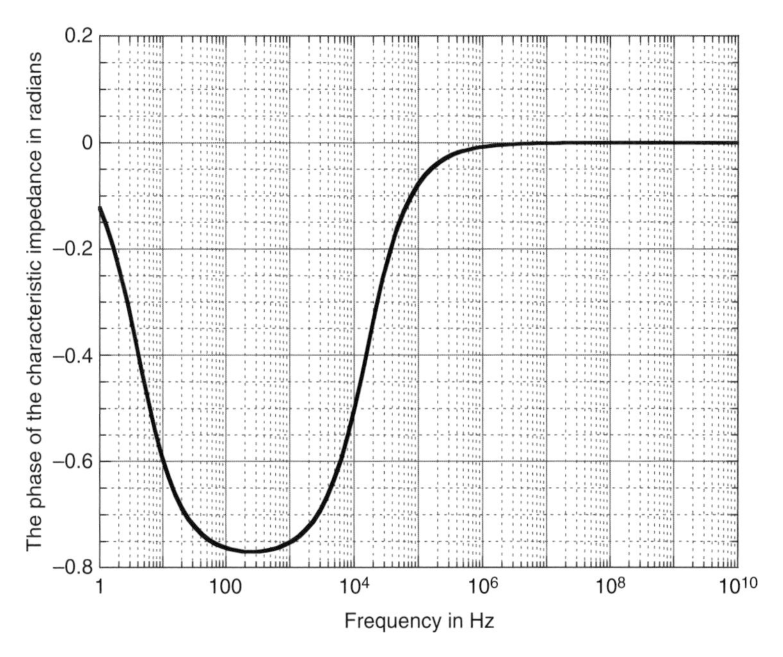

Figure 6.5 The phase of the characteristic impedance in radians against frequency for the same parameters as in Example 6.1.

$$Z_0 = \sqrt{\frac{R}{G}}. \tag{6.16}$$

The phase of the characteristic impedance is given by

$$\frac{1}{2}\left(\arctan\left(\frac{\omega L}{R}\right) - \arctan\left(\frac{\omega C}{G}\right)\right), \tag{6.17}$$

which has a lowest value of $-\pi/4$ radians between the frequencies of 4 Hz and 16 kHz. This is shown in Figure 6.5.

These two figures show that measurements of the reflection coefficient at these low frequencies are not straightforward and a detailed knowledge of Z_0 is needed before they can be interpreted. It is for this reason that most wave measurements stop below about 10 kHz. However, the reader may remember the examples in Chapter 1 which had batteries as sources. For the most part, the transients discussed in those examples would contain frequencies well above this 10 kHz limit. Obviously, if the final result was a steady current, the value of Z_0 at 0 Hz would be considerably higher than the traditional 50 Ω. This change of impedance would only be detected if the changes at the source were sufficiently slow, that is longer than 0.1 ms in duration. This does raise the question of the electricity supply at 50 Hz, where the characteristic impedance may well be well above the value expected. However, due to the fact that the distribution circuits rarely extend for a fraction of a wavelength – taking a low velocity of $1.5.10^7\,\text{ms}^{-1}$ from Figure 6.2 gives a wavelength of 300 km – these wave effects will be largely undetected. Since the resistive losses are usually greater than the conductive losses, the phase of Z_0 is a negative angle as the second term in Equation (6.17) is larger than the first. An approximate expression for this can be obtained by putting $G = 0$ and assuming $\omega L > R$:

$$Z_0(\text{lossy}) = \sqrt{\frac{R + \text{j}\omega L}{\text{j}\omega C}} = \sqrt{\frac{L}{C} - \text{j}\frac{R}{\omega C}} = \sqrt{\frac{L}{C}}\left(1 - \text{j}\frac{R}{\omega L}\right)^{\frac{1}{2}} = \sqrt{\frac{L}{C}}\angle - \arctan\left(\frac{R}{2\omega L}\right)$$

$$\cong \sqrt{\frac{L}{C}} - \text{j}\sqrt{\frac{L}{C}}\left(\frac{R}{2\omega L}\right),$$

or using the loss-less Z_0:

$$Z_0(\text{lossy}) \cong Z_0 - \text{j}Z_0\left(\frac{R}{2\omega L}\right). \tag{6.18}$$

A problem can arise in measurements concerning matched loads and power. If a matched load is defined as an impedance which does not cause a reflection, then from Equation (1.13), which will still apply, the impedance must be equal to the characteristic impedance of the line. So in a classical power measurement, the power detector would be matched in this way. However, the power on the line as defined in Section 2.3 now needs modifying:

$$\frac{V_1^2}{2Z_0\left(1 + \left(\frac{R}{2\omega L}\right)^2\right)} = \frac{V_1^2\ \text{Re}(Z_0(\text{lossy}))}{2|Z_0(\text{lossy})|^2}. \tag{6.19}$$

The implication of this is that less power arrives at the detector and so precise power measurements require an accurate value of the characteristic impedance and the input impedance of the power detector before any estimations of the power can be made. For the rest of this chapter, the focus will be on the use of transmission lines above this lower frequency limit.

6.3 The input impedance of a length of lossy line

The input impedance of a length of line, D, was considered in Section 2.5. This section will modify the equations in that section so as to include the attenuation. As in the loss-less case, one end of the line is terminated with an impedance Z_L. The voltage at the other end can be represented as

$$V = V_1 \exp(\alpha D) \exp(\text{j}\beta D) + |\rho| V_1 \exp(-\alpha D) \exp(\text{j}(\angle\rho - \beta D)). \tag{6.20}$$

This is the same as the equation leading up to (2.20) but with the addition of the attenuation terms. The equation for the current will be

$$I = I_1 \exp(\alpha D) \exp(\text{j}\beta D) - |\rho| I_1 \exp(-\alpha D) \exp(\text{j}(\angle\rho - \beta D)). \tag{6.21}$$

The ratio of these two will give the input impedance of the length of lossy transmission line:

$$Z_{\mathrm{IN}}(\text{lossy}) = Z_0\left(\frac{\exp(\gamma D) + \rho \exp(-\gamma D)}{\exp(\gamma D) - \rho \exp(-\gamma D)}\right), \tag{6.22}$$

where $V_1/I_1 = Z_0$; $\gamma = \alpha + \mathrm{j}\beta$ and $\rho = |\rho| \exp(\mathrm{j}\angle\rho)$.

Using Equation (1.13) for the reflection coefficient, i.e. $\rho = (Z_{\mathrm{L}} - Z_0)/(Z_{\mathrm{L}} + Z_0)$ gives

$$Z_{\mathrm{IN}}(\text{lossy}) = Z_0\left(\frac{(Z_{\mathrm{L}} + Z_0)\exp(\gamma D) + (Z_{\mathrm{L}} - Z_0)\exp(-\gamma D)}{(Z_{\mathrm{L}} + Z_0)\exp(\gamma D) - (Z_{\mathrm{L}} - Z_0)\exp(-\gamma D)}\right).$$

This can be re-expressed in terms of hyperbolic functions as

$$Z_{\mathrm{IN}}(\text{lossy}) = Z_0\left(\frac{Z_{\mathrm{L}} + Z_0 \tanh(\gamma D)}{Z_0 + Z_{\mathrm{L}} \tanh(\gamma D)}\right). \tag{6.23}$$

This equation is similar to the loss-less Equation (2.21), if $\alpha = 0$ and using

$$\tanh(\mathrm{j}\beta D) = \mathrm{j}\tan(\beta D).$$

Many of the features described in Section 2.5 still exist in the line with losses but in a modified form. So if $Z_{\mathrm{L}} = 0$, i.e. a short circuit termination, then

$$Z_{\mathrm{IN}}\ (\text{lossy}) = Z_0 \tanh(\gamma D). \tag{6.24}$$

For an open circuit termination:

$$Z_{\mathrm{IN}}(\text{lossy}) = Z_0 \coth(\gamma D). \tag{6.25}$$

For $D = \frac{\lambda}{4} + \frac{n\lambda}{2}$ or $\beta D = \frac{\pi}{2} + n\pi$, where $n = 0, 1, 2, \ldots$,

$$Z_{\mathrm{IN}}\ (\text{lossy}) = Z_0\left(\frac{Z_{\mathrm{L}} + Z_0 \coth(\alpha D)}{Z_0 + Z_{\mathrm{L}} \coth(\alpha D)}\right) \approx Z_0\left(\frac{\alpha D Z_{\mathrm{L}} + Z_0}{\alpha D Z_0 + Z_{\mathrm{L}}}\right) \text{ for } \alpha D < 0.4. \tag{6.26}$$

Finally, for $D = \frac{\lambda}{2}$ or $\beta d = n\pi$, where $n = 0, 1, 2, \ldots$,

$$Z_{\mathrm{IN}}\ (\text{lossy}) = Z_0\left(\frac{Z_{\mathrm{L}} + Z_0 \tanh(\alpha D)}{Z_0 + Z_{\mathrm{L}} \tanh(\alpha D)}\right) \approx Z_0\left(\frac{Z_{\mathrm{L}} + \alpha D Z_0}{Z_0 + \alpha D Z_{\mathrm{L}}}\right) \text{ for } \alpha D < 0.4. \tag{6.27}$$

Example 6.2 A low-loss transmission line is 10 m long and at one end is a terminating impedance. When the *VSWR* was measured near this impedance it had a value of 4, but 10 m away, when the measurement was repeated, the *VSWR* was only 3. Find the attenuation constant of the transmission line.

Solution to Example 6.2

The *VSWR* is defined in Equation (2.18) as

$$VSWR = \frac{V_{\mathrm{MAX}}}{V_{\mathrm{MIN}}} = \frac{|V_+| + |V_-|}{|V_+| - |V_-|}.$$

So, near the terminating impedance, assuming the losses are small and that the measurement is made close to the termination,

$$VSWR = 4 = \frac{|V_+| + |V_-|}{|V_+| - |V_-|} \quad \text{so} \quad 3|V_+| = 5|V_-|.$$

At 10 m away, the losses need to be taken into account:

$$VSWR(+\text{losses}) = \frac{|V_+| \exp(10\alpha) + |V_-| \exp(-10\alpha)}{|V_+| \exp(10\alpha) - |V_-| \exp(-10\alpha)} = 3,$$

$$\text{so} \quad 5 \exp(10\alpha) + 3 \exp(-10\alpha) = 15 \exp(10\alpha) - 9 \exp(-10\alpha)$$

$$\text{or } 10 \exp(10\alpha) = 12 \exp(-10\alpha) \text{ and } \exp(20\alpha) = 1.2 \text{ so } \alpha = 0.08 \text{ dBm}^{-1}.$$

Example 6.3 A 2 m length of low-loss transmission line has a characteristic impedance of 50 Ω. The attenuation constant is 0.752 dBm^{-1} and the wavelength is one metre. A resistive termination is connected at one end of the line and measured at the other. If the measured value is 100 Ω, what is the actual value? What would have been measured at the end of a 20 m line?

Solution to Example 6.3

The measurement is also a pure resistance, since the line is four half-wavelengths long. Starting at the measurement end of the line, the reflection coefficient there is

$$\rho_{2\text{m}} = \frac{V_-}{V_+} = \frac{100 - 50}{100 + 50} = \frac{1}{3}.$$

At the unknown impedance, the reflection coefficient is

$$\rho = \frac{V_- \exp(2\alpha)}{V_+ \exp(-2\alpha)} = \frac{R - 50}{R + 50},$$

where R is the unknown resistive impedance. Rearranging the equations and using the value of the attenuation constant:

$$\rho = \frac{\sqrt{2}}{3} = \frac{R - 50}{R + 50}, \quad \text{which gives } R = 139\ \Omega.$$

For 20 m of line, the reflection coefficient will be

$$\rho_{20\text{m}} = \frac{\sqrt{2} \exp(-20\alpha)}{3 \exp(20\alpha)} = 0.0147$$

which gives a measured resistance of 51.5 Ω.

In Equation (6.27) as $\alpha D \to \infty$, $\tanh(\alpha D) \to 1$ and Z_0 (lossy) $\to Z_0$. This is illustrated in this example. The 139 Ω resistance has reduced to 100 Ω 2 m away and is almost 50 Ω when the distance is 20 m. Although *modern* instrumentation is able to eliminate or 'de-embed' the attenuation of a cable, clearly it is easier to measure the difference between 139 Ω and 50 Ω than the much smaller difference between 51.5 Ω and 50 Ω. If the frequency response of these measurements were plotted on a Smith chart, the result would be a spiral going inwards towards the centre.

6.4 The conductance, *G*

In Chapter 4, the nature of both L and C was investigated using the theory of electromagnetism. Using the same theory, an equation for G can be obtained by exchanging the dielectric between the plates for a conducting medium. Then, in each case, G can be obtained from the equation for C by exchanging ε and σ. To illustrate this Example 4.1 on the coaxial cable is used, where in Equation (4.4) from Gauss' Theorem the electric flux density, D, was given by

$$D = \frac{\Phi_e}{2\pi rl} = \frac{Q}{2\pi rl}\ \text{Cm}^{-2}\ \text{(Equation (4.4))}.$$

Differentiating this with respect to time gives

$$\frac{\partial D}{\partial t} = J = \frac{1}{2\pi rl}\frac{\partial Q}{\partial t} = \frac{I}{2\pi rl}\ \text{Am}^{-2}, \tag{6.28}$$

where I is the current flowing from one conductor to the other through the conducting medium. Now the current density, J, is linked to the electric field by

$$J = \sigma E. \tag{6.29}$$

Hence the electric field in the conducting medium produced by the flow of the current, I, is given by

$$E = \frac{I}{2\pi rl\sigma}\ \text{Vm}^{-1}.$$

Integrating to obtain the voltage gives

$$V = \int_a^b \frac{I}{2\pi rl\sigma}\text{d}r = \frac{I}{2\pi l\sigma}\ln\left(\frac{b}{a}\right).$$

$$\text{Hence } G = \frac{I}{V} = \frac{2\pi l\sigma}{\ln\left(\frac{b}{a}\right)}\ \text{Sm}^{-1}. \tag{6.30}$$

Comparing this with Equation (4.7):

$$C = \frac{2\pi l\varepsilon}{\ln\left(\frac{b}{a}\right)}$$

it shows that the σ term has replaced the ε in Equation (6.30).

Although this result is for zero frequency, it will also apply for TEM modes. So for a wide range of frequencies, for the lowest order mode in two conductor transmission lines:

$$\sigma C = \varepsilon G. \tag{6.31}$$

Now an alternative way of representing the losses in a dielectric material is to use a complex permittivity which gives the same results as Equation (6.31). The reactance of a parallel combination of C and G is given by

$$X = G + \mathrm{j}\omega\, C. \tag{6.32}$$

If the ε is now replaced by $\varepsilon' - \mathrm{j}\varepsilon''$, the following results:

$$C = \frac{2\pi l \varepsilon'}{\ln\left(\frac{b}{a}\right)} - \mathrm{j}\frac{2\pi l \varepsilon''}{\ln\left(\frac{b}{a}\right)}.$$

The reactance of this parallel combination gives

$$X = \mathrm{j}\omega C = \mathrm{j}\omega \frac{2\pi l \varepsilon'}{\ln\left(\frac{b}{a}\right)} + \frac{2\pi l \omega \varepsilon''}{\ln\left(\frac{b}{a}\right)}. \tag{6.33}$$

Comparing Equations (6.32) and (6.33) shows that the real part of ε gives the usual equation for capacitance and the imaginary part gives the equation for the conductance. So the analogy is complete if

$$\sigma = \omega \varepsilon''. \tag{6.34}$$

Now the ratio of the two parts of the dielectric constant is sometimes called the loss tangent for the dielectric and the angle between the two is called δ:

$$\tan\,\delta = \frac{\varepsilon''}{\varepsilon'}. \tag{6.35}$$

For a low-loss dielectric the value of δ is very small and often reasonably constant with frequency. If ε' is also constant with frequency, then Equation (6.34) implies that the conductivity is increasing with frequency. Now the high frequency condition used in Equation (6.7):

$$G < \omega C$$

can now be combined with a modified version of Equation (6.31):

$$\omega \varepsilon'' C = \varepsilon' G,$$

which makes the high frequency condition

$$\frac{\varepsilon''}{\varepsilon'} < 1 \text{ or } \tan\,\delta < 1 \quad \text{or } \delta < \frac{\pi}{4}. \tag{6.36}$$

In practice ε' and ε'' do vary with frequency. The real part of the dielectric constant often decreases with increasing frequency – called dielectric relaxation – and the imaginary part can have a maximum at these relaxation frequencies. The most extreme example of this is water which changes its real part from about 80 down to 2.25 at optical frequencies with a peak value of $\tan\delta = 1$ in the middle of the microwave frequencies. For most dielectrics, the changes are far smaller, and although this is a cause for some dispersion in transmission lines it is not a major obstacle for most applications. Some typical values:

$$\text{alumina, } \varepsilon_{\mathrm{R}} = 10.8, \quad \tan\delta = 5\times10^{-4};$$

$$\text{Perspex, } \varepsilon_{\mathrm{R}} = 2.6, \quad \tan\delta = 6\times10^{-3};$$

$$\text{PVC, } \varepsilon_{\mathrm{R}} = 2.8, \quad \tan\delta = 1\times10^{-2};$$

$$\text{quartz, } \varepsilon_{\mathrm{R}} = 3.8, \quad \tan\delta = 5\times10^{-4};$$

$$\text{sapphire, } \varepsilon_{\mathrm{R}} = 9.4, \quad \tan\delta = 2\times10^{-4}.$$

Finally, the cross link to plane waves can be made using the second term in the expression for the attenuation constant, (6.10):

$$\alpha = \frac{GZ_0}{2} = \frac{\omega\varepsilon'' C Z_0}{2\varepsilon'} = \frac{\omega\varepsilon'' C}{2\varepsilon'}\sqrt{\frac{L}{C}} = \frac{\omega\varepsilon''\sqrt{LC}}{2\varepsilon'} = \frac{\omega\varepsilon''\sqrt{\mu\varepsilon'}}{2\varepsilon'} = \frac{\omega\varepsilon''}{2}\sqrt{\frac{\mu}{\varepsilon'}} = \frac{\sigma\eta}{2}. \quad (6.37)$$

The last of these expressions gives the attenuation as a plane wave moves through a lossy dielectric, the first is for a TEM wave on a two conductor transmission line. There are obvious similarities.

6.5 The resistance *R* and the skin effect

It might be expected that the same theory could be used for R, but the reason why this can not be employed is the skin effect. At quite low frequencies, the distribution of current in a conductor is not uniform as it would be at zero frequency. This distribution changes as frequency increases and at radio and microwave frequencies the currents only flow in the surface or the 'skin' of the conductors. This has two effects: one is to restrict the flow of current and thus raise the value of R and the other is to reduce almost completely the internal or self inductance of the conductor to zero. A starting point for describing the skin effect is to begin with the complex form of the dielectric constant used in the previous section. The definition of a good dielectric with low losses is

$$\varepsilon' \gg \varepsilon''.$$

If the losses are much greater than the dielectric effect, then this becomes the definition of a conductor:

$$\varepsilon'' \gg \varepsilon'.$$

The equations for wave propagation in conductors can be obtained by using the complex form of the dielectric constant as follows:

$$\varepsilon = \varepsilon' - \mathrm{j}\varepsilon'', \quad \text{which for high conductivity } \varepsilon \to -\mathrm{j}\varepsilon'' = -\mathrm{j}\frac{\sigma}{\omega}.$$

Using this in Equation (5.6), the electromagnetic wave equation for a conductor becomes

$$\nabla^2 E = -\frac{\mathrm{j}\mu\sigma}{\omega}\frac{\partial^2 E}{\partial t^2}, \quad \text{which becomes}$$

$$\nabla^2 E = \mathrm{j}\omega\mu\sigma E \quad \text{for sinusoidal waves.}$$

Limiting this wave equation to plane waves where $\nabla^2_{x,y}E = 0$ gives

$$\frac{\partial^2 E}{\partial z^2} = \mathrm{j}\omega\mu\sigma E. \tag{6.38}$$

Solutions to this wave equation for plane waves in a conductor are of the form

$$E = E_0 \exp\,(-\gamma z).$$

Substituting this into Equation (6.38) gives

$$\gamma^2 = \mathrm{j}\omega\mu\sigma,$$

and taking the square root,

$$\gamma = \alpha + \mathrm{j}\beta = \frac{1}{\sqrt{2}}(1+\mathrm{j})\sqrt{\omega\mu\sigma}. \tag{6.39}$$

This solution shows that waves both decay and propagate in a conductor at the same rate:

$$\alpha = \beta = \sqrt{\frac{\omega\mu\sigma}{2}}. \tag{6.40}$$

The distance a wave needs to travel in order to decay to exp(−1) of its original value or 36.8% is called the 'skin depth' and is given the symbol δ. This is identical to the symbol given to the dielectric loss tangent in the previous section. Such an overlap is unfortunate, but the Greek alphabet is a little short of easy symbols, ξ for instance is not over-used. In practice, the specific areas of the two usually remove any ambiguity and the two uses of δ are now so widespread that it would be difficult to change. The decay of the fields is given by

$$E = E_0 \exp(-\alpha\delta) = E_0 \exp\,(-1) \text{ so } \delta = \frac{1}{\alpha} = \sqrt{\frac{2}{\omega\mu\sigma}} = \frac{1}{\sqrt{\pi f \mu\sigma}}. \tag{6.41}$$

The phase constant β can be used to find the velocity of propagation as

$$v_{\mathrm{p}} = \frac{\omega}{\beta} = \omega\sqrt{\frac{2}{\omega\mu\sigma}} = \sqrt{\frac{2\omega}{\mu\sigma}}. \tag{6.42}$$

Finally, the wave impedance can be found by substituting $-\mathrm{j}\sigma/\omega$ for ε, so

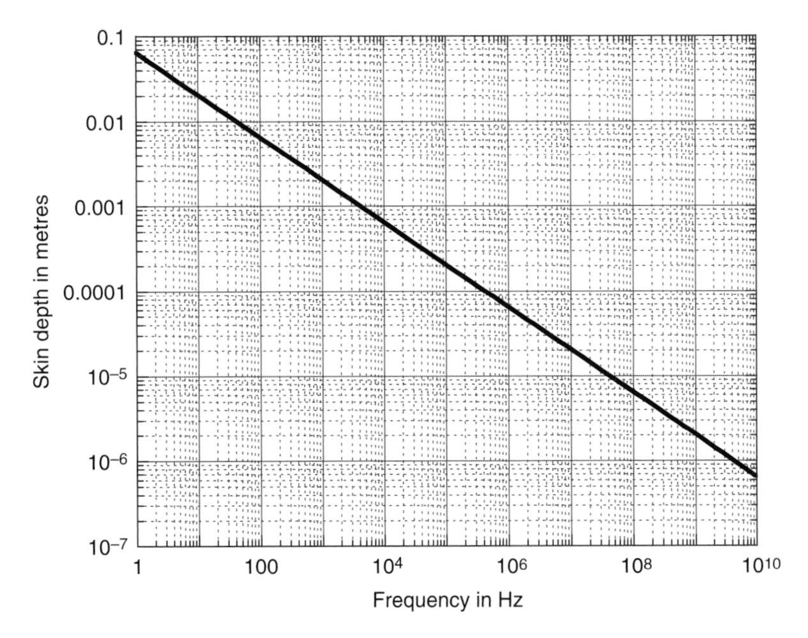

Figure 6.6 The skin depth in metres against frequency for a conductor with a conductivity of 6×10^7 Sm^{-1}.

$$\eta = \sqrt{\frac{\mu}{\varepsilon}} = \sqrt{\frac{\mathrm{j}\omega\mu}{\sigma}} = (1+\mathrm{j})\sqrt{\frac{\omega\mu}{2\sigma}}. \tag{6.43}$$

This wave impedance has a resistive and an inductive component which are equal in amplitude. Some typical values of these three parameters δ, v_{p} and η against frequency are shown in Figures 6.6, 6.7 and 6.8.

The conductivities of common metals at room temperature are:

aluminium, 4.00×10^7 Sm^{-1};

copper, 6.45×10^7 Sm^{-1};

gold, 4.88×10^7 Sm^{-1};

silver, 6.80×10^7 Sm^{-1}.

The relationship between the circuit parameter, *R*, and the skin effect is established for higher frequencies by using the concept of a sheet resistance. If the currents only travel in the skin of the conductor, then the surface of the conductor is the important region to consider. First of all consider the sheet resistance or Ω/square, R_{sheet}. Taking a square resistive sheet, and placing two good conducting strips along opposite edges, the resistance can now be measured, between the strips, see Figure 6.9.

To make it more general, if the width of the sheet is *w* and the length is *l* then the resistance is given by

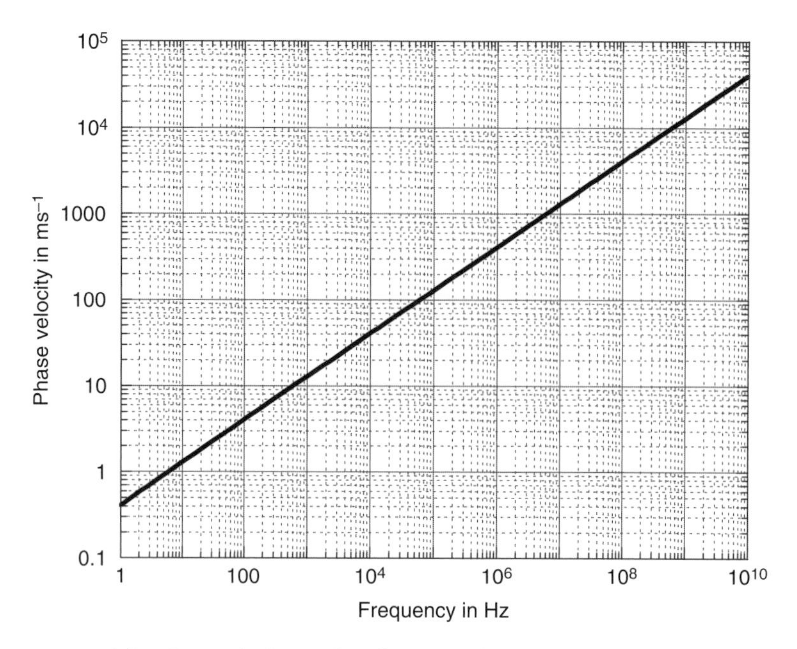

Figure 6.7 The phase velocity against frequency in a conductor with a conductivity of 6×10^7 Sm^{-1}.

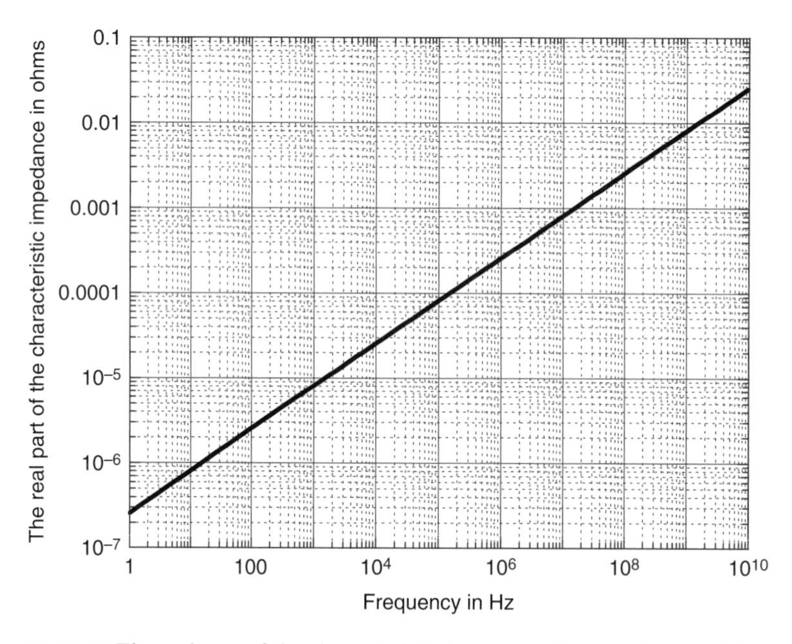

Figure 6.8 The real part of the characteristic impedance in a conductor with a conductivity of 6×10^7 Sm^{-1} against frequency. The imaginary part is equal to the real part.

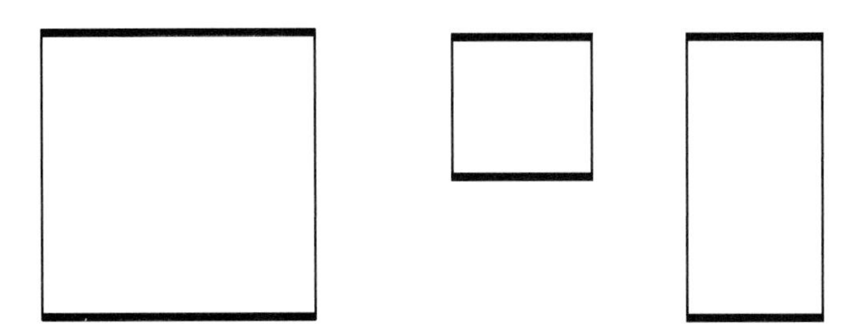

Figure 6.9 Various sheets of resistive material, with conducting strips at the top and bottom. The first two are both squares and their resistances are both R_{sheet}. The right-hand sheet is twice as long as the middle sheet and its resistance is $2R_{\text{sheet}}$.

$$R = \frac{l}{w} R_{\text{sheet}}. \tag{6.44}$$

Now the solution for a plane wave propagating in a conductor is given in Equation (6.39) as

$$E = E_0 \exp(-\gamma z).$$

The current density in the conductor is then

$$J = \sigma E_0 \exp(-\gamma z).$$

By integrating this current density over a width, w, and into the conductor to infinity, the total current, I_{total}, can be found as

$$I_{\text{total}} = \int_0^\infty J \mathrm{d}z = \frac{\sigma E_0 w}{\gamma}. \tag{6.45}$$

Now taking a length l, the voltage at the surface across this length will be

$$V = lE_0. \tag{6.46}$$

Hence the surface impedance, using Equations (6.44), (6.45) and (6.46), is given by

$$\frac{V}{I_{\text{total}}} = lE_0 \times \frac{\gamma}{\sigma E_0 w} = \frac{l\gamma}{\sigma w}. \tag{6.47}$$

Now by considering a square, i.e. putting $l = w$, the surface impedance per square is called Z_{S}, given by

$$Z_{\text{S}} = \frac{\gamma}{\sigma} = \sqrt{\frac{\omega\mu}{2\sigma}}(1 + \mathrm{j}). \tag{6.48}$$

The real part of Equation (6.47) is usually called the skin resistance, R_{S}, and the imaginary part the skin reactance, X_{S}. So

$$Z_{\text{S}} = R_{\text{S}} + \mathrm{j}X_{\text{S}} \text{ and } R_{\text{S}} = X_{\text{S}} = \sqrt{\frac{\omega\mu}{2\sigma}} = R_{\text{sheet}}. \tag{6.49}$$

In other words, the sheet resistance is equal to the real part of the wave impedance given in Equation (6.49). These equations apply in round wires where the skin depth is much less than the radius of the wire or the thickness of the conductor. For typical wires this usually occurs at frequencies above 100 kHz, as can be seen in Figure 6.6. In the next example, this will be applied to some of the two-wire transmission lines discussed so far. The sheet inductance is often much smaller than the main inductance, L, of a transmission line, but in some configurations it can become a significant factor in both increasing the characteristic impedance and producing a slower velocity of propagation.

Example 6.4 The attenuation of some two-wire transmission lines: The first example will be the coaxial cable (see Example 4.1), followed by the parallel plate line (see Example 4.2) and the two parallel wires (see Example 4.3).

---

1 – Attenuation in a coaxial cable

Use Equation (6.10),

$$\alpha = 8.686\left(\frac{R}{2Z_0} + \frac{GZ_0}{2} + \cdots\right) \text{ dBm}^{-1},$$

and take the first part which concerns just the resistive loss. In the high frequency region, i.e. well above 100 kHz, the skin effect is the dominant source of resistive loss. Now consider the surface of the inner and outer conductors as sheets (by a suitable cut in the z direction), as shown in Figure 6.10.

The left-hand rectangle represents the surface of the outer conductor which has a radius b, and the right-hand rectangle represents the inner conductor. The total resistance due to the skin effect is between A and B in the figure, i.e.

$$R = \frac{R_S}{2\pi b} + \frac{R_S}{2\pi a}. \tag{6.50}$$

Hence the attenuation constant due to the skin effect is

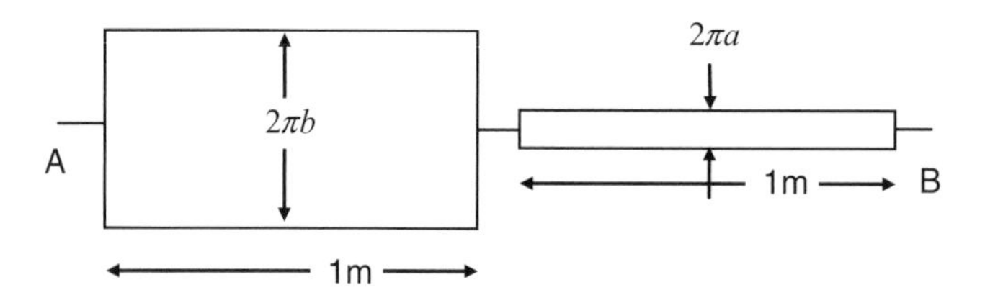

Figure 6.10 The two surfaces of a 1 m length of a coaxial cable ‘opened out’ to show the derivation of the total sheet resistance.

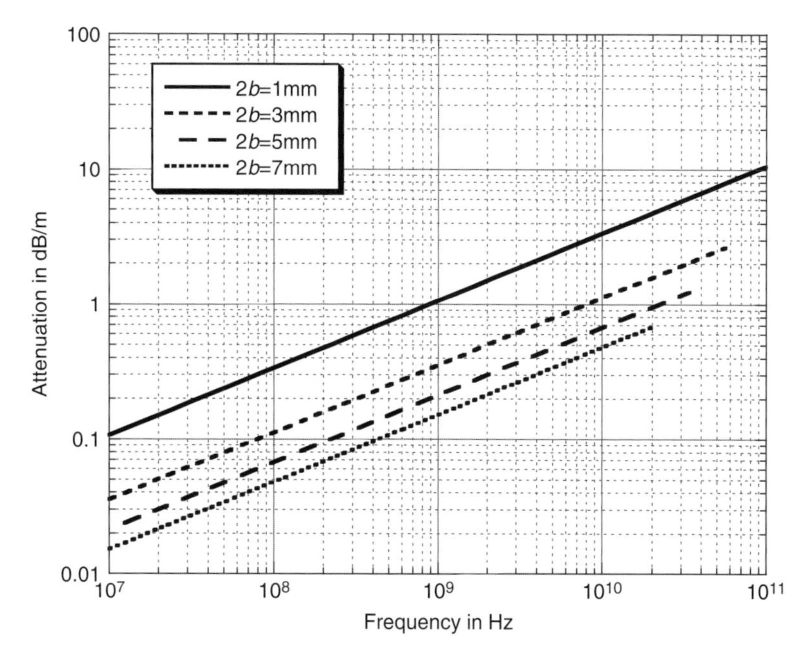

Figure 6.11 The theoretical attenuation due to the skin resistance in coaxial cables against frequency, for various diameters of the outer conductor. The relative dielectric constant of the dielectric in the cable is 2.5 and the conductivity on the conductors is $6.10^7\ \mathrm{Sm}^{-1}$. The upper limit is the onset of the first higher order mode in the cable.

$$\alpha = 8.686\left(\frac{R_S}{2Z_0}\right)\left(\frac{1}{2\pi b}+\frac{1}{2\pi a}\right) = 8.686\sqrt{\frac{\omega\mu}{2\sigma}}\frac{1}{4\pi Z_0}\left(\frac{1}{b}+\frac{1}{a}\right)\ \mathrm{dBm}^{-1}. \tag{6.51}$$

Now Equation (6.51) is used to compute the attenuation constant variation with frequency and in Figure 6.11 are shown the theoretical attenuations in various commonly used coaxial cables. The attenuation is shown from 10 MHz to the frequency at which the first higher order mode starts, in other words the working bandwidth of the cable.

It is worth noting that in loss-less coaxial cable the electric fields are orthogonal to the conducting surfaces. In this lossy case, the losses give rise to a small electric field in the z direction, i.e. the direction of propagation. Now this field, E_z, is linked to the other fields by

$$\sigma|E_z| = |J_z|,\ |J_z| = |H_\theta| \text{ and } \frac{E_r}{H_\theta} = \eta \text{ so } |E_z| = \frac{|E_r|}{\sigma\eta} \approx \frac{|E_r|}{3.10^9}.$$

This small E_z electric field means that the propagation is no longer a pure TEM mode. However, if the losses are small, this deviation from the TEM mode will usually have a negligible effect on either the characteristic impedance or the velocity. This does remove the perfect orthogonality of the modes and in some cases can give rise to mode coupling.

2 – Attenuation in a parallel plate line

The same procedure can be used for the parallel plate line. The width of each plate in figure 4.2 was w and so the total skin resistance for one metre length is given by

$$R = R_S \times \frac{2}{w}, \tag{6.52}$$

so the attenuation coefficient due to the skin effect is

$$\alpha = 8.686 \times \frac{R}{2Z_0} = 8.686 \times \frac{R_S}{wZ_0} = 8.686 \times \frac{R_S}{\eta d} \quad \text{dBm}^{-1}. \tag{6.53}$$

This has used the equation for Z_0 from Example 4.12. The attenuation is a function of not the width of the plates but the separation, d, which is a surprising result.

Another way of deriving the attenuation for small losses is to use the power equation,

$$P = P_0 \exp(-2\alpha z) \approx P_0(1 - 2\alpha z + \cdots). \tag{6.54}$$

If the length is taken as 1 m, i.e. $z = 1$, then rearranging the terms:

$$\alpha = \frac{P_0 - P}{2P_0} = \frac{P_L}{2P_0} \quad \text{nepers m}^{-1}, \tag{6.55}$$

where P_0 is the power travelling down the line and P_L is the small amount of power lost per metre. Using the circuit variables:

$$P_0 = \frac{|V|^2}{2Z_0} = \frac{|I|^2 Z_0}{2} \quad \text{and} \quad P_L = \frac{|I|^2 R}{2} = \frac{|I|^2 R_S}{w}.$$

Hence:

$$\alpha = 8.686 \times \frac{R_S}{wZ_0} = 8.686 \times \frac{R_S}{\eta d} \quad \text{dBm}^{-1} \text{as before.}$$

3 – Attenuation in a pair of parallel wires

The attenuation in a pair of parallel wires can be found by a similar approach to the one used for the coaxial cable. The radius of each wire is a so the resistance per metre is given by

$$R = \frac{R_S}{\pi a},$$

putting $a = b$ in Equation (6.50).

So the attenuation coefficient is given by

$$\alpha = 8.686 \times \frac{R_S}{2\pi a Z_0} \quad \text{dBm}^{-1}. \tag{6.56}$$

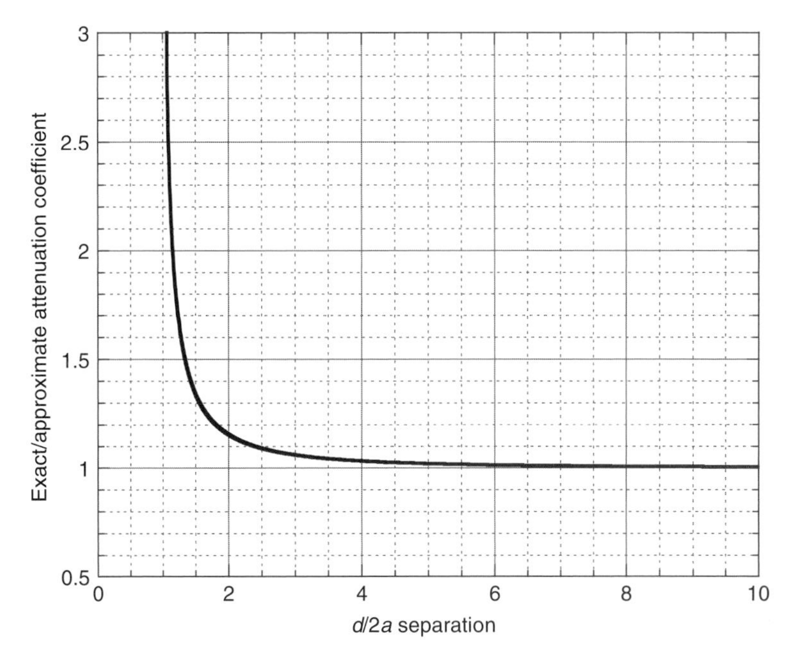

Figure 6.12 The ratio of the exact to the approximate attenuation coefficient for two parallel wires. Once $d/2a$ is greater than 2, the values from the approximate equation are only a few percent less than those from the exact equation.

Using the exact equations, which take into account the non-uniform distribution of current in the conductors, particularly when they are close together, the attenuation coefficient can be obtained from conformal mapping as

$$\alpha = 8.686 \times \frac{R_S}{2\pi a Z_0} \times \left(\frac{\frac{d}{2a}}{\sqrt{\left(\frac{d}{2a}\right)^2 - 1}} \right) \text{ dBm}^{-1}. \tag{6.57}$$

The ratio of these two equations is shown in Figure 6.12.

4 – Attenuation in microstrip

There are numerous empirical equations for the attenuation constant due to conductor losses in microstrip and, rather than repeat them all, a simple one has been selected to act as a 'rule of thumb'. The reason for the complication is that the current distribution in both conductors is not uniform and also varies slowly with frequency. An approximate equation is

$$\alpha_C \approx 0.0434 \times \frac{R_S\sqrt{\varepsilon_{\text{eff}}\ (f = 0)}}{h} \text{ dBm}^{-1}, \tag{6.58}$$

where h is the thickness of the substrate. The ratio of the width of the top conductor, w, to h is approximately unity. For other values of this ratio and different conductor thicknesses, the number 0.0434 in the above equation varies. A full range of values for this number is given in the books listed at the end of the chapter. As frequency increases, the currents are more concentrated in the ground plane and so the attenuation is higher than the above equation predicts. An example of the calculation of a typical value is as follows.

Let $w/h = 1$ and $\varepsilon_R = 10$, then Equation (4.109) gives ε_{eff} $(f = 0) = 6.85$. See also Figure 4.28.

Using Equation (6.49), $R_S = 0.025\,\Omega$ at 10 GHz. See also Figure 6.8.

Finally, choosing $h = 0.5$ mm gives $\alpha_C \approx 5.7\,\mathrm{dBm}^{-1}$. So a rough figure is $6\,\mathrm{dBm}^{-1}$ at 10 GHz.

5 – Attenuation in coplanar waveguide

Just like microstrip, there are numerous equations for the attenuation constant due to conduction losses. However, many engineers claim that the losses are similar to microstrip and often a bit smaller. So an approximate equation is

$$\alpha_C \approx 0.0434 \times \frac{R_S\sqrt{\varepsilon_{eff}(f=0)}}{w}\ \mathrm{dBm}^{-1}, \tag{6.59}$$

where w is the width of the centre conductor. Since this transmission line is less dispersive than microstrip, this value changes with frequency with the normal $\sqrt{f}$ dependency. An example of the calculation of a typical value follows.

Let $w = 0.5$ mm and $\varepsilon_R = 10$, then Equation (4.113) gives ε_{eff} $(f = 0)$ as 5.5. See also Figure 4.30. Using the value of R_S at 10 GHz gives a value of α_C as $5.1\ \mathrm{dBm}^{-1}$. So, in summary, at 10 GHz, for microstrip the attenuation is about $6\,\mathrm{dBm}^{-1}$ and for coplanar waveguide it is about $5\,\mathrm{dBm}^{-1}$. Both figures, in practice, will be higher than this due to surface roughness, by a factor between 1 and 2. The other source of attenuation, that due to dielectric loss, is nearly always much smaller than the attenuation caused by conductor loss.

6 – Attenuation in stripline

The attenuation constant due to conductor losses in stripline has an approximate equation as

$$\alpha_C = 8.686 \times \frac{R_S}{2\eta d}\left(\frac{\dfrac{\pi w}{2d} + \ln\left(\dfrac{8d}{\pi t}\right)}{\ln 2 + \dfrac{\pi w}{4d}}\right)\ \mathrm{dBm}^{-1}, \tag{6.60}$$

which is valid if $w > 4d$ and $t < 0.2d$ and t is the thickness of the centre conductor.

A typical value can be found for 10 GHz and the following.

Take $w/d = 0.8$ and $\varepsilon_R = 5$, then from Equation (4.104) or Figure 4.22 this gives $Z_0 = 50\,\Omega$. If t is taken to be 0.05 mm and $d = 0.5$ mm, then $\alpha_C \approx 4.4\,\mathrm{dBm}^{-1}$.

So a 'rule of thumb' for all three of these transmission lines, the value for α_C at 10 GHz and for similar dimensions and $Z_0 = 50\,\Omega$ is

microstrip	≈ 6 dBm^{-1},
coplanar waveguide	≈ 5 dBm^{-1},
stripline	≈ 4.5 dBm^{-1}.

None of these three transmission lines are likely to be used at anything like one metre in length. At the free space wave length of 3 cm, i.e. 10 GHz, the attenuation will be only 0.18 dB/wave length for microstrip etc. In other words, this is not a major problem to circuit design, where the lengths of the lines are usually in centimetres or less, rather than metres.

6.6 Overall attenuation

The overall attenuation for transmission lines is the sum of the two attenuation coefficients. For all transmission lines with one dielectric the attenuation due to dielectric losses is given by Equation (6.37) as

$$\alpha_D = 8.686 \times \frac{GZ_0}{2} = 8.686 \times \frac{\omega \varepsilon'' \eta}{2} \quad \text{dBm}^{-1}. \tag{6.61}$$

For all transmission lines with a dielectric and air combination, i.e. microstrip, coplanar waveguide and dielectric waveguide, a filling factor, q, is used:

$$\alpha_D = 8.686 \times q \times \frac{\omega \varepsilon'' \eta}{2} \quad \text{dBm}^{-1}, \quad \text{where } q = \frac{\varepsilon_{\text{eff}} - 1}{\varepsilon_R - 1}, \quad 0 < q < 1, \tag{6.62}$$

the assumption being that the dielectric losses only occur in the dielectric substrate and not in the air above the substrate.

The attenuation due to the conductor losses, α_C, has been discussed for several lines so far and the overall attenuation constant is given by

$$\alpha_{\text{overall}} = \alpha_C + \alpha_D. \tag{6.63}$$

In all of the above transmission lines, the practical applications usually have $\alpha_D \ll \alpha_C$.

6.7 Attenuation in waveguides

This next section uses electromagnetic theory to find the attenuation in various waveguides. Exact solutions can be obtained for both rectangular and circular metallic waveguides.

Example 6.5 Attenuation of the TE_{10} mode in rectangular metallic waveguides: This discussion will be limited to just the main mode in rectangular waveguide, i.e. the TE_{10} mode. The technique can be extended to all higher order waveguide modes and this is covered in most texts. However, during the normal operation of rectangular waveguide these higher order modes are below their individual cut-off frequencies, so they will be attenuated far more by this than by any attenuation due to the skin resistance. Also, it is difficult to measure the attenuation of higher order modes with other modes present, except in a resonant structure or some special mode restricting waveguide. Using the field solutions from Chapter 5, Equations (5.47) and (5.48) for the TE_{10} mode:

$$E_y = E_0 \sin\left(\frac{\pi x}{a}\right) \exp \mathrm{j}(\omega t - k_z z),$$

$$H_x = -\frac{k_z}{\omega\mu} E_0 \sin\left(\frac{\pi x}{a}\right) \exp \mathrm{j}(\omega t - k_z z),$$

$$H_z = \frac{\mathrm{j}\pi}{\omega\mu a} E_0 \cos\left(\frac{\pi x}{a}\right) \exp \mathrm{j}(\omega t - k_z z),$$

the currents induced in the conducting walls of the waveguide can be found from the magnetic fields at the walls. The two walls at $x = 0$ and $x = a$, that is the walls of width b, have just the H_z present. So the amplitude of the current per metre in these walls, $|J_y|$, is given by

$$|J_y| = |H_z| = \frac{\pi E_0}{\omega\mu a}. \tag{6.64}$$

However, in the walls at $y = 0$ and $y = b$, that is the broader walls of width a, both H_x and H_z are present. So the current per metre in these walls is given by

$$|J_x| = |H_z| = \frac{\pi E_0}{\omega\mu a} \cos\left(\frac{\pi x}{a}\right),$$

$$|J_z| = |H_x| = \frac{k_z E_0}{\omega\mu} \sin\left(\frac{\pi x}{a}\right).$$

So the losses per metre in each pair of are walls due to the skin resistance are given by

$$P_\mathrm{L} = 2 \times \frac{R_\mathrm{S}}{2}\left(b\left(\frac{\pi E_0}{\omega\mu a}\right)^2 + \left(\frac{\pi E_0}{\omega\mu a}\right)^2 \int_0^a \cos^2\left(\frac{\pi x}{a}\right)\mathrm{d}x + \left(\frac{k_z E_0}{\omega\mu}\right)^2 \int_0^a \sin^2\left(\frac{\pi x}{a}\right)\mathrm{d}x\right),$$

$$P_\mathrm{L} = R_\mathrm{S}\left(\left(\frac{\pi E_0}{\omega\mu a}\right)^2\left(b + \left(\frac{a}{2}\right)\right) + \left(\frac{k_z E_0}{\omega\mu}\right)^2\left(\frac{a}{2}\right)\right).$$

Now from Equation (5.53):

$$P_\mathrm{AV} = P_0 = \frac{k_z ab E_0^2}{4\ \omega\mu}.$$

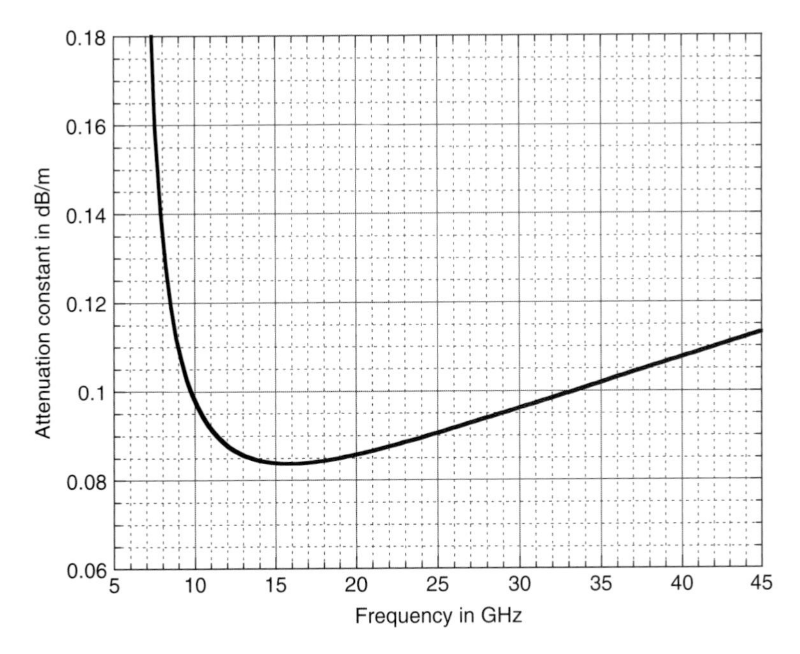

Figure 6.13 The attenuation constant for the TE_{10} metallic rectangular waveguide due to conductor losses against frequency. The 'a' dimension was taken as 2.286 cm and $a = 2b$ for this graph (although in practice this is not quite the case for X-band). The conductivity was 6.10^7 Sm^{-1}, and the waveguide was air filled.

Finally, using Equation (6.55), assuming only small losses:

$$\alpha = 8.686 \times \left(\frac{P_L}{2P_0}\right) = 8.686 \times 2R_S \left(\frac{\left(\frac{\pi}{a}\right)^2 \left(b + \left(\frac{a}{2}\right)\right) + k_z^2 \left(\frac{a}{2}\right)}{k_z ab\omega\mu}\right).$$

Now $(\pi/a)^2 + k_z^2 = k_0^2 = \omega^2\mu\varepsilon$, so

$$\alpha = 8.686 \times R_S \left(\frac{2b\left(\frac{\pi}{a}\right)^2 + \omega^2\mu\varepsilon a}{k_z ab\omega\mu}\right),$$

and since $k_z = \omega\sqrt{\mu\varepsilon}\left(1 - (f_c/f)^2\right)^{\frac{1}{2}}$ and $(\pi/a) = \omega_c\sqrt{\mu\varepsilon}$, we have

$$\alpha = 8.686 \times \frac{R_S}{b\eta\left(1 - \left(\frac{f_c}{f}\right)^2\right)^{\frac{1}{2}}} \left(1 + \frac{2b}{a}\left(\frac{f_c}{f}\right)^2\right) \quad \mathrm{dBm}^{-1}. \tag{6.65}$$

The graph of the attenuation coefficient for an X-band guide is shown in Figure 6.13. For very high frequencies, Equation (6.65) becomes

$$\alpha = 8.686 \times \frac{R_S}{b\eta} = 8.686 \times \frac{\sqrt{\frac{\omega\mu}{2\sigma}}}{b\eta} = 8.686 \times \frac{1}{b}\sqrt{\frac{\pi f \varepsilon}{\sigma}} \quad \mathrm{dBm}^{-1}. \tag{6.66}$$

This shows that the attenuation rises with $\sqrt{f}$ due to the skin resistance at frequencies well above the cut-off frequency. In Figure 6.13, the section of the graph above about 20 GHz is due to the skin resistance. At the cut-off frequency (for X-band this is at 6.56 GHz) the attenuation is infinite. This will be discussed more in the next chapter. The minimum attenuation occurs at 15.83 GHz or $(1+\sqrt{2})f_{\mathrm{c}}$, and has a value of

$$\alpha_{\mathrm{MIN}} = 8.686 \times 2 \times \frac{\sqrt{\frac{\pi f_{\mathrm{c}} \mu}{\sigma}}}{b\eta} = 8.686 \times \frac{2}{b}\sqrt{\frac{\pi f_{\mathrm{c}}\, \varepsilon}{\sigma}} \quad \mathrm{dBm}^{-1}. \tag{6.67}$$

This can be found by differentiating Equation (6.65). Since the operating range for this waveguide is 8.4 to 12.5 GHz, as discussed in Chapter 5, the attenuation is not a minimum in this range. It is worth noting that all higher order modes start with an infinite attenuation at their respective cut-off frequencies, and so propagation above 12.5 GHz is not unreasonable up to $2f_{\mathrm{c}}$ or 13.12 GHz. This attenuation is the minimum attenuation due to the conductivity of the waveguide walls. There may be in addition an attenuation due to surface roughness which can cause reflections as well. This additional attenuation increases with frequency and can increase the attenuation above the minimum by as much as a factor of two at 100 GHz.

In addition to the conductor losses, some waveguides may be filled with a dielectric which may also have losses. Modifying Equation (5.36) to include these dielectric losses gives

$$\gamma = \left(\left(\frac{\pi}{a}\right)^2 - \omega^2\mu(\varepsilon' - \mathrm{j}\varepsilon'')\right)^{\frac{1}{2}}. \tag{6.68}$$

Then by rearranging:

$$\gamma = \left(\left(\frac{\pi}{a}\right)^2 - \omega^2\mu\varepsilon'\right)^{\frac{1}{2}}\left(1 + \frac{\mathrm{j}\omega^2\mu\varepsilon''}{\left(\left(\frac{\pi}{a}\right)^2 - \omega^2\mu\varepsilon\right)}\right)^{\frac{1}{2}}.$$

If the dielectric losses are small, then the last term in the second bracket is small and this bracket can be expanded by the binomial theorem:

$$\gamma = \left(\left(\frac{\pi}{a}\right)^2 - \omega^2\mu\varepsilon'\right)^{\frac{1}{2}}\left(1 + \frac{\mathrm{j}\omega^2\mu\varepsilon''}{2\left(\left(\frac{\pi}{a}\right)^2 - \omega^2\mu\varepsilon'\right)} + \cdots\right).$$

So above the cut-off frequency when $\omega^2\mu\varepsilon > (\pi/a)^2$:

$$\gamma = \alpha + \mathrm{j}\beta \approx \frac{\mathrm{j}\omega^2\mu\varepsilon''}{2\mathrm{j}\omega\sqrt{\mu\varepsilon'}\left(1 - \left(\frac{f_{\mathrm{c}}}{f}\right)^2\right)^{\frac{1}{2}}} + \mathrm{j}\omega\sqrt{\mu\varepsilon'}\left(1 - \left(\frac{f_{\mathrm{c}}}{f}\right)^2\right)^{\frac{1}{2}}. \tag{6.69}$$

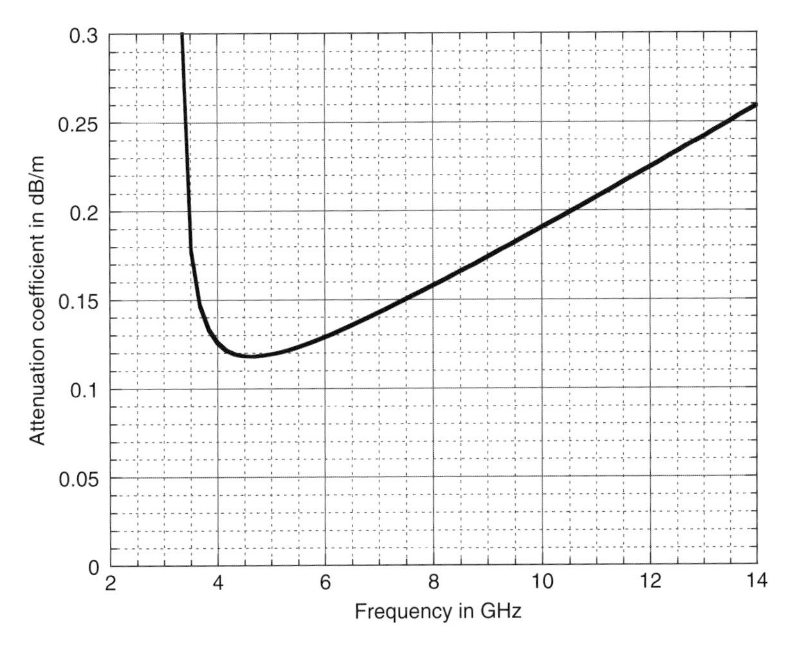

Figure 6.14 The attenuation constant due to dielectric losses in a rectangular metallic waveguide against frequency. The dielectric constant of the dielectric filling the waveguide was $4(1 - j10^{-4})\varepsilon_0$ and the waveguide dimensions were the same as in Figure 6.13, i.e. a was equal to 2.286 cm and $a = 2b$.

Hence:

$$\gamma = \alpha + j\beta \approx \frac{j\omega^2\mu\varepsilon''}{2j\omega\sqrt{\mu\varepsilon'}\left(1 - \left(\frac{f_c}{f}\right)^2\right)^{\frac{1}{2}}} + j\omega\sqrt{\mu\varepsilon'}\left(1 - \left(\frac{f_c}{f}\right)^2\right)^{\frac{1}{2}}; \tag{6.70}$$

this is just the same form as Equation (6.37). The graph of this equation is shown in Figure 6.14. The minimum attenuation is given by

$$\alpha_{\text{MIN}} = 8.686\omega_c\varepsilon''\eta \quad \text{dBm}^{-1}, \text{ which occurs when } f = \sqrt{2}f_c. \tag{6.71}$$

At higher frequencies the attenuation rises in proportion to frequency. It might be assumed from this that any dielectric losses will eventually exceed the conductor losses. However, the waveguides are also reduced in size as the frequency increases to maintain a mono-mode propagation. So in practice the waveguide attenuation due to the skin effect losses increases with $f^{3/2}$ but dielectric losses increase with just frequency. This is significant when dielectric waveguides are compared with metallic waveguides later on in this chapter.

Example 6.6 Attenuation of the TE_{11} mode in circular metallic waveguide: The attenuation in circular metallic waveguide can be calculated in a very similar way to the method used in the previous section. Since this involves some difficult manipulations of Bessel functions, only the results will be reproduced here so as to complete this survey of attenuation. The attenuation due to the skin effect losses in the conducting walls is given by

$$\alpha = 8.686 \times \frac{R_S}{a\eta} \frac{1}{\left(1 - \left(\frac{f_c}{f}\right)^2\right)^{\frac{1}{2}}} \left(0.420 + \left(\frac{f_c}{f}\right)^2\right) \text{ dBm}^{-1}. \tag{6.72}$$

The attenuation due to dielectric losses is given by

$$\alpha = 8.686 \times \frac{\omega\eta\varepsilon''}{2\left(1 - \left(\frac{f_c}{f}\right)^2\right)^{\frac{1}{2}}} = 8.686 \times \frac{\omega\varepsilon'' Z_{TE11}}{2} \text{ dBm}^{-1}. \tag{6.73}$$

The equations are very similar to those of rectangular waveguide. In most applications using metallic waveguide, the lengths employed would usually be only a few metres, so that this attenuation would not normally be a limitation. In power applications, though, the attenuation would result in heating of the waveguide walls which could be the main design criterion. The author has experienced 1 kW going down a coaxial cable to a mismatched termination, where the standing waves inside the transmission line could be detected from the 'hot-spots' on the outside! Obviously, if precision measurements are being made, the characteristic impedance, or wave impedance, would be complex as a result of this attenuation. In most modern network analysers corrections for this would be made automatically. However, in view of the complexity of this correction, it is advisable to verify what exactly the computer program is doing to the raw data in such measurements. Most network analysers have either coaxial or rectangular waveguide outputs and assume that the same transmission line is used to connect to the measurand. This is often not the case in practice.

Example 6.7 Attenuation in optical fibre and rectangular dielectric waveguide: In optical fibre and rectangular dielectric waveguides, only the dielectric losses and not the skin effect losses are present. In optical fibre these losses are remarkably small due to the exceptional purity of the materials used. Attenuations much lower than 3 dB km^{-1} have been easily achieved. Details of how this phenomenonal attenuation has been achieved are given in the books listed at the end of the chapter.

For a rectangular dielectric guide, losses can be smaller than expected if the dielectric is surrounded by air which can normally be assumed to be loss-less. Equation (6.37) gives the general form for the attenuation due to dielectric losses:

$$\alpha = 8.686 \times \frac{\omega\varepsilon'' Z}{2} \ \text{dBm}^{-1}.$$

So the attenuation in a rectangular dielectric guide is of the form

$$\alpha \approx 8.686 \times q \times \frac{\omega\varepsilon'' Z_{E^y_{11}}}{2} \ \text{dBm}^{-1}, \quad \text{where} \quad q \approx \frac{\varepsilon_{\text{EFF}} - 1}{\varepsilon_{\text{R}} - 1}, \tag{6.74}$$

where q is the ratio of the power in the dielectric to the total travelling power. At frequencies near the cut-off frequency, nearly all the power is travelling outside the guide, whereas at high frequencies, all the power is travelling inside the dielectric. So in theory, $0 < q < 1$, however, in the discussion on bandwidth, the cut-off frequency for practical guides was set at the point where at least half the power is travelling in the guide. In this case $0.5 < q < 1$, which means that Equation (6.74) gives a reasonable estimate of the attenuation constant with, say, $q = 0.75$. The net effect is that the attenuation increases in proportion to frequency, with an extra increase over a wide frequency range due to the filling factor, q, also increasing with frequency.

Example 6.8 Some notes on the comparison of coaxial cable, metallic and dielectric rectangular waveguide: Clearly, at low frequencies, of these three transmission lines, the coaxial cable is the one to use. Although coaxial cable does have some problems below about 10 kHz which are discussed in Example 6.1, above this frequency the line performs well, right up to 100 GHz or more. The main limitation on coaxial cable at very high frequencies is the small cross-section and not the attenuation. In practice, at 100 GHz, the centre conductor could be less than 0.2 mm in diameter. So the connectors become harder to make accurately and for low uncertainty measurements, this can often be the limiting factor.

Metallic rectangular waveguide has been used for a long time in the frequency range 20 –100 GHz. In the future it may be used up to 1 THz. It has a much smaller bandwidth than coaxial cable and is more dispersive. In addition, at the top of the frequency range, the connectors become the main problem. It is increasingly difficult to align a rectangular aperture inside a waveguide flange, and this is the main factor which determines the uncertainty of measurements using this guide. Although the attenuation is high, this is not the limiting factor. Dielectric waveguide does not have the huge bandwidth of coaxial cable, but does have lower losses than both metallic waveguide and coaxial cable, particularly above 100 GHz, due to absence of skin effect losses. It also has the advantage that connections can be made relatively easily if the guide is surrounded by air. So for measurements above 100 GHz, their ease of connection may mean they could be used for accurate measurements in preference to the other two.

When a microwave cryogenic application is being considered, the attenuation is often not the main consideration, especially as the lower temperatures can increase the conductivity and thus lower the attenuation. The main factors are the thermal conductivity and the bandwidth. In many cases, a coaxial cable is the best solution, except at very high frequencies where dielectric waveguide may be easier to use. In some applications, however, an optical fibre carrying an optical signal modulated with the appropriate microwave signal may meet all the design criteria.

6.8 The *Q* factor of a length of line

When a length of transmission line is terminated with two short circuits at either end and is n half wavelengths long it forms a resonant circuit, with a Q factor. The definitions of the Q factor are as follows:

$$Q = \frac{\omega_0 \times \text{stored energy}}{\text{power dissipated}}, \tag{6.75}$$

where ω_0 is the resonant angular frequency.

$$Q = \frac{\text{resonant frequency}}{\text{bandwidth}}. \tag{6.76}$$

The bandwidth is defined as the 3 dB bandwidth, i.e. where the power falls to half its value at the resonant frequency:

$$Q = \frac{\omega_0}{2R} \times \frac{\partial X}{\partial \omega} \quad (\omega = \omega_0) \quad \text{or} \quad Q = \frac{\omega}{2G} \times \frac{\partial B}{\partial \omega} \quad (\omega = \omega_0). \tag{6.77}$$

All these definitions give the same value of Q and, to illustrate this, the first three equations will be applied to the series resonant circuit shown in Figure 6.15. (A parallel resonant circuit would require the last equation in (6.77).)

Starting with Equation (6.75):

$$Q = \frac{\omega_0 \times \text{stored energy}}{\text{power dissipated}} = \frac{\omega_0 \times \frac{1}{2}LI^2}{\frac{1}{2}I^2R} = \frac{\omega_0 L}{R}. \tag{6.78}$$

Now the impedance of the circuit is given by

$$Z_{\text{series}} = R + \mathrm{j}\left(\omega L - \frac{1}{\omega C}\right).$$

The resonant frequency, ω_0, is where the imaginary part of this impedance goes to zero, i.e.

Figure 6.15 A series resonant circuit. The current I is common to all the elements, L, R and C.

$$\omega_0 = \frac{1}{\sqrt{LC}}.$$

The 3 dB point, or half power point, is when the current falls to $\frac{1}{\sqrt{2}}$ of its value at the resonant frequency, i.e. when

$$R = \left(\omega L - \frac{1}{\omega C}\right).$$

This gives a quadratic equation for ω:

$$\omega^2 - \frac{\omega R}{L} - \frac{1}{LC} = 0 \text{ which gives } \omega = \frac{R}{2L} + \omega_0\left(1 + \frac{1}{4Q^2}\right)^{\frac{1}{2}} \approx \frac{R}{2L} + \omega_0, \quad \text{if } Q \gg 1.$$

So the angular bandwidth is R/L and using these results in Equation (6.76) gives

$$Q = \frac{\text{resonant frequency}}{\text{bandwidth}} = \frac{\omega_0 L}{R}. \tag{6.79}$$

Finally, Equation (6.77) gives

$$Q = \frac{\omega_0}{2R} \times \frac{\partial X}{\partial \omega} = \frac{\omega_0}{2R} \times \frac{\partial}{\partial \omega}\left(\omega L - \frac{1}{\omega C}\right) = \frac{\omega_0}{2R} \times \left(L + \frac{1}{\omega_0^2 C}\right) = \frac{\omega_0 L}{R}. \tag{6.80}$$

Now all the equations, (6.78), (6.79) and (6.80), do give the same equations for the Q factor. Apply these equations to the transmission line circuit shown in Figure 6.16 starting with

$$Q = \frac{\omega_0 \times \text{stored energy}}{\text{power dissipated}}.$$

The energy stored in the line can be found by imagining some power, P_0, travelling in one direction and back again. If the velocity of the wave is v, then the total energy is given by the product

$$\text{stored energy} = 2 \times P_0 \times \text{transit time},$$

where the transit time is given by

$$\text{transit time} = \frac{\frac{n\lambda_0}{2}}{v} = \frac{n\lambda_0}{2v}.$$

So the stored energy is given by

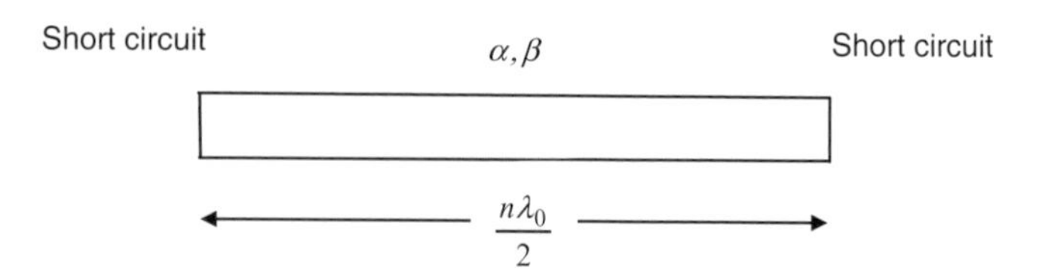

Figure 6.16 A length of transmission line, terminated at both ends with short circuits. The length of the line is a number n of half wavelengths at the resonant frequency.

$$\text{stored energy} = \frac{n\lambda_0 P_0}{v}.$$

Now if we assume the losses in the circuit are small, then as this power goes back and forth in the cavity there will be some power lost:

$$\text{power dissipated} = P_0 - P_0 \exp\left(-2\alpha n\lambda_0\right) \approx 2\alpha n\lambda_0 P_0 \quad \text{if} \quad 2\alpha n\lambda_0 \ll 1.$$

Hence the Q factor is given by

$$Q = \frac{\omega_0 \times \text{stored energy}}{\text{power dissipated}} = \frac{\omega_0 \times n\lambda_0 P_0}{v} \times \frac{1}{2\alpha n\lambda_0 P_0} = \frac{\beta}{2\alpha}. \tag{6.81}$$

This simple equation for the Q factor is applicable to most transmission lines where there is negligible dispersion. It is independent of the length of the line and gives an easy way to compare the resonant structures made with different transmission lines. The following table shows some typical examples.

	α nepers cm^{-1}	β radians cm^{-1}	Q
metallic waveguide at 10 GHz	$1.15.10^{-4}$	1.57	6830
metallic waveguide at 100 GHz	$3.64.10^{-3}$	15.7	2160
coaxial cable at 10 GHz	$5.76.10^{-4}$	2.96	2180
coaxial cable at 100 GHz	$1.15.10^{-2}$	29.6	1290
dielectric waveguide at 10 GHz	$1.05.10^{-4}$	2.50	11 900
dielectric waveguide at 100 GHz	$1.05.10^{-3}$	25.0	11 900
microstrip at 10 GHz	$1.51.10^{-2}$	6.28	208
microstrip at 100 GHz	$4.77.10^{-1}$	62.8	66

Coplanar waveguide and stripline have similar values to microstrip.

The values for α and β are taken from practical examples.

Finally, there is a dispersive effect on the Q factor, which can be shown using the third equation of (6.77). Consider the transmission line shown in Figure 6.16 without the short circuit on the left-hand side. The impedance looking into the line from the left-hand side is

$$Z_{IN} = Z_0 \tanh\left(\frac{\gamma n\lambda_0}{2}\right)$$

from Equation (6.23) with $Z_L = 0$,

$$Z_{IN} = Z_0\left(\frac{\tanh \alpha D + \mathrm{j} \tan \beta D}{1 + \mathrm{j} \tanh \alpha D \tan \beta D}\right),$$

where $\gamma = \alpha + \mathrm{j}\beta$ and $D = n\lambda_0/2$.

If $\alpha D \ll 1$ and the frequency is near to resonance, then $\tanh \alpha D \tan \beta D \ll 1$, so

$$Z_{IN} \approx Z_0(\tanh \alpha D + \mathrm{j} \tan \beta D). \tag{6.82}$$

Now the third equation of (6.77) is

$$Q = \frac{\omega_0}{2R} \times \frac{\partial X}{\partial \omega} \quad (\omega = \omega_0).$$

Applying this to Equation (6.82) gives

$$Q = \frac{\omega_0}{2\alpha D} \times \sec^2(\beta D) \times \frac{\partial \beta D}{\partial \omega},$$

where $\tanh \alpha D \approx \alpha D$ as $\alpha D \ll 1$.

At the resonant frequency, $\sec \beta D = 1$ so

$$Q = \frac{\omega_0}{2\alpha} \times \frac{\partial}{\partial \omega}\left(\omega\sqrt{\varepsilon_R \varepsilon_0 \mu_0}\right).$$

Assuming ε_R is a function of frequency:

$$Q = \frac{\omega_0}{2\alpha}\left(\sqrt{\varepsilon_R \varepsilon_0 \mu_0} + \frac{\omega}{2}\sqrt{\frac{\varepsilon_0 \mu_0}{\varepsilon_R}} \times \frac{\partial \varepsilon_R}{\partial \omega}\right),$$

$$Q = \frac{\beta}{2\alpha}\left(1 + \frac{\omega_0}{2\varepsilon_R} \times \frac{\partial \varepsilon_R}{\partial \omega}\right) \quad \text{at} \quad \omega = \omega_0. \tag{6.83}$$

In practice, the extra term in Equation (6.83) compared with Equation (6.81) is often very much less than unity, so Equation (6.81) is the more useful one for estimating the Q factor of a resonant length of line. The relative dielectric constant can also be the effective dielectric constant, and this second term can be positive in, for example, dielectric waveguides. So far, the losses discussed in the resonant length have been the conductor or skin effect losses and the dielectric losses. In order to get energy into a resonant length of line there must be some form of coupling and this will also involve some losses and a reduction of the Q factor. Finally, the short circuits at the ends of the resonant lengths may not be perfect reflectors and so some further losses will occur here. In the case of open transmission lines, there may be losses due to radiation at the open circuits. The topic of radiation is not discussed in this book, but it is worth noting that for linear transmission lines supporting proper modes there will be no radiation losses. However, at bends, discontinuities and mismatches, there may well be radiation losses in open lines. In coaxial cable and metallic waveguide these are sufficiently enclosed not to have radiation losses, except at open circuits where the waves are free to leave the transmission line and form expanding plane waves or, in other words, radiation.

6.9 Phase and group velocity

When a sinusoidal signal on a transmission line is modulated there are signals produced at different frequencies. A classic case is amplitude modulation, which results in two signals of different frequency. In order to illustrate this, these two signals might be of the form

$$\begin{aligned} V_1 &= V_0 \exp \mathrm{j}\Big((\omega + \Delta\omega)t - (\beta + \Delta\beta)z\Big), \\ V_2 &= V_0 \exp \mathrm{j}\Big((\omega - \Delta\omega)t - (\beta - \Delta\beta)z\Big). \end{aligned} \tag{6.84}$$

These two signals are shifted from the centre frequency, ω, by an amount $\Delta\omega$, which is the modulation frequency. If the line is dispersive, the phase constant is also changed at different frequencies. Adding these two together gives

$$V_{\mathrm{TOTAL}} = V_0 \exp \mathrm{j}(\omega t - \beta z)(\exp \mathrm{j}(\Delta\omega t - \Delta\beta z) + \exp\ -\mathrm{j}(\Delta\omega t - \Delta\beta z)),$$

$$V_{\mathrm{TOTAL}} = 2V_0 \cos\ (\Delta\omega t - \Delta\beta z) \exp \mathrm{j}(\omega t - \beta z). \tag{6.85}$$

Equation (6.85) shows a travelling wave with a cosine amplitude modulation. The original wave is still there in the exponential term and it has a phase velocity given by

$$\omega t = \beta z \text{ and thus } v_{\mathrm{p}} = \frac{z}{t} = \frac{\omega}{\beta}. \tag{6.86}$$

However, the modulation envelope has a different velocity, v_{g}, given by $\Delta\omega t = \Delta\beta z$ and thus

$$v_{\mathrm{g}} = \frac{z}{t} = \frac{\Delta\omega}{\Delta\beta}.$$

In the limit as $\Delta\beta \to 0$, this becomes

$$v_{\mathrm{g}} = \frac{\partial\omega}{\partial\beta}. \tag{6.87}$$

This velocity is called the group velocity and hence has the subscript g. It describes the velocity of the modulation envelope which in some cases will be different from the phase velocity. This is the consequence of adding the two travelling waves in Equation (6.84). Now the link between v_{p} and v_{g} is given by the following:

$$v_{\mathrm{p}} = \frac{\omega}{\beta} \quad \text{so} \quad \beta = \frac{\omega}{v_{\mathrm{p}}}$$

and differentiating gives

$$\frac{\partial\beta}{\partial\omega} = \frac{1}{v_{\mathrm{p}}} - \frac{\omega}{v_{\mathrm{p}}^2}\frac{\partial v_{\mathrm{p}}}{\partial\omega}.$$

Inverting these equations gives

$$v_{\mathrm{g}} = \frac{\partial \omega}{\partial \beta} = \frac{v_{\mathrm{p}}}{\left(1 - \dfrac{\omega}{v_{\mathrm{p}}} \dfrac{\partial v_{\mathrm{p}}}{\partial \omega}\right)}. \tag{6.88}$$

Now Equation (6.88) gives an insight to the various types of dispersion. The normal dispersion is when $\partial v_{\mathrm{p}}/\partial \omega < 0$ and the anomalous dispersion is when $\partial v_{\mathrm{p}}/\partial \omega > 0$. In the normal dispersion $v_{\mathrm{g}} < v_{\mathrm{p}}$, whereas for the anomalous dispersion $v_{\mathrm{g}} > v_{\mathrm{p}}$. In between these two definitions is the case where the phase velocity is independent of frequency, so

$$\frac{\partial v_{\mathrm{p}}}{\partial \omega} = 0 \text{ and } v_{\mathrm{g}} = v_{\mathrm{p}}. \tag{6.89}$$

This occurs in air-filled loss-less transmission lines where

$$v_{\mathrm{p}} = v_{\mathrm{g}} = \frac{1}{\sqrt{\mu_0 \varepsilon_0}}.$$

If the transmission line is filled with a dielectric, then

$$v_{\mathrm{p}} = \frac{1}{\sqrt{\mu_0 \varepsilon_{\mathrm{R}} \varepsilon_0}} \text{ and } \frac{\partial v_{\mathrm{p}}}{\partial \omega} = \frac{1}{\sqrt{\mu_0 \varepsilon_0}} \times -\frac{1}{2} \varepsilon_{\mathrm{R}}^{-3/2} \frac{\partial \varepsilon_{\mathrm{R}}}{\partial \omega} = -\frac{v_{\mathrm{p}}}{2\varepsilon_{\mathrm{R}}} \frac{\partial \varepsilon_{\mathrm{R}}}{\partial \omega}$$

and the group velocity is

$$v_{\mathrm{g}} = \frac{v_{\mathrm{p}}}{\left(1 + \dfrac{\omega}{2\varepsilon_{\mathrm{R}}} \dfrac{\partial \varepsilon_{\mathrm{R}}}{\partial \omega}\right)}. \tag{6.90}$$

It is quite common for the relative dielectric constant to decrease with frequency, by a process called relaxation. This means that the group velocity is faster than the phase velocity and so the dispersion is anomalous. A similar situation exists for the low frequency response of transmission lines. In Figure 6.3 it can be seen that the velocity increases with frequency in this region which also gives rise to anomalous dispersion. For propagation in a conductor the phase velocity is given by Equation (6.42) as

$$v_{\mathrm{p}} = \sqrt{\frac{2\omega}{\mu\sigma}} \text{ so } \frac{\partial v_{\mathrm{p}}}{\partial \omega} = \sqrt{\frac{2}{\mu\sigma}} \times \frac{1}{2} \omega^{-\frac{1}{2}}, \quad \text{which gives}$$

$$v_{\mathrm{g}} = \frac{v_{\mathrm{p}}}{1 - \dfrac{\omega}{v_{\mathrm{p}}} \sqrt{\dfrac{2}{\mu\sigma}} \times \dfrac{1}{2} \omega^{-\frac{1}{2}}} = 2 v_{\mathrm{p}}. \tag{6.91}$$

This is rather a surprising result, where the group velocity is twice the phase velocity and again the dispersion is anomalous.

Turning to metallic rectangular waveguide, this reveals an example of normal dispersion:

$$v_{\mathrm{p}} = \frac{\dfrac{1}{\sqrt{\mu_0 \varepsilon_0}}}{\left(1 - \left(\dfrac{\omega_{\mathrm{c}}}{\omega}\right)^2\right)^{\frac{1}{2}}} \quad \text{so} \quad \frac{\partial v_{\mathrm{p}}}{\partial \omega} = \frac{1}{\sqrt{\mu_0 \varepsilon_0}} \times -\frac{1}{2} \left(1 - \left(\frac{\omega_{\mathrm{c}}}{\omega}\right)^2\right)^{-\frac{3}{2}} \times \frac{2\omega_{\mathrm{c}}^2}{\omega^3}$$

$$\text{and } \frac{\omega}{v_p}\frac{\partial v_p}{\partial \omega} = -\frac{\omega}{v_p} \times \frac{v_p \omega_c^2}{\omega^3} \times \left(1 - \left(\frac{\omega_c}{\omega}\right)^2\right)^{-1} = -\left(\frac{\omega_c}{\omega}\right)^2 \left(1 - \left(\frac{\omega_c}{\omega}\right)^2\right)^{-1},$$

$$\text{so } v_g = \frac{v_p}{\left(1 - \dfrac{\omega}{v_p}\dfrac{\partial v_p}{\partial \omega}\right)} = v_p\left(1 - \left(\frac{\omega_c}{\omega}\right)^2\right), \quad \text{so} \quad v_g = \frac{1}{\sqrt{\mu_0 \varepsilon_0}}\left(1 - \left(\frac{\omega_c}{\omega}\right)^2\right)^{\frac{1}{2}},$$

$$\text{and finally } v_p \times v_g = \frac{1}{\mu_0 \varepsilon_0}. \tag{6.92}$$

In this case, the group velocity is always less than the phase velocity and actually goes to zero at the cut-off frequency.

In all the dielectric guides, i.e. slab guide, rectangular dielectric guide and optical fibre, there are two changes with frequency. The first is the change of the dielectric constant with frequency, which is discussed after Equation (6.90) and is called *dielectric* or *material* dispersion. The second is the change of shape of the modes with frequency, usually called *waveguide* or *modal* dispersion. In modal dispersion, the tendency is always to move the energy into the higher dielectric constant as frequency increases. Thus the waves move slower at higher frequency, which makes this dispersion normal. The two opposite types of dispersion can cancel each other to a certain extent over a limited frequency range.

In all these cases, if $\Delta\omega \ll \omega$ so that $\partial v_p/\partial\omega$ = a constant for all the modulation frequencies, the group velocity is also a constant. Under these conditions, all the modulation components arrive at the end of a transmission line at the same time, which is called the group delay, τ_g. If L is the length of the line then

$$\tau_g = \frac{L}{v_g} = L\frac{\partial \beta}{\partial \omega}, \tag{6.93}$$

so although the transit time for the information is not the same as the transit time for the phase, there is no distortion of the signal under these circumstances. The distortion occurs when $\partial v_p/\partial\omega \neq$ a constant and this is called *group* dispersion. This usually results in pulse broadening and distortion and these will be discussed in the next section. There is an ongoing discussion about the velocity of the energy on a transmission line and several authors relate this to the group velocity. A discussion of this is given in the books listed at the end of this chapter.

6.10 Pulse broadening and distortion

When $\partial v_p/\partial\omega$ is not equal to a constant, the various components of a modulated signal do not arrive at the end of a transmission line at the same time. This clearly results in distortion as the components spread out or disperse in time. An approximate measure of this pulse broadening is to expand the term $\partial\beta/\partial\omega$ using the Taylor series:

$$\frac{\partial \beta}{\partial \omega}(at\ \omega + \Delta\omega) = \frac{\partial \beta}{\partial \omega}(at\ \omega) + \Delta\omega \frac{\partial^2 \beta}{\partial \omega^2}(at\ \omega) + \frac{\Delta\omega^2}{2!}\frac{\partial^3 \beta}{\partial \omega^3}(at\ \omega) + \cdots \qquad (6.94)$$

Assuming the third and subsequent terms in the series are small:

$$\tau_g + \Delta\tau_g \approx L\frac{\partial \beta}{\partial \omega} + L\Delta\omega\frac{\partial^2 \beta}{\partial \omega^2},$$

so the spread in group delay is

$$\Delta\tau_g \approx L\Delta\omega\frac{\partial^2 \beta}{\partial \omega^2}. \qquad (6.95)$$

This expression is often used in optical fibre calculations to find the spread in an optical pulse over a distance L.

6.11 Pulse distortion on transmission lines caused by the skin effect

When a pulse propagates down a transmission line, there are two main effects that can occur due to attenuation. The first is a simple reduction in the pulse amplitude. The second is a change of shape which usually involves a broadening of the pulse. If the attenuation coefficient is constant with frequency, then the first effect is the only one that takes place. However, at microwave frequencies, the main cause of attenuation is the skin effect and it is the dominant mechanism for attenuation in those transmission lines involving metallic conductors. The losses in any dielectric material which is also present are usually much smaller than the skin effect losses. At frequencies well above 100 GHz and up to optical frequencies, the main loss mechanism in dielectric guides is just the dielectric loss. However, simple pulses are not usually used in these guides. Instead, at such frequencies, a pulse may be used to modulate a sine-wave carrier. The resultant frequency spectrum of such a 'pulse of sine-waves' is sufficiently small for the attenuation constant to be considered independent of frequency. So, in this section, the change of pulse shape at only microwave frequencies will be considered. It is worth commenting that nowadays, unbelievable numbers of pulses are sent down transmission lines in both the internet and in other computing networks. So this section has widespread implications for their use. In order to obtain the shape of the pulses, it is necessary to use Laplace transforms. The reason for this is that the propagation constant is defined in the frequency domain and any solution for pulses needs to be in the time domain. In a similar way to the treatment in Chapter 1, a step wave will be analysed first and then the pulse response can be found from the addition of a second reverse step delayed in time. So, starting with the propagation constant for the skin effect from Equation (6.39):

$$\gamma = \alpha + j\beta = \frac{1}{\sqrt{2}}(1 + j)\sqrt{\omega\mu\sigma} = \sqrt{j\omega\mu\sigma} = k\sqrt{j\omega}, \qquad (6.96)$$

where k is a constant for the skin effect. If the inductive part of the skin effect causes a small change in the phase velocity, then for waves on a transmission line, the propagation constant can be written as

$$\gamma = \alpha + \mathrm{j}\beta = \sqrt{\frac{\omega\mu\sigma}{2}} + \mathrm{j}\left(\sqrt{\frac{\omega\mu\sigma}{2}} + \omega\sqrt{\mu\varepsilon}\right) \approx k\sqrt{\mathrm{j}\omega} + \mathrm{j}\omega\sqrt{\mu\varepsilon}. \quad (6.97)$$

Putting these equations in the form of a Laplace transform with variable p gives

$$\gamma = k\sqrt{p} + p\sqrt{\mu\varepsilon}. \quad (6.98)$$

So a wave propagating down a line in the z direction would have a Laplace transform given by

$$E(p) = E_0 \exp\ -z\left(k\sqrt{p} + p\sqrt{\mu\varepsilon}\right) = E_0 \exp\left(-zk\sqrt{p}\right) \exp\left(-zp\sqrt{\mu\varepsilon}\right). \quad (6.99)$$

If the unit step, or Heaviside function, is introduced as the input waveform, the wave becomes

$$E(p) = \frac{E_0}{p} \exp\left(-zk\sqrt{p}\right) \exp\left(-zp\sqrt{\mu\varepsilon}\right). \quad (6.100)$$

When this is transformed back into the time domain, the second exponential term gives a delay in time as the wave propagates at the usual velocity down the line. The first exponential has an inverse transform

$$L^{-1}\left(\frac{1}{p} \exp\left(-zk\sqrt{p}\right)\right) = \mathrm{erfc}\left(\frac{zk}{2\sqrt{t}}\right), \text{ where } \mathrm{erfc}(x) = 1 - \frac{2}{\sqrt{\pi}}\int_0^x \exp\left(-t^2\right)\mathrm{d}t. \quad (6.101)$$

If the delay in time is now added to this result:

$$E(t) = E_0 \mathrm{erfc}\left(\frac{zk}{2\sqrt{t - z\sqrt{\mu\varepsilon}}}\right). \quad (6.102)$$

Example 6.9 Pulses in coaxial cables: In order to illustrate this result, consider a pulse of unit amplitude and pulse width of 100 ns which is sent down a coaxial cable. The cable has a characteristic impedance of 50 Ω and an outer diameter of 7 mm, and the dielectric between the conductors in the cable has a relative permittivity of 2.5. Take the conductivity of the conductors in the cable as $6.10^7\ \mathrm{Sm}^{-1}$. Find the shape of the pulses at distances down the cable of 203 m, 642 m and 2030 m.

Solution to Example 6.9

Use Equation (4.71), which is

$$Z_0 = \frac{1}{2\pi}\sqrt{\frac{\mu}{\varepsilon}} \ln\left(\frac{b}{a}\right) \text{ and } \sqrt{\frac{\mu_0}{\varepsilon_0}} = 120\pi.$$

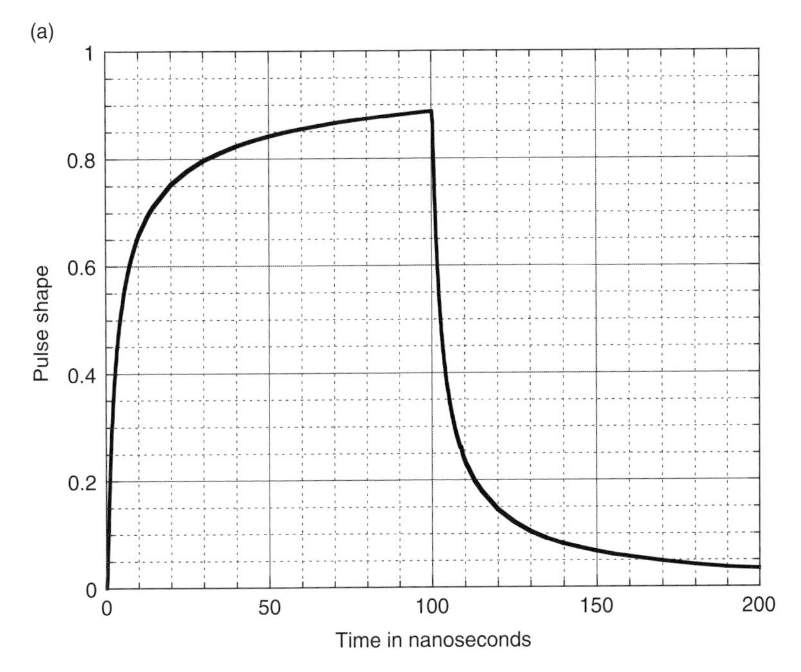

Figure 6.17(a) Graph 1: the distortion of a 100 ns pulse with unity amplitude as it travels down a 7 mm diameter coaxial cable for 203 m. The cable has a dielectric with a relative permittivity of 2.5. The delay in time is 11 ns and the pulse has a leading edge at the origin, i.e. it is travelling to the left of the picture.

$$\text{Then } 50 = \frac{60}{\sqrt{2.5}} \ln\left(\frac{b}{a}\right) \text{ and } \frac{b}{a} = 3.7345.$$

Now use this result in Equation (6.51), which is

$$\alpha = 8.686\sqrt{\frac{\omega\mu}{2\sigma}}\frac{1}{4\pi Z_0}\left(\frac{1}{b} + \frac{1}{a}\right) \text{ dBm}^{-1}.$$

Rewriting this equation gives

$$\frac{\alpha}{\sqrt{\omega}} = \sqrt{\frac{\mu}{2\sigma}}\frac{1}{4\pi b Z_0}\left(1 + \frac{b}{a}\right) = 2.2031.10^{-7} \text{ nepers m}^{-1}.$$

For this coaxial cable, the constant $k = \alpha\sqrt{2/\omega} = 3.118.10^{-7}\text{nepers m}^{-1}/\sqrt{\omega}$.

The first graph, shown in Figure 6.17(a), is constructed by choosing to make $t = 100$ to represent the pulse width of 100 ns. Ignoring for the moment the delay in time in Equation (6.102), the variables other than time can be made equal to unity, i.e.

$$\frac{kz}{2} = 1.$$

Now taking into account the units of time, i.e. nano-seconds,

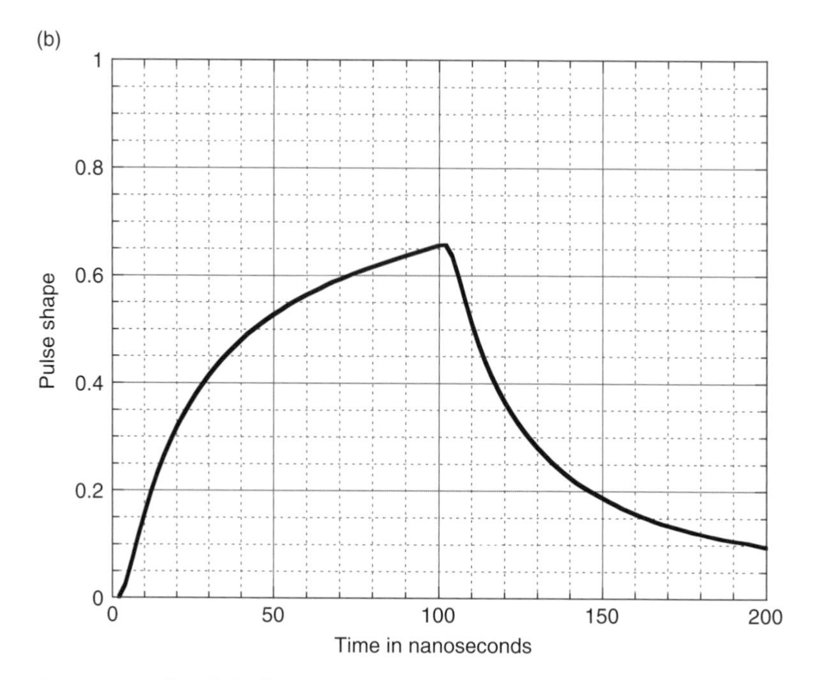

Figure 6.17(b) Graph 2: the same pulse and cable as Figure 6.17(a) but now at a distance of 624 m. Note the extensive trailing edge of the pulse which goes well beyond 200 ns after the pulse began. This pulse is delayed by about 34 ns.

$$z = \frac{2 \times 3.1622 \times 10^{-5}}{3.118 \times 10^{-7}} = 203 \text{ m},$$

where the term 3.1622×10^{-5} is the adjustment from seconds to nano-seconds taking into account the $\sqrt{t}$ relationship. The velocity is $1.9 \times 10^{8}\,\text{ms}^{-1}$, so the pulse will be delayed in time by approximately 11 ns.

The second graph, shown in Figure 6.17(b), is constructed by choosing to make $t = 10$ to represent the pulse width of 100 ns. This increases the distance and delay by a factor of $\sqrt{10}$, i.e. to 642 m and about 34 ns. Finally, the third graph, shown in Figure 6.17(c), uses $t = 1$ to represent a pulse of width 100 ns. This gives the distance as 2030 m and the delay as about 110 ns.

It is also worth noting that very similar shapes to that shown in graph 3 could arise from different initial pulse shapes. The shape is the remains of an almost completely dispersed packet of energy.

Finally, these results can be used to predict the results for a variety of distances and pulse widths or duration, as shown in Figure 6.18. Below the solid line for graph 1 in Figure 6.18 there will be some small distortion. Above the dotted line for graph 3 there will be increasing spreading of the pulse. Similar figures can easily be produced for transmission lines with higher attenuation. For instance, a parallel wire transmission line, or a twisted pair, will have an attenuation about 100 times higher than 7 mm coaxial cable. As a result, the lines in Figure 6.18 will be shifted

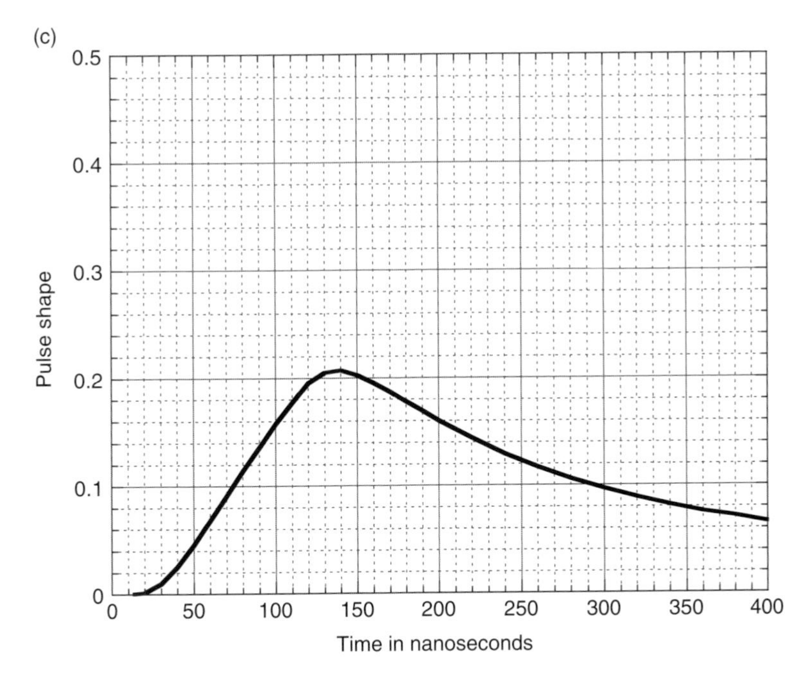

Figure 6.17(c) Graph 3: this is the same pulse as in the other two graphs. However, the scale has been changed to show the extensive spreading of the pulse as well as the reduction in amplitude. This pulse will take about 110 ns to travel the 2030 m.

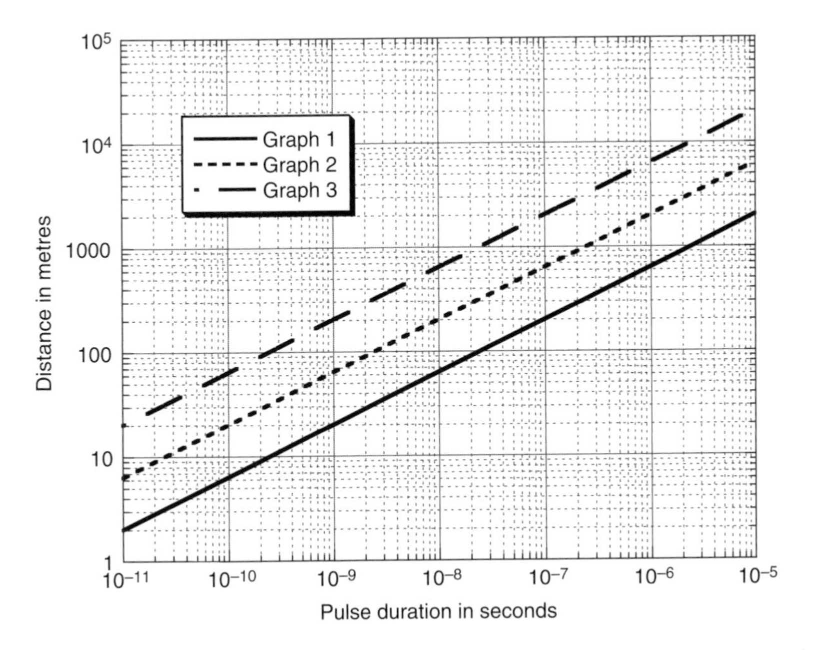

Figure 6.18 The graph of the distance in metres down a 7 mm coaxial cable which pulses of various durations can travel before the distortions described in this section occur. Graph 1 refers to Figure 6.17(a), graph 2 to Figure 6.17(b) and graph 3 to Figure 6.17(c).

downwards by a factor of 100. Not surprisingly, the bandwidth of copper telephone lines supplying digital data to many villages is limited to around 1 MHz, as the author knows well. Similarly, the shift downwards for microstrip, coplanar waveguide and stripline would be a factor of about 400. So a pulse of duration only 0.1 ns would become badly distorted after about a couple of centimetres.

6.12 Conclusion

This chapter has discussed many aspects of attenuation in transmission lines. For most textbooks, this chapter would complete the coverage of the study of transmission lines. In this book, the first three chapters investigated the subject using only equivalent circuits. The next three chapters extended the topic by using electromagnetic waves as well. In the final part of the book, an approach will be made using photons. All three aspects are useful for solving many applications. However, all three are needed to obtain a more complete picture of the propagation of waves in transmission lines.

6.13 Further reading

Attenuation

S. Ramo, J. R. Whinnery and T. Van Duzer *Fields and Waves in Communication Electronics*, Third edition, New York, Wiley, 1993. Chapter 5, section 5.11.

The skin effect

S. Ramo, J. R. Whinnery and T. Van Duzer *Fields and Waves in Communication Electronics*, Third edition, New York, Wiley, 1993. Chapter 6, section 6.4 and chapter 8, sections 8.5, 8.6 (for stripline, microstrip and coplanar waveguide), 8.7, 8.9.

P. C. Magnusson, G. C. Alexander and V. K. Tripathi *Transmission Lines and Wave Propagation*, New York, CRC Press, 1992. Chapter 14: Skin effect in coaxial conductors.

R. E. Collin *Field Theory of Guided Waves*, New York, IEEE Press, 1991. Chapter 5, section 5.4 on mode coupling.

C. R. Paul *Transmission Lines in Digital and Analogue Electronic Systems*, New York, Wiley, 2010. Section 3.11.

R. E. Collin *Foundations for Microwave Engineering*, New York, McGraw-Hill, 1992. Parts of Chapter 3.

Attenuation in microstrip, coplanar waveguide and stripline

S. Ramo, J. R. Whinnery and T. Van Duzer *Fields and Waves in Communication Electronics*, Third edition, New York, Wiley, 1993. Chapter 8, sections 8.5, 8.6.

B. C. Wadell *Transmission Line Design Handbook*, Norwood, MA, Artech House, 1991. Chapter 3, sections 3.4 to 3.6 (also includes coupled striplines).

7 Transmission lines and photons

For many scientists and engineers working at microwave frequencies or below, the word 'photon' is rarely mentioned. It seems to be used only by physicists and engineers who are involved with optical frequencies. In this chapter and the next, a brief introduction to photons will be given to familiarise those readers who do not regularly use them. The purpose of this discussion is to develop a photonic theory of transmission lines which will reveal some additional aspects to those already covered in the previous six chapters.

7.1 Properties of photons – energy and rectilinear propagation

Photons are the smallest quantity of electromagnetic energy, just as electrons are the smallest quantity of charge. This smallest quantity of energy is related to its frequency, via Planck's Law:

$$\text{energy} = hf \quad \text{J}, \tag{7.1}$$

where h is Planck's constant and is equal to $6.626\,0755 \times 10^{-34}$ Js. So at optical frequencies, say 10^{15} Hz, the number of photons a second, n, to make 1 mW is $1.51.10^{15}$. However, in an optical pulse of duration 10 ps at the same power level there will be only $1.51.10^{4}$ photons. At lower frequencies, in the microwave range, at say 10 GHz, the number per second becomes even larger at $1.51.10^{20}$ and if the frequency is reduced to, say, only 50 Hz, then the number is $3.02.10^{28}$. These large numbers do not mean that the characteristics of the individual photons are hidden. Photons are fundamental particles within the group called bosons, which means that they can all co-exist in the same state. This is not true of the other type of fundamental particles, fermions, which have to have separate quantum states. Now the photon is an unusual particle in that it has neither a rest mass nor a charge. It always moves at the speed of light in a vacuum and in the absence of strong gravitational fields, its motion is rectilinear, i.e. it travels in straight lines. Astronomy would be a very difficult subject without this last property! The electromagnetic fields of a photon will be discussed later, but when large numbers of photons are present, their fields add together to form the familiar fields present in the various electromagnetic waves discussed so far.

7.2 Detecting photons

In many of the measurements made to detect photons at frequencies well below optical frequencies, very low temperatures are needed to reduce the unwanted electromagnetic signals generated typically by the thermal agitation of neighbouring atoms. This thermal background is familiar to most electronic engineers as thermal or Johnson noise and in a bandwidth, B, this noise power is given by

$$\text{thermal noise power} = kTB \text{ watts,} \tag{7.2}$$

where k is Boltzmann's constant and is equal to $1.38.10^{-23}$ J K^{-1} and T is the absolute temperature. This expression is derived from the Rayleigh–Jeans Law for the energy density, dU, in the range f to $f + \mathrm{d}f$:

$$\mathrm{d}U = \frac{8\pi f^2 kT\mathrm{d}f}{c^3} \text{ Jm}^{-3}. \tag{7.3}$$

However, this law is only valid for $hf \ll kT$. The complete law is the Planck's Radiation Law for the energy density, dU, in the range f to $f + \mathrm{d}f$:

$$\mathrm{d}U = \frac{8\pi hf^3 \mathrm{d}f}{c^3\left(\exp\left(\frac{hf}{kT}\right) - 1\right)} \text{ Jm}^{-3}, \tag{7.4}$$

and deriving the thermal noise power available from a resistor from this law gives

$$\text{thermal noise power} = \frac{hfB}{\left(\exp\left(\frac{hf}{kT}\right) - 1\right)} \text{ watts.} \tag{7.5}$$

The transition back to the familiar Equation (7.2) takes place when

$$f \ll \frac{kT}{h} \text{ or } f \ll 20.8T \text{ GHz.} \tag{7.6}$$

For most electronic frequencies, including high microwave frequencies, the condition (7.6) is easily met at room temperature and so Equation (7.2) is used extensively. However, if the temperature is lowered to 10^{-3} K using cryogenic techniques, then the transition frequency is reduced to 20 MHz. Equation (7.6) is plotted in Figure 7.1 for three temperatures: 290, 1 and 0.001 K.

For all three curves there is a region from 0 Hz upwards where Equation (7.2) applies and, in this region, individual photons would not be easily detected. However, above the transition frequency, this source of noise is greatly reduced. For 290 K, this will be at frequencies above 6 THz, but as the temperature is lowered to 1 K, this will be for frequencies above 20 GHz. So that in the microwave range it should be possible to see individual photons at these low temperatures. The other main source of noise in electronic circuits – shot noise – is not reduced by lowering the temperature or by raising the frequency. This source of noise is usually the limit to very sensitive measurements of individual photons.

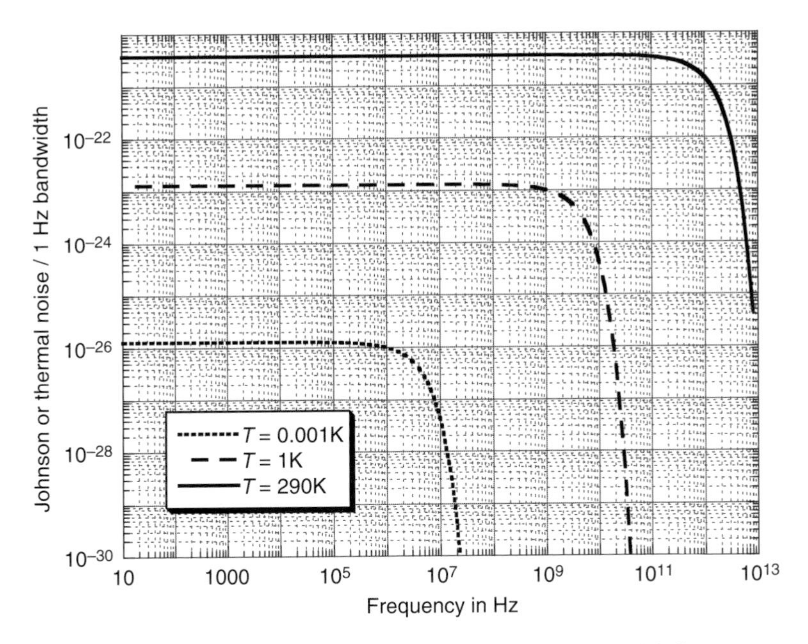

Figure 7.1 Johnson or thermal noise in watts per 1 Hz bandwidth for temperatures of 290, 1 and 0.001 K.

7.3 Plane wave analysis of transmission lines

The photon property described in the rest of this chapter is concerned with the rectilinear propagation at the speed of light. All the major textbooks contain a little on plane wave analysis, but in many cases it is only as an afterthought or even as a curiosity. If photons travel in straight lines and together can form an electromagnetic wave, then the use of plane waves in guided wave analysis may be more fundamental than is argued generally. To illustrate their use, one or two of the structures that have already been analysed using Maxwell's equations in previous chapters will be re-examined to show that the plane wave solutions produce the same results, but often with more insight into the physical processes. The simplest starting point is the analysis of a metallic rectangular waveguide. In order to analyse some structures, further properties of plane waves will be derived in this chapter.

Example 7.1 Analysis of a metallic rectangular waveguide using plane waves: Metallic rectangular waveguide has been analysed in Chapters 5 and 6 using Maxwell's equations. The analysis in this example will use the plane waves described in Group 1 of Section 5.2. The starting point is to imagine two plane waves propagating in the waveguide in just the x and z directions, and with their electric field components in the y direction. Such an arrangement is shown in Figure 7.2.

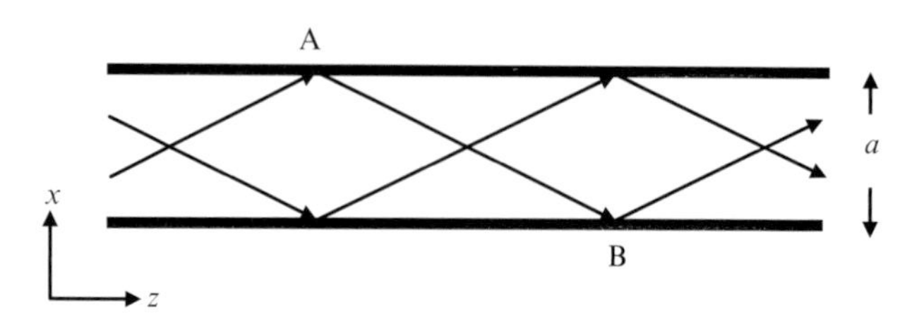

Figure 7.2 A metallic rectangular waveguide with two plane waves propagating in both the x and z directions. The electric fields are in the y direction, which is normal to this page.

For these two plane waves to form a waveguide mode, there has to be a phase relation between the two of them which effectively couples them together. The wave approaching point A is reflected at both A and B and then leaves B. The phase relationship requires the wave approaching A and the wave leaving B to have the same phase within $2m\pi$, where $m = 1, 2, 3, \ldots$ The effect of this is that there will be just one wave going towards the side of the waveguide at A and one going towards the other side at B. This is illustrated in Figure 7.3, where the phase fronts for one of the waves have been added. These phase fronts are lines of constant phase. In order to emphasise the link between the waves, the phase fronts have been extended outside the guide.

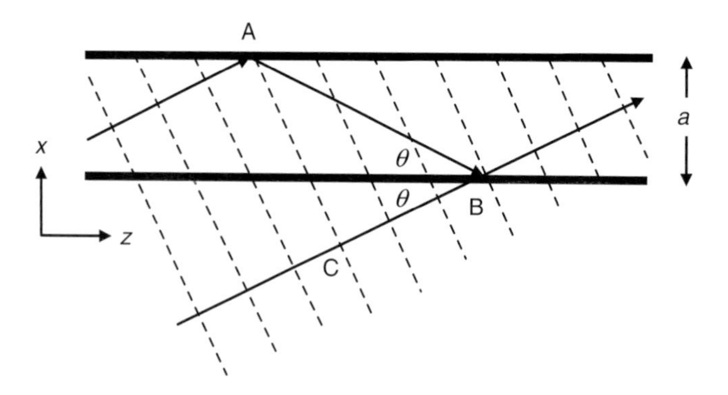

Figure 7.3 This shows the phase relationship between the waves in the waveguide. The phase fronts are shown in dotted lines and have been extended outside the guide to show the relationship between the wave approaching A and the wave leaving B. They are one and the same wave in a waveguide mode.

If the direction of propagation of the waves makes an angle of θ with the z-axis, then the length AB is related to the broad dimension of the guide a by

$$\mathrm{AB}\ \sin\theta = a. \tag{7.7}$$

As the wave propagates along AB, the phase delay is given by

$$k_0\ \mathrm{AB} = \frac{k_0\, a}{\sin\theta}. \tag{7.8}$$

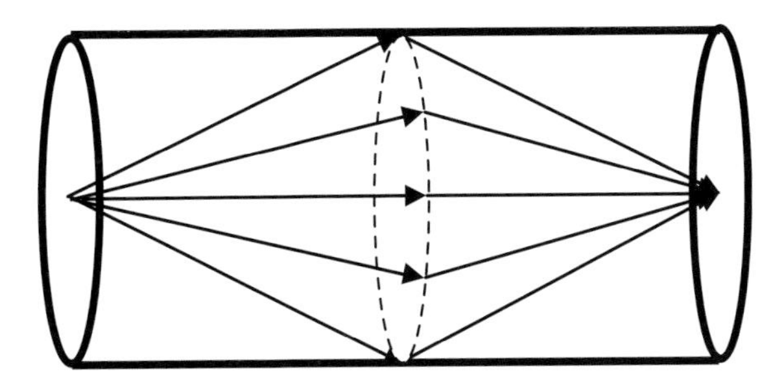

Figure 7.5 A cone of plane waves in a metallic circular waveguide which form the modes of propagation. Only the wave vectors are shown in the diagram.

integral can be formed which is identical to the integral form of a Bessel function. These Bessel functions were discussed in Chapter 5, Section 5.5. The cone of plane waves is shown in Figure 7.5, where only the directions of the wave vectors are shown.

To see the link to the solutions in Chapter 5, only the TE_{11} mode will be discussed. In this mode the electric fields are all transverse to the direction of propagation. If only the magnetic fields in the plane waves are considered, then Figure 7.6 shows the orientation of these fields for just one pair of waves.

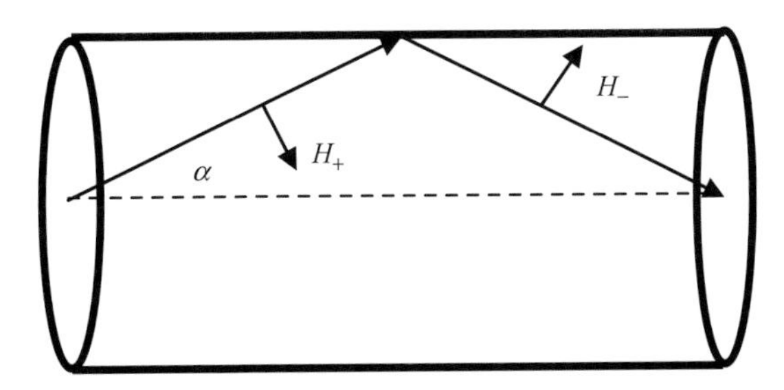

Figure 7.6 The orientation of the magnetic fields of just one pair of plane waves in the circular waveguide shown in Figure 7.5.

In Figure 7.7 are shown two wave vectors, the vector $\overline{P}$ is transverse to the z direction and at an angle of θ to the x-axis. The other vector, $\overline{\zeta}$, is a unit vector at angle β to the x-axis and an angle α to the z-axis, as shown in Figure 7.6. The two vectors are given by

$$\overline{P} = r \cos(\theta)\overline{x} + r \sin(\theta)\overline{y},$$

$$\overline{\zeta} = \sin(\alpha)\cos(\beta)\overline{x} + \sin(\alpha)\sin(\beta)\overline{y} + \cos(\alpha)\overline{z}. \tag{7.31}$$

Now each plane wave is travelling in the z and r directions. Resolving the wavenumber in these directions gives

$$k_z = k\cos(\alpha) \text{ and } k_r = k\sin(\alpha). \tag{7.32}$$

An equation for the magnetic field, H_+, shown in Figure 7.6, could be

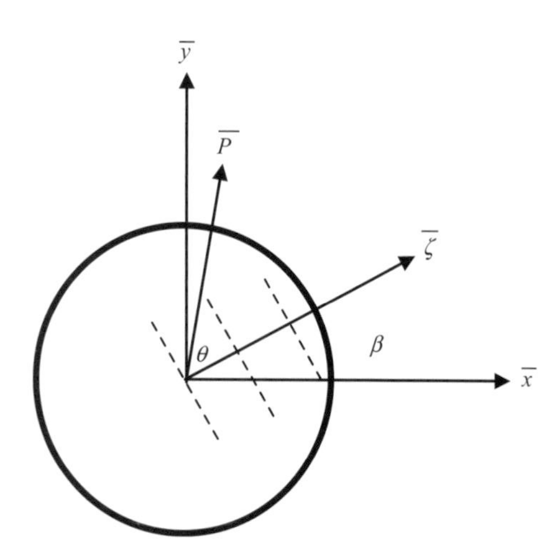

Figure 7.7 This shows a cross-section through the circular guide. The dotted lines show the lines of constant phase for a wave moving along the vector direction $\overline{\zeta}$.

$$H_+ = H_0 \exp \mathrm{j}(\omega t - k_z z) \exp(-\mathrm{j}k_r(\cos(\beta)x + \sin(\beta)y)),$$

$$H_+ = H_0 \exp \mathrm{j}(\omega t - k_z z) \exp(-\mathrm{j}k_r r),$$

where the radius, r, is at an angle β to the x-axis. Now the component of this field along the vector $\overline{P}$, $H_{\overline{P}}$, is given by

$$H_{\overline{P}} = H_0 \exp \mathrm{j}(\omega t - k_z z) \exp\left(-\mathrm{j}k\left(\overline{\zeta}.\overline{P}\right)\right),$$

$$H_{\overline{P}} = H_0 \exp \mathrm{j}(\omega t - k_z z) \exp(-\mathrm{j}k_r(\cos(\beta)\cos(\theta) + \sin(\beta)\sin(\theta))),$$

$$H_{\overline{P}} = H_0 \exp \mathrm{j}(\omega t - k_z z) \exp(-\mathrm{j}k_r \cos(\theta - \beta)). \tag{7.33}$$

Now as β varies from 0 to 2π, each plane wave's contribution to the field on $\overline{P}$ can be found using the following integral:

$$H = H_0 \exp \mathrm{j}(\omega t - k_z z) \int_0^{2\pi} g(\beta) \exp(-\mathrm{j}k_r \cos(\theta - \beta))\mathrm{d}\beta. \tag{7.34}$$

The function $g(\beta)$ is needed to prevent the integral becoming zero as it would if all the field vectors formed a circle. Now a mathematical dodge is now needed to expand Equation (7.34) to give the required result. Since both θ and β are independent variables, let $\varphi = \beta - \theta$. Then, by the separation of variables:

$$g(\beta) = g(\varphi + \theta) = g_1(\varphi)g_2(\theta).$$

Using this in Equation (7.34) gives

$$H = H_0 \exp \mathrm{j}(\omega t - k_z z) g_2(\theta) \int_0^{2\pi} g_1(\varphi) \exp \mathrm{j}k_r \cos(\varphi)\mathrm{d}\varphi.$$

Now Stratton, in the book listed at the end of the chapter, shows that the integral is a solution of Bessel's equation as a long as $g_1(\varphi)$ is a solution of the simple equation

$$\frac{d^2 g}{d\varphi^2} + n^2 g = 0$$

and for the TE_{11} mode $g_1 = \cos(\varphi)$ and $n = 1$.

Now $J_1(k_r r) = \frac{1}{2\pi j}\int_0^{2\pi} \exp(jk_r r \cos(\varphi)) \cos(\varphi) d\varphi$, so by a suitable change of constant:

$$H = H_1 \exp j(\omega t - k_z z) g_2(\theta) J_1(k_r r).$$

Since g_2 also has simple solutions, for this mode $g_2 = \cos(\theta)$ and hence

$$H = H_1 \exp j(\omega t - k_z z) \cos(\theta) J_1(k_r r). \tag{7.35}$$

This is the same form as the equation for H_z in Equation (5.67) for the TE_{11} mode.

Now this has shown that it is possible, with some difficulty, to resolve complex functions into plane waves. However, there is a difficulty here for photons. The number of photons is finite and the number of plane waves in the above analysis is infinite and their individual amplitudes are zero. So a conclusion must be that an individual photon does not become an infinite number of plane waves. It may become a conical wave as described in Figure 7.5, propagating at the speed of light and with a special shape for the cylindrical boundaries.

7.4 Oblique incidence of plane waves on a dielectric interface

Before slab guides can be analysed using plane waves, some of the basic theory of oblique incidence needs to be established. This is described in great detail by many textbooks, so only an outline will be given here. In Figure 7.8 is shown a diagram of a plane wave obliquely incident on a flat dielectric surface. The boundary conditions at the interface are that the tangential electric and magnetic fields should be continuous. Since the waves are partly travelling along this surface, then the components of the wavenumbers must also equate, i.e. the rate of change of phase must be the same along the interface for all three waves. Using the angles in the diagram:

$$\sqrt{\varepsilon_1} k_0 \sin\theta = \sqrt{\varepsilon_1} k_0 \sin\theta' = \sqrt{\varepsilon_2} k_0 \sin\theta'', \tag{7.36}$$

$$\text{which implies } \theta = \theta' \text{ and } \sqrt{\varepsilon_1}\sin\theta = \sqrt{\varepsilon_2}\sin\theta'' \text{ (Snell's Law).} \tag{7.37}$$

The TM incidence

Taking the surface components of both the electric and magnetic fields along the surface of the interface, using Figure 7.8:

$$E_+ \cos\theta + E_- \cos\theta = E_T \cos\theta'' \text{ and } H_+ + H_- = H_T, \tag{7.38}$$

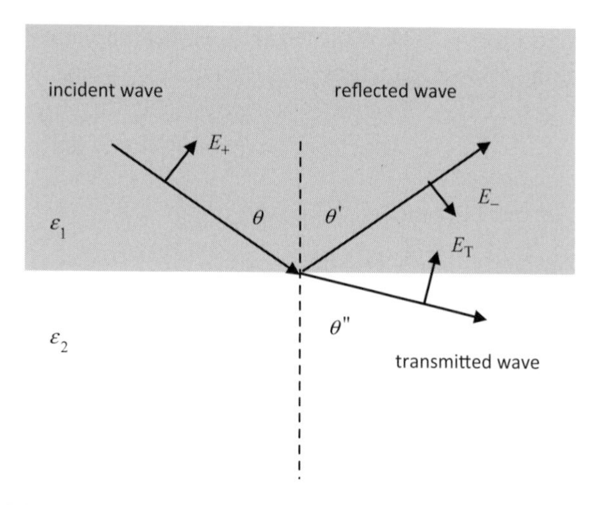

Figure 7.8 The oblique incidence of a plane wave on a flat dielectric interface where $\varepsilon_1 > \varepsilon_2$. The polarisation of the incident wave's magnetic field is along the surface and normal to this page. This is called TM incidence.

where θ' has been assumed equal to θ using Equation (7.36). Now for each wave, the relationships between the electric and magnetic fields are

$$\frac{E_+}{H_+} = \frac{\eta_0}{\sqrt{\varepsilon_1}} = -\frac{E_-}{H_-} \text{ and } \frac{E_T}{H_T} = \frac{\eta_0}{\sqrt{\varepsilon_2}}. \tag{7.39}$$

Using Equation (7.38), the ratio of the electric and magnetic fields becomes

$$\frac{(E_+ + E_-)\cos\theta}{(H_+ + H_-)} = \frac{E_T \cos\theta''}{H_T}. \tag{7.40}$$

Now this polarisation has the magnetic field transverse to or along the interface and is called the TM or transverse magnetic polarisation. A reflection coefficient can be defined as follows:

$$\rho_{TM} = \frac{E_-}{E_+} = -\frac{H_-}{H_+}. \tag{7.41}$$

Using Equations (7.39), (7.40) and (7.41), the following expression for ρ_{TM} can be derived:

$$\frac{E_+ \cos\theta(1+\rho_{TM})}{H_+(1-\rho_{TM})} = \frac{E_T \cos\theta''}{H_T} \text{ and so} \sqrt{\varepsilon_2}\cos\theta\frac{(1+\rho_{TM})}{(1-\rho_{TM})} = \sqrt{\varepsilon_1}\cos\theta''. \tag{7.42}$$

Hence

$$\rho_{TM} = \frac{\sqrt{\varepsilon_1}\cos\theta'' - \sqrt{\varepsilon_2}\cos\theta}{\sqrt{\varepsilon_1}\cos\theta'' + \sqrt{\varepsilon_2}\cos\theta}. \tag{7.43}$$

Equation (7.43) is the first Fresnel reflection coefficient for the oblique incidence.

$$\left(\frac{\sqrt{\varepsilon_1}\left(\frac{\varepsilon_1}{\varepsilon_2}\sin^2\theta - 1\right)^{\frac{1}{2}}}{\sqrt{\varepsilon_2}\cos\theta} \right) = \tan(m\pi - k_x a) = -\tan k_x a.$$

Now $\cos\theta = k_x/\sqrt{\varepsilon_1}k_0$, and eliminating θ by squaring the above gives

$$\frac{\varepsilon_1^2}{\varepsilon_2}k_0^2\left(\frac{\varepsilon_1}{\varepsilon_2} - \frac{k_x^2}{\varepsilon_2 k_0^2} - 1\right) = k_x^2\tan^2 k_x a \quad \text{or} \quad k_0^2(\varepsilon_1 - \varepsilon_2) = k_x^2\left(1 + \left(\frac{\varepsilon_2}{\varepsilon_1}\right)^2\tan^2 k_x a\right), \tag{7.56}$$

which is the same as Equation (5.103) for TM waves in a slab guide. So the solution is obtained in a different route by assuming that there are two plane waves inside the slab. The Maxwell's equations used in Chapter 5 to find this solution give the fields without any reference to the plane waves. What this analysis shows is that these plane waves exist and that they lie behind the shapes of the fields discussed extensively in Chapter 5. Finally, to complete the slab guide picture, the same method can be used for TE waves. Starting with the phase condition:

$$2k_x a + 2\tan^{-1}\left(\frac{\sqrt{\varepsilon_2}\left(\frac{\varepsilon_1}{\varepsilon_2}\sin^2\theta - 1\right)^{\frac{1}{2}}}{\sqrt{\varepsilon_1}\cos\theta} \right) = 2m\pi, \qquad m = 0, 1, 2, \ldots, \tag{7.57}$$

and so

$$\left(\frac{\sqrt{\varepsilon_2}\left(\frac{\varepsilon_1}{\varepsilon_2}\sin^2\theta - 1\right)^{\frac{1}{2}}}{\sqrt{\varepsilon_1}\cos\theta} \right) = \tan(m\pi - k_x a) = -\tan k_x a.$$

By squaring both sides and using $\cos\theta = k_x/\sqrt{\varepsilon_1}k_0$:

$$\frac{k_x^2 a^2}{\cos^2 k_x a} = (\varepsilon_1 - \varepsilon_2)k_0^2 a^2, \tag{7.58}$$

which is the same as Equation (5.88) (taking the square root). Thus, again in these examples, the analysis with plane waves gives the same results as those obtained from Maxwell's equations. In general, this is true for all transmission lines because the electromagnetic waves themselves are made up of plane waves. These plane waves move in straight lines as they are made from individual photons. The various modes in a transmission line are the result of the allowable criss-cross paths of plane waves or photons. What is fascinating is the solution to these problems when the number of photons is small. Then the rules of quantum mechanics dictate that these modes are the only possible ones and that there is no limit on the number photons in each mode. Indeed, the theory suggests that once a number of photons have entered a mode, other photons then preferentially join their number, thus forming the waves familiar in electromagnetic theory.

Before this discussion of plane waves is finished, a brief look at attenuation is included to show that the same equations result from both approaches.

7.5 Oblique incidence of plane waves on a conductor

The theory of the skin effect has already been discussed in Chapter 6, and the technique for analysing oblique incidence on dielectrics has been described in Example 7.3. Combining these accounts will give the two reflection coefficients for the oblique reflection from a conductor. First of all the wave vector, k_0, is for most frequencies much larger than the wave vector in a conductor, $k = \sqrt{\omega\mu\sigma/2}$ (see Equation (6.40)). So the equivalent equation to Equation (7.36) for the component of the wavevector along the surface of a conductor is

$$k_0 \sin\theta = \sqrt{\frac{\omega\mu\sigma}{2}} \sin\theta'' \quad \text{or} \quad \omega\sqrt{\mu_0\varepsilon_0} \sin\theta = \sqrt{\frac{\omega\mu_0\sigma}{2}} \sin\theta''. \tag{7.59}$$

This assumes that the permeability of the conductor is the same as that of free space, i.e. μ_0.

Simplifying Equation (7.59) gives

$$\sin\theta = \sqrt{\frac{\sigma}{2\omega\varepsilon_0}} \sin\theta'', \tag{7.60}$$

which is the equivalent of Snell's Law for conductors. However, for a wide range of frequencies the term in the square root is very large. For example, if $\sigma = 5.10^7$ and the frequency is as high as 1 THz, then

$$\sin\theta \approx 670 \times \sin\theta''. \tag{7.61}$$

Since the maximum value of $\sin\theta = 1$, then $\sin\theta''$ must always be very small and so the angle of refraction for most conductors is very nearly equal to zero. This argument begins to break down at optical frequencies, as the conductivity begins to fall with frequency as the frequency approaches the plasma frequency for conduction electrons. However, it is a reasonable approximation if the conductivity of the conductor is still at its low frequency value. With this value for θ'', the equations for the two reflection coefficients are easy to derive. Firstly, for the TM reflection, the surface fields are given by

$$\frac{E_+ \cos\theta + E_- \cos\theta}{H_+ + H_-} = \frac{E_T \cos\theta''}{H_T} = \frac{E_T}{H_T} = \sqrt{\frac{\omega\mu}{2\sigma}}(1+\mathrm{j}). \tag{7.62}$$

This is taking $\theta'' = 0$ and so $\cos\theta'' = 1$, and using Equation (6.43). By defining the reflection coefficient for the TM incidence as

$$\rho_{TM} = \frac{E_-}{E_+}, \tag{7.63}$$

Equation (7.62) can now be rearranged to give

$$\rho_{TM} = \frac{\sqrt{\frac{\omega\mu_0}{2\sigma}}(1+\mathrm{j}) - \eta_0 \cos\theta}{\sqrt{\frac{\omega\mu_0}{2\sigma}}(1+\mathrm{j}) + \eta_0 \cos\theta} = \frac{\sqrt{\frac{\omega\varepsilon_0}{2\sigma}}(1+\mathrm{j}) - \cos\theta}{\sqrt{\frac{\omega\varepsilon_0}{2\sigma}}(1+\mathrm{j}) + \cos\theta}. \tag{7.64}$$

Secondly, for the TE reflection, the surface fields are given by

$$\frac{E_+ + E_-}{H_+ \cos\theta + H_- \cos\theta} = \frac{E_T}{H_T \cos\theta''} = \frac{E_T}{H_T} = \sqrt{\frac{\omega\mu}{2\sigma}}. \tag{7.65}$$

Defining the TE reflection coefficient as

$$\rho_{TE} = \frac{E_-}{E_+}, \tag{7.66}$$

again, Equation (7.65) can be rearranged to give

$$\rho_{TE} = \frac{\sqrt{\frac{\omega\mu}{2\sigma}}(1+\mathrm{j}) \cos\theta - \eta_0}{\sqrt{\frac{\omega\mu}{2\sigma}}(1+\mathrm{j}) \cos\theta + \eta_0} = \frac{\sqrt{\frac{\omega\varepsilon_0}{2\sigma}}(1+\mathrm{j}) \cos\theta - 1}{\sqrt{\frac{\omega\varepsilon}{2\sigma}}(1+\mathrm{j}) \cos\theta + 1}. \tag{7.67}$$

In Figure 7.12, the fraction of the incident power absorbed in the conductor is shown for both the TM and TE waves. In both cases the fractions are

$$1 - |\rho_{TM}|^2 \text{ and } 1 - |\rho_{TE}|^2.$$

The factor $\sqrt{\omega\varepsilon/2\sigma}$ is called A in Figure 7.12 and is given an arbitrary value of 0.001 which corresponds to a frequency of about 2 THz if the conductivity is 6.10^7 Sm^{-1}.

It can be seen from Figure 7.12 that the fraction of absorbed power for TE polarised waves decreases to zero as the angle of incidence increases. However, a different phenomenon occurs with the waves with TM polarisation. Near to an angle of incidence approaching $\pi/2$, the fraction of power absorbed reaches a maximum, as shown in Figure 7.13. Now assuming that this theory is still valid for these large angles of incidence, Equation (7.61) suggests that even if $\sin\theta = 1$, $\sin\theta''$ is still very small. The equation becomes

$$2A \sin\theta = 2A = \sin\theta''. \tag{7.68}$$

In the above analysis, it was assumed that $\theta'' = 0$, but from Equation (7.68):

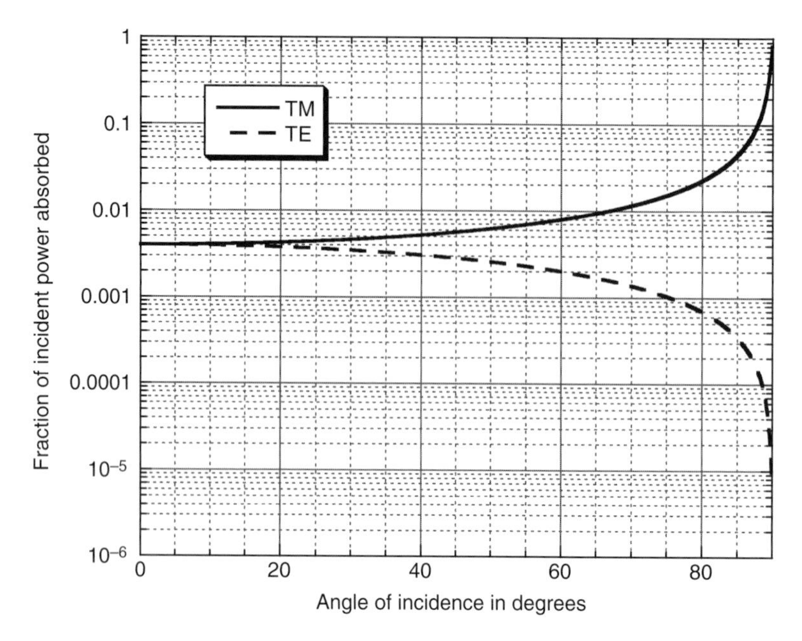

Figure 7.12 The fraction of the incident power absorbed for both the TM and the TE waves for a value of $A = 0.001$.

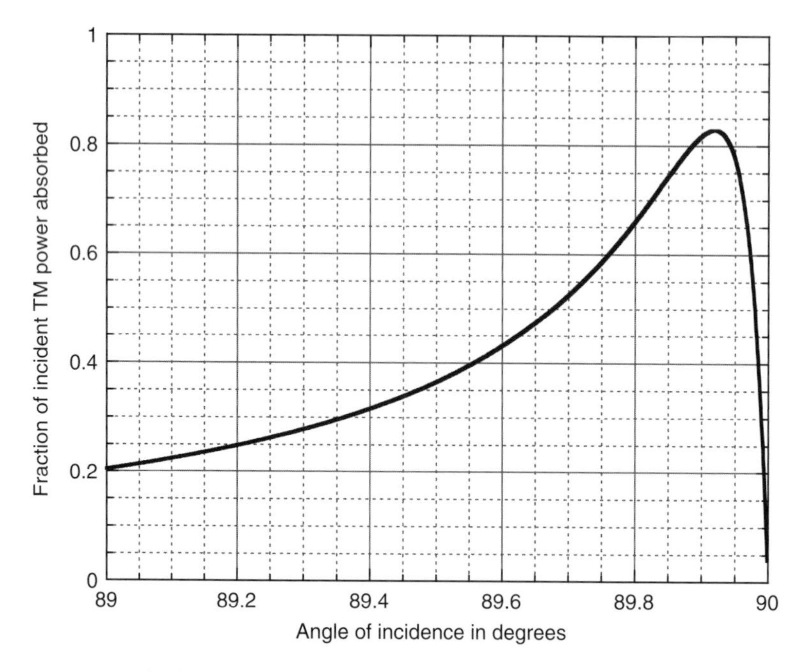

Figure 7.13 The fraction of the incident power absorbed for the TM polarisation near an angle of incidence of $\pi/2$. The value of A is 0.001 and the peak occurs when $\sqrt{2}A = \cos\theta$ and the peak value is $2(\sqrt{2} - 1) = 0.828$ at $\theta = 89.92°$.

$$\cos\theta'' = \left(1 - (2A\sin\theta)^2\right)^{\frac{1}{2}},$$

which gives the error in $\cos\theta''$ as $\approx A$ or 0.001. The effect on Equations (7.64) and (7.67) if the exact $\cos\theta''$ term had been used rather than unity would be negligible.

These results will now be used to examine plane wave solutions to the attenuation in some simple transmission lines.

Example 7.4 Attenuation in a parallel plate guide:

1 TEM modes

The parallel plate guide has been discussed in both Chapters 4 and 6. Equation (6.55) is the starting point for this discussion:

$$\alpha = \frac{P_L}{2P_0}, \tag{7.69}$$

where α is the attenuation constant, P_L is the power lost per metre and P_0 is the travelling power. The equation is valid for small attenuation. Now if the lowest order mode in the parallel plate guide is a pure TEM mode, the electric fields are normal to the conductors and there is no attenuation according to Figure 7.13. However, if the mode is divided into two plane waves which are obliquely incident at an angle near $\pi/2$, then there will be some attenuation. Figure 7.14 shows such an arrangement.

The wave approaching the region AB loses a fraction of its power and then travels on. Now the length AB $= d\tan\theta$, so the fraction of the incident power lost per metre is given by

$$\frac{\left(1 - |\rho_{TM}|^2\right)}{d\tan\theta} \approx \frac{\left(1 - |\rho_{TM}|^2\right)\cos\theta}{d} \quad \text{since } \theta \to \frac{\pi}{2}. \tag{7.70}$$

Now the attenuation constant for a parallel plate guide is given in Equation (6.53) as:

$$\alpha = 8.686\frac{R_S}{\eta d} = 8.686\frac{R_S}{Z_0 w} = 8.686\frac{A}{d} \quad \text{dBm}^{-1}. \tag{7.71}$$

Now $|\rho_{TM}|^2$ can be written as

$$|\rho_{TM}|^2 = \frac{(A - \cos\theta)^2 + A^2}{(A + \cos\theta)^2 + A^2} \text{ and } 1 - |\rho_{TM}|^2 = \frac{4A\cos\theta}{(A + \cos\theta)^2 + A^2}. \tag{7.72}$$

From Equations (7.69), (7.70) and (7.72), the attenuation coefficient becomes

$$\alpha = 8.686\frac{2A\cos^2\theta}{\left((A + \cos\theta)^2 + A^2\right)d}, \tag{7.73}$$

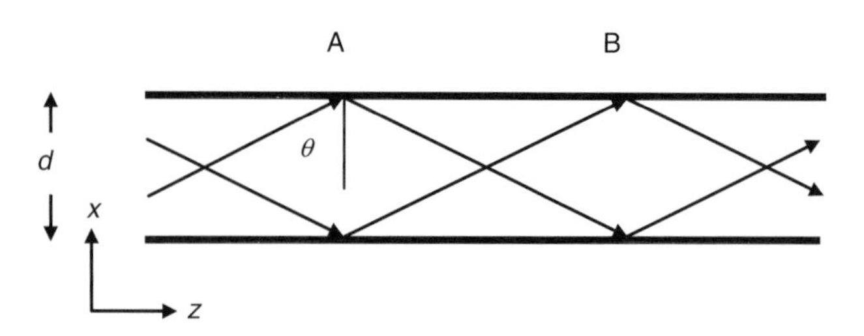

Figure 7.14 Two plane waves obliquely incident in a parallel plate guide. The separation between the plates is d.

where the losses in both conductors of the parallel plate guide in Figure 7.14 have been taken into account.

Now Equations (7.71) and (7.73) can be reconciled if

$$\frac{2\cos^2\theta}{(A+\cos\theta)^2+A^2}=1. \tag{7.74}$$

This is valid if $\cos\theta = 2A/(\sqrt{3}-1)$. If $A = 0.001$, then $\theta = 89.84°$. So the model works if the waves are obliquely incident at an angle very close to $\pi/2$. Another way of looking at this is that there will be a component of the electric field along the surface of the plates to compensate for the skin effect losses. This field will combine with the main electric field to tilt the electric field vector and the wave vector a small amount – only 0.16° – to account for the attenuation. Also, the energy flow in the transmission line must be partly towards the plates to compensate for the losses and this must also change the direction of the wave vectors. There must be two wave vectors to account for the flow into both of the plates. It is worth noting that if the frequency were lowered to the microwave range, the angle would be much smaller and even at optical frequencies it could be much less than 1°.

2 TM modes

The higher order modes between two parallel plates can also be discussed here. For the TM modes, there will be two plane waves incident at an angle θ similarly to Figure 7.14. Since the angle will in this case be much less than $\pi/2$, there can be an approximation which simplifies the derivation of the attenuation coefficient. From the second equation in (7.70), if $\cos\theta \gg A$ then

$$1-|\rho_{\mathrm{TM}}|^2=\frac{4A\cos\theta}{(A+\cos\theta)^2+A^2}\approx\frac{4A}{\cos\theta}. \tag{7.75}$$

Now the attenuation coefficient is given by

$$\alpha_{\mathrm{TM}}=8.686\times\frac{2A}{\cos\theta}\times\frac{1}{d\tan\theta}=8.686\times\frac{2A}{d\sin\theta}. \tag{7.76}$$

In a similar way to waveguides, the wavenumber in the transverse direction, k_x, is also the cut-off wavenumber, k_c. For the TM_m mode this will be

$$k_x = k_c = \frac{m\pi}{d} \text{ and } k_0^2 = k_x^2 + k_z^2. \tag{7.77}$$

From the geometry of Figure 7.14:

$$\sin\theta = \frac{k_z}{k_0} = \frac{\sqrt{k_0^2 - k_c^2}}{k_0} = \sqrt{1 - \left(\frac{f_c}{f}\right)^2}. \tag{7.78}$$

Hence the attenuation constant is

$$\alpha_{TM} = 8.686 \times \frac{2A}{d\sqrt{1 - \left(\frac{f_c}{f}\right)^2}} = 8.686 \times \frac{2R_S}{\eta d\sqrt{1 - \left(\frac{f_c}{f}\right)^2}} \text{ dBm}^{-1}. \tag{7.79}$$

3 TE modes

Finally, for TE modes, the attenuation constant can be found from Equations (7.67) and (7.70) as

$$\alpha_{TE} = 8.686 \times \frac{\left(1 - |\rho_{TE}|^2\right)}{2d\tan\theta} = 8.686 \times \frac{4A\cos\theta}{(A+1)^2 + A\cos^2\theta} \times \frac{1}{2d\tan\theta}.$$

If $A \ll 1$, then this becomes

$$\alpha_{TE} = 8.686 \times \frac{2A\cos\theta}{d\tan\theta} = 8.686 \times \frac{2A\cos^2\theta}{d\sin\theta} = 8.686 \times \frac{2R_S\left(\frac{f_c}{f}\right)^2}{\eta d\sqrt{1 - \left(\frac{f_c}{f}\right)^2}} \text{ dBm}^{-1}. \tag{7.80}$$

These results are given in most textbooks, but usually they are derived by integrating the losses in the conducting surfaces caused by the induced surface currents. The above derivations, by comparison, do yield the same results fairly easily and show that the two plane waves can be used to model the attenuation of these modes in parallel plate guides.

Example 7.5 Attenuation in the TE_{10} mode in rectangular waveguide: For the TE_{10} mode, the attenuation was derived in Chapter 6, Equation (6.65) as

$$\alpha = 8.686 \times \frac{R_S}{b\eta\left(1 - \left(\frac{f_c}{f}\right)^2\right)^{\frac{1}{2}}}\left(1 + \frac{2b}{a}\left(\frac{f_c}{f}\right)^2\right) \text{ dBm}^{-1}.$$

By considering two waves reflecting off the vertical walls separated by a distance a, the attenuation coefficient is given by Equation (7.80)

$$\alpha_{\mathrm{TE}} = 8.686 \times \frac{2A\left(\frac{f_{\mathrm{c}}}{f}\right)^2}{a\sqrt{1-\left(\frac{f_{\mathrm{c}}}{f}\right)^2}}. \tag{7.81}$$

The same two waves have losses in the walls at the top and bottom, now separated by a distance b, in a similar fashion to the TEM waves discussed in Example 7.6. The only difference is that the waves are not propagating straight down the guide in the z direction, but at a small angle to that direction. So modifying Equation (7.79) to take this into account gives

$$\alpha = 8.686\frac{A}{b\cos\left(\frac{\pi}{2}-\theta\right)} = \frac{A}{b\sin\theta} = \frac{A}{b\sqrt{1-\left(\frac{f_{\mathrm{c}}}{f}\right)^2}}. \tag{7.82}$$

Now the overall attenuation will be the sum of these two:

$$\alpha_{\mathrm{TE10}} = 8.686 \times \frac{A}{b\sqrt{1-\left(\frac{f_{\mathrm{c}}}{f}\right)^2}} \times \left(\frac{2b}{a}\left(\frac{f_{\mathrm{c}}}{f}\right)^2 + 1\right), \tag{7.83}$$

which is the same as Equation (6.65) as $A = R_{\mathrm{S}}/\eta$. Comparing the two derivations reveals how much easier the above method is compared with the one used in Chapter 6. Also, the method shows that the losses in the vertical walls are not the same as those in the horizontal walls. If the power in the waveguide is reduced to a few photons per second, the probability of a photon being absorbed by the walls per unit length is 2α, using Equation (7.69) (which is only valid for small losses). The remaining photons continue down the waveguide. So the number of photons is being reduced by the attenuation. For a single photon transmission, the probability $1 - 2\alpha$ determines whether or not the photon will arrive at the end of a one metre length of transmission line, assuming α is small.

Example 7.6 Attenuation in a coaxial cable: The wave propagating in a coaxial cable is not easily resolved into plane waves moving in different directions, which has been a feature of both the TE and TM modes in the waveguides discussed so far. It is tempting to use the conformal transformation technique to change the coaxial cable into a parallel plate guide. This transformation does not change the capacitance, inductance and characteristic impedance as these all involve ratios of the dimensions of the transmission lines. The attenuation does involve these dimensions directly and so some care has to be taken in transforming the structures and keeping the relevant part having the correct dimension. When the coaxial cable is transformed using the equation $w = \ln z$ the circumferences of both the inner and outer conductor become just 2π, as shown in Chapter 4, Example 4.4 and Figure 4.5. Now, if both surfaces are treated independently, a transformation of $w = a \ln z$, where a is the radius of the inner conductor, will give a parallel plate guide with both plate widths of $2\pi a$. Now the parallel plate guide has an attenuation constant given by Equation (7.71) as

$$\alpha = 8.68 \frac{R_S}{Z_0 w}.$$

This can be rewritten as two terms, each representing the losses in one of the plates. So the equation of the attenuation coefficient becomes

$$\alpha = 8.686 \frac{R_S}{2Z_0} \left(\frac{1}{w} + \frac{1}{w} \right). \tag{7.84}$$

Now, under the new conformal transformation, the width of one of the plates is now correct and so the first term for the attenuation coefficient of the coaxial cable is given by

$$\alpha = 8.686 \frac{R_S}{2Z_0} \left(\frac{1}{2\pi a} + \cdots \right).$$

Now changing the conformal transformation to $w = b \ln z$ gives the correct dimension to the other plate and so the full expression for the attenuation constant becomes

$$\alpha = 8.686 \frac{R_S}{4\pi Z_0} \left(\frac{1}{a} + \frac{1}{b} \right),$$

which is the same as Equation (6.51).

Although this route is not too mathematically rigorous, since the wave is mainly a TEM wave at least it shows how conformal transformation can be used to find the attenuation.

7.6 Plane waves and thin resistive films

Example 7.7 Normal incidence of plane waves on thin resistive films: There is currently a considerable interest in the use of thin resistive films in many applications. These films can be made much thinner than the skin depth, and it is only these very thin films which will be considered here. Under these conditions the skin depth is no longer relevant and the bulk conductivity of the film, σ, is the key parameter. However, if the film is very thin, such that the thickness is approaching the mean free path of electrons, then this bulk conductivity will be reduced. Since the thickness is less than the skin depth it is also less than the wavelength in the film.

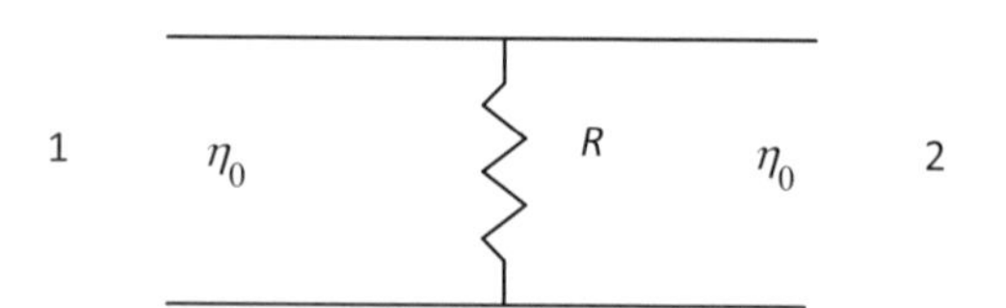

Figure 7.15 The equivalent circuit for a plane wave in air incident normal to a thin resistive film. The resistive film is very thin so that the thickness is much less than the wavelength in the resistive medium.

So it is possible to assume that the tangential electric field is continuous through the film. This will not, however, be true for the tangential magnetic field. So for a plane wave in air incident on a very thin film, the analysis can follow the one dimensional analysis of the first three chapters. The diagram in Figure 7.15, which is the same as Example 2.5 circuit a, is the starting point.

Now if the plane wave in air has a cross-section of $1\,\mathrm{m}^2$, then the resistance of the film is given by

$$R = \frac{1}{\sigma d}, \tag{7.85}$$

where d is the thickness of the film and $d \ll \delta$ and $d \ll \lambda$. The reflection and transmission coefficients are given by

$$\rho = \frac{-\eta_0}{2R + \eta_0} \text{ and } \tau = \frac{2R}{2R + \eta_0}. \tag{7.86}$$

In terms of power, the fraction of the incident power reflected is

$$|\rho|^2 = \frac{\eta_0^2}{(2R + \eta_0)^2},$$

the fraction of the incident power transmitted is

$$|\tau|^2 = \frac{4R^2}{(2R + \eta_0)^2},$$

and so the fraction of incident power absorbed in the resistive film is

$$1 - |\rho|^2 - |\tau|^2 = \frac{4R\eta_0}{(2R + \eta_0)^2}. \tag{7.87}$$

These three terms are also the probabilities of a photon being reflected, transmitted or absorbed in the resistive film. If the incident electric field is E_+, then the field across the resistive film, E_{R}, and the transmitted electric field, E_{T}, are given by

$$E_{\mathrm{R}} = E_{\mathrm{T}} = E_+(1 + \rho) = E_+\frac{2R}{(2R + \eta_0)}. \tag{7.88}$$

However, the magnetic field is not continuous across the resistive film as a result of the currents in the film itself. Using the notation H_{R1} and H_{R2} for the magnetic fields on each side of the film, the following equations link them together:

$$H_{R1} = \frac{E_+}{\eta_0}(1-\rho) = \frac{E_+}{\eta_0} \times \frac{2(R+\eta_0)}{(2R+\eta_0)}, \tag{7.89}$$

$$H_{R2} = \frac{E_T}{\eta_0} = \frac{E_+}{\eta_0} \times \frac{2R}{(2R+\eta_0)}. \tag{7.90}$$

Now the current in the film is given by

$$I = E_+ \sigma d \frac{2R}{(2R+\eta_0)} = \frac{2E_+}{(2R+\eta_0)} \text{ (using Equation (7.85)).} \tag{7.91}$$

Integrating the magnetic field round the resistive film should give this current, by using Equations (7.89), (7.90) and (7.91):

$$\oint H.dl = H_{R1} - H_{R2} = \frac{2E_+}{(2R+\eta_0)} = I = \frac{E_R}{R}$$

$$\text{and } H_{R1} = \frac{E_R}{R} + H_{R2} = \frac{E_R}{R} + H_T. \tag{7.92}$$

Example 7.8 Oblique incidence of plane waves on resistive films: In the case of oblique incidence, the surface or tangential electric field can again be assumed to be continuous through the film. So for a TM incidence:

$$E_R = E_+ \cos\theta + E_- \cos\theta = E_T \cos\theta, \tag{7.93}$$

$$H_{R1} = H_+ + H_- = \frac{E_R}{R} + H_T, \tag{7.94}$$

where θ is the angle of incidence.

So if $\rho_{TM} = E_-/E_+$ then

$$E_+(1+\rho_{TM}) = E_T, \quad \frac{E_+}{\eta_0}(1-\rho_{TM}) = \frac{E_T \cos\theta}{R} + \frac{E_T}{\eta_0}$$

$$\text{and } \rho_{TM} = -\frac{\eta_0 \cos\theta}{2R + \eta_0 \cos\theta}, \tag{7.95}$$

which for $\theta = 0$ is the same as Equation (7.86) for normal incidence. For the case of $\theta = \pi/2$, the reflection coefficient goes to zero. The transmission coefficient is then given by

$$\tau_{TM} = \frac{E_T}{E_+} = (1+\rho_{TM}) = \frac{2R}{2R + \eta_0 \cos\theta}. \tag{7.96}$$

The equations for TE incidence are very similar:

$$E_{\mathrm{R}} = E_+ + E_- = E_+(1+\rho_{\mathrm{TE}}) = E_{\mathrm{T}}, \tag{7.97}$$

$$H_{\mathrm{R1}} = H_+ \cos\theta + H_- \cos\theta = H_+ \cos\theta(1-\rho_{\mathrm{TE}}) = \frac{E_{\mathrm{R}}}{R} + H_{\mathrm{T}} \cos\theta, \tag{7.98}$$

which gives

$$\rho_{\mathrm{TE}} = -\frac{\eta_0}{2R\cos\theta + \eta_0} \quad \text{and} \quad \tau_{\mathrm{TE}} = \frac{2R\cos\theta}{2R\cos\theta + \eta_0}. \tag{7.99}$$

Again the equations become identical to Equations (7.86) for normal incidence.

Example 7.9 The wave impedance of rectangular waveguide using plane waves and resistive films: The wave impedance of rectangular metallic waveguide has been discussed in Chapter 5, and for the TE modes Equation (5.51) gives

$$Z_{\mathrm{TE}} = \frac{\eta}{\sqrt{1-\left(\frac{f_{\mathrm{c}}}{f}\right)^2}}.$$

Using plane waves in this chapter, Equation (7.25) gives the same result:

$$Z_{\mathrm{TE}} = \frac{E_y}{H_x} = \frac{E_0}{H_0\cos\theta} = \frac{\eta}{\sqrt{1-\left(\frac{f_{\mathrm{c}}}{f}\right)^2}}.$$

The result for TM modes can also be derived from plane waves as

$$Z_{\mathrm{TM}} = \frac{E_x}{H_y} = \frac{E_0\cos\theta}{H_0} = \eta\sqrt{1-\left(\frac{f_{\mathrm{c}}}{f}\right)^2}. \tag{7.100}$$

Now imagine a resistive film placed transversely across a rectangular waveguide. This can be analysed in two ways. The first is using the equivalent circuit shown in Figure 7.16.

It is fairly simple to show that the reflection and transmission coefficients are given by

$$\rho_{\mathrm{TE}} = -\frac{Z_{\mathrm{TE}}}{2R+Z_{\mathrm{TE}}} = -\frac{\eta_0}{2R\cos\theta+\eta_0}; \quad \tau_{\mathrm{TE}} = 1+\rho_{\mathrm{TE}} = \frac{2R}{2R+Z_{\mathrm{TE}}} = \frac{2R\cos\theta}{2R\cos\theta+\eta_0};$$

$$\rho_{\mathrm{TM}} = -\frac{Z_{\mathrm{TM}}}{2R+Z_{\mathrm{TM}}} = -\frac{\eta_0\cos\theta}{2R+\eta_0\cos\theta}; \quad \tau_{\mathrm{TM}} = 1+\rho_{\mathrm{TM}} = \frac{2R}{2R+Z_{\mathrm{TM}}} = \frac{2R}{2R+\eta_0\cos\theta}. \tag{7.101}$$

On the other hand, if the models using plane waves with TE and TM incidence are considered then the same equations result, as shown in Equations (7.95), (7.96)

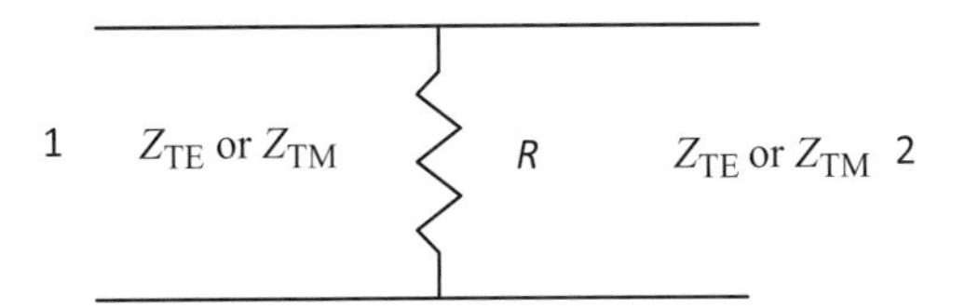

Figure 7.16 The equivalent circuit of a rectangular waveguide with a thin resistive film mounted transversely across the guide.

and (7.99). This nicely demonstrates that the plane wave analysis can be used to complement the more formal techniques of analysis using Maxwell's equations.

Example 7.10 Normal incidence of plane waves on resistive films on a dielectric substrate: The purpose of this example and the next one is partly to bring together much of the theory developed in this chapter and partly to show a surprising broad-band condition that can arise. In many practical cases, a very thin film may not be self supporting and a dielectric substrate will be needed. In order to analyse the combination of these two, the scattering matrix of each of them is required. Using the same techniques that were described in Section 2.8 and Equation (7.86), the scattering matrices are

$$S_R = \frac{1}{2R + \eta_0}\begin{bmatrix} -\eta_0 & 2R \\ 2R & -\eta_0 \end{bmatrix} \quad \text{for the resistive film}$$

and for the dielectric substrate:

$$S_{DS} = \frac{1}{2\eta_1\eta_0 + \mathrm{j}\left(\eta_1^2 + \eta_0^2\right)\tan kd}\begin{bmatrix} -\mathrm{j}\left(\eta_0^2 - \eta_1^2\right)\tan kd & 2\eta_1\eta_0 \sec kd \\ 2\eta_1\eta_0 \sec kd & -\mathrm{j}\left(\eta_0^2 - \eta_1^2\right)\tan kd \end{bmatrix}, \tag{7.102}$$

where the dielectric substrate has an intrinsic impedance of η_1 and a thickness of d.

By changing these into transmission matrices, which are described in Chapter 2, these can be combined and converted back to an overall scattering matrix given by

$$S_{R+DS} = \frac{1}{A}\begin{bmatrix} -\dfrac{\eta_1\eta_0^2}{R} + \mathrm{j}\tan kd\left(\eta_1^2 - \eta_0^2 - \dfrac{\eta_0\eta_1^2}{R}\right) & 2\eta_1\eta_0 \sec kd \\ 2\eta_1\eta_0 \sec kd & -\dfrac{\eta_1\eta_0^2}{R} + \mathrm{j}\tan kd\left(\eta_1^2 - \eta_0^2 + \dfrac{\eta_0\eta_1^2}{R}\right) \end{bmatrix},$$

$$\text{where} \quad A = 2\eta_0\eta_1\left(1 + \frac{\eta_0}{2R}\right) + \mathrm{j}\tan kd\left(\eta_1^2 + \eta_0^2 + \frac{\eta_0\eta_1^2}{R}\right). \tag{7.103}$$

Now this rather cumbersome matrix has an unusual property. The terms for S_{12} and S_{21} are equal and the square of the modulus of each of them is given by

$$|S_{12}|^2 = |S_{21}|^2 = \frac{4\eta_1^2\eta_0^2}{4\eta_1^2\eta_0^2\left(1+\frac{\eta_0}{2R}\right)^2\cos^2 kd + \left(\eta_1^2+\eta_0^2+\frac{\eta_0\eta_1^2}{R}\right)^2\sin^2 kd}. \tag{7.104}$$

Now if the coefficients of $\cos^2 kd$ and $\sin^2 kd$ are equated, then the frequency dependent terms are removed and the square of the modulus is constant with frequency. Equating these coefficients and taking a square root gives

$$2\eta_1\eta_0\left(1+\frac{\eta_0}{2R}\right) = \left(\eta_1^2+\eta_0^2+\frac{\eta_0\eta_1^2}{R}\right),$$

and solving for R gives

$$R = \frac{\eta_1\eta_0}{\eta_0-\eta_1} = \frac{\eta_0}{\sqrt{\varepsilon_R}-1}. \tag{7.105}$$

The fraction of the incident power transmitted under the special conditions of Equation (7.105) is given by

$$|S_{12}|^2 = |S_{21}|^2 = \frac{1}{\left(1+\frac{\eta_0}{2R}\right)^2} = \frac{4}{\left(1+\sqrt{\varepsilon_R}\right)^2}. \tag{7.106}$$

Under this condition, the scattering matrix becomes

$$S_{R+DS} = \frac{1}{A}\begin{bmatrix} -(\eta_0-\eta_1)+\mathrm{j}\left(\frac{2\eta_1^2}{\eta_0}-(\eta_0+\eta_1)\right)\tan kd & 2\eta_1\sec kd \\ 2\eta_1\sec kd & -(\eta_0-\eta_1)(1-\mathrm{j}\tan kd) \end{bmatrix},$$

$$A = (\eta_0+\eta_1)(1+\mathrm{j}\tan kd). \tag{7.107}$$

The asymmetry in the scattering matrix results from the asymmetry in the arrangement. When a wave is incident on the dielectric interface, the combination of the resistive film and the air beyond gives an impedance equal to that of the dielectric. So the reflection coefficient is simply

$$|S_{22}| = \frac{\eta_1-\eta_0}{\eta_1+\eta_0}, \tag{7.108}$$

where the fraction of the incident power in the resistive sheet is given by

$$\frac{4\eta_1(\eta_0-\eta)}{(\eta_0+\eta_1)^2}. \tag{7.109}$$

In the reverse case, when the wave is incident on the resistive sheet first, there is a frequency dependent power reflected and a complementary frequency dependent power in the resistive sheet.

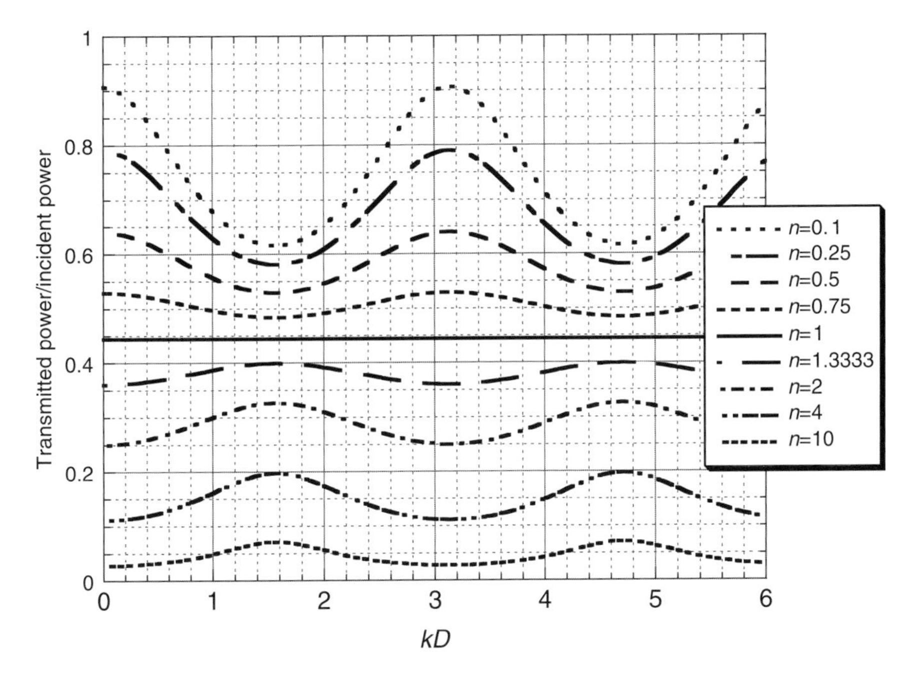

Figure 7.17 The frequency response for the transmitted power through a resistive film and a dielectric substrate. The dielectric substrate has a relative dielectric constant of 4 and the resistance of the film is η_0/n.

Figure 7.17 shows the variation with frequency, or dielectric thickness, of the transmitted power discussed above. The relative dielectric constant is 4, so, using Equation (7.101), the special value of R for the broad-band condition is η_0.

So, with this arrangement, it is possible to remove any frequency dependence from the transmission through a dielectric slab, window or a length of transmission line containing a higher dielectric material. This would be modified if the dielectric has some loss. However, if the loss were small, the only modification would be a reduction in the transmitted power rather than a new frequency dependence.

Example 7.11 Oblique incidence of plane waves on a resistive film on a dielectric substrate: In this example, the previous example will be extended to show that for obliquely incident plane waves, the frequency independent properties are still available, but in a modified form. In order to obtain a fairly simple solution, the thickness of the dielectric substrate is going to be made much smaller than the beam-width of the plane waves involved. This means that the resultant reflected and transmitted waves will have approximately the same width as the incident wave despite multiple reflections in the substrate. This simplification is not a serious limit in practice. Figure 7.18 shows the arrangement for these multiple reflections in just the dielectric substrate. Once a transmission matrix for this

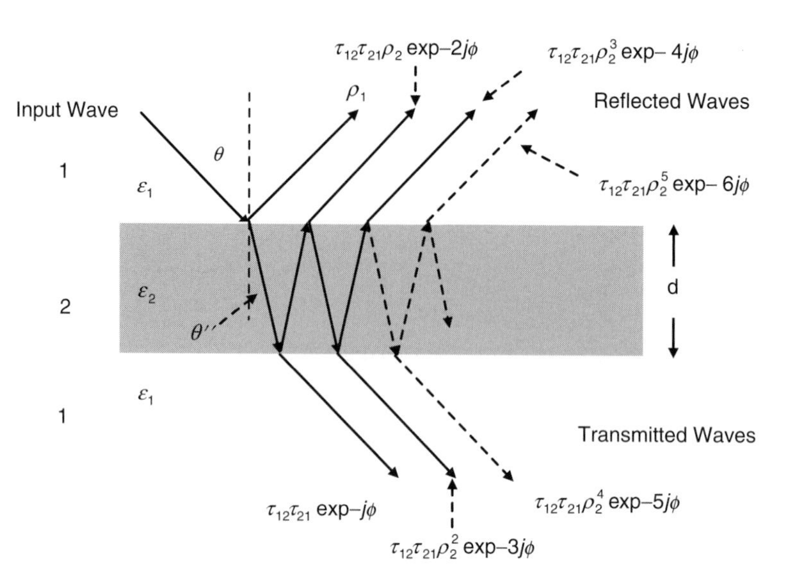

Figure 7.18 A plane wave obliquely incident on a resistive film on a dielectric substrate. The subsequent reflected and transmitted waves are shown. The dotted lines symbolise that the waves decay away eventually to zero.

substrate has been derived it will be combined with the transmission matrix for the resistive film and then converted back to a scattering matrix in the same way that was used in the previous example.

The phase delay through medium 2 is given by

$$\phi = k_2 d \sec \theta'', \tag{7.110}$$

where d is the thickness of the dielectric substrate.

Since both the reflected and transmitted waves form geometric series and $|\rho_2| < 1$, the sum of all the waves can be found using the usual formula for the sum:

$$\text{reflected wave} \qquad \rho' = \rho_1 + \frac{\tau_{12}\tau_{21} \exp - 2j\phi}{1 - \rho_2^2 \exp - 2j\phi}, \tag{7.111}$$

$$\text{transmitted wave} \qquad \tau' = \frac{\tau_{12}\tau_{21} \exp - j\phi}{1 - \rho_2^2 \exp - 2j\phi}, \tag{7.112}$$

where ρ_1 is the reflection coefficient of a wave in medium 1 with a dielectric constant ε_1 at the interface with medium 2 with a dielectric constant ε_2. The other reflection coefficient is for waves inside medium 2. The transmission coefficients are similar; τ_{12} is for waves going from medium 1 to 2 and τ_{21} is for the opposite direction. For TM incidence, the following equations can be found from Equation (7.43):

$$\rho_1 = \frac{\sqrt{\varepsilon_1} \cos \theta'' - \sqrt{\varepsilon_2} \cos \theta}{\sqrt{\varepsilon_1} \cos \theta'' + \sqrt{\varepsilon_2} \cos \theta} \text{ and } \tau_{12} = \frac{2\sqrt{\varepsilon_1} \cos \theta}{\sqrt{\varepsilon_1} \cos \theta'' + \sqrt{\varepsilon_2} \cos \theta}, \tag{7.113}$$

$$\rho_2 = \frac{\sqrt{\varepsilon_2} \cos \theta - \sqrt{\varepsilon_1} \cos \theta''}{\sqrt{\varepsilon_2} \cos \theta + \sqrt{\varepsilon_1} \cos \theta''} \text{ and } \tau_{21} = \frac{2\sqrt{\varepsilon_2} \cos \theta''}{\sqrt{\varepsilon_2} \cos \theta + \sqrt{\varepsilon_1} \cos \theta''}, \tag{7.114}$$

where the angles θ and θ'' are shown in Figure 7.18.

Putting ρ_2, τ_{12} and τ_{21} in terms of just ρ_1 gives

$$\rho_2 = -\rho_1, \ \tau_{12} = \frac{(1+\rho_1) \cos \theta}{\cos \theta''} \text{ and } \tau_{21} = \frac{(1-\rho_1) \cos \theta''}{\cos \theta}. \tag{7.115}$$

The reflected wave, ρ', and the transmitted wave, τ', can now be written as

$$\rho' = \frac{\rho_1(1 - \exp\ -2j\phi)}{1 - \rho_1^2 \exp\ -2j\phi} \text{ and } \tau' = \frac{(1-\rho_1^2) \exp\ -j\phi}{1 - \rho_1^2 \exp\ -2j\phi}. \tag{7.116}$$

These can be combined into a scattering matrix for oblique waves:

$$[S_{\text{DSoblique}}] = \begin{bmatrix} \rho' & \tau' \\ \tau' & \rho' \end{bmatrix}. \tag{7.117}$$

Using the usual transformations discussed in Chapter 2, Section 2.9, a transmission matrix can be found using

$$[T_{\text{DSoblique}}] = \frac{1}{\tau'} \begin{bmatrix} 1 & -\rho' \\ \rho' & (\tau')^2 - (\rho')^2 \end{bmatrix}. \tag{7.118}$$

The result is obtained after a little manipulation as

$$[T_{\text{DSoblique}}] = \begin{bmatrix} \cos \phi + j(A+B) \sin \phi & -j(A-B) \sin \phi \\ j(A-B) \sin \phi & \cos \phi - j[A+B] \sin \phi \end{bmatrix},$$

$$\text{where } A = \frac{1}{2}\sqrt{\frac{\varepsilon_1 \cos \theta''}{\varepsilon_2 \cos \theta}} \text{ and } B = \frac{1}{2}\sqrt{\frac{\varepsilon_2 \cos \theta}{\varepsilon_1 \cos \theta''}}. \tag{7.119}$$

Now the transmission matrix for the resistive film is found by using Equation (7.101) modified for the surrounding medium having a relative dielectric constant of ε_1. The result is

$$[T_{\text{Roblique}}] = \begin{bmatrix} 1+D & D \\ -D & 1-D \end{bmatrix}, \quad \text{where} \quad D = \frac{\eta_0 \cos \theta}{2R\sqrt{\varepsilon_1}}. \tag{7.120}$$

Multiplying these two transmission matrices together and transforming them back to a scattering matrix gives:

$$[S_{\text{R+DSoblique}}] = \frac{1}{F}\begin{bmatrix} -D \cos \phi - j(A(2D-1)+B) \sin \phi & 1 \\ 1 & -D \cos \phi + j(A(2D+1)-B) \sin \phi \end{bmatrix},$$

$$\text{where} \quad F = (1+D) \cos \phi + j(A(2D+1)+B) \sin \phi. \tag{7.121}$$

The two transmission coefficients are equal and their amplitudes can be made frequency independent if the coefficients of $\cos \phi$ and $\sin \phi$ are made equal, i.e.

$$1 + D = (A(2D+1)) + B. \tag{7.122}$$

This gives the condition for the resistive film as

$$R_{\text{TM}} = \frac{\eta_0 \cos\theta \cos\theta''}{\sqrt{\varepsilon_2}\cos\theta - \sqrt{\varepsilon_1}\cos\theta''}. \tag{7.123}$$

This is a modified form of Equation (7.116). To complete the picture, the TE incidence follows very similar lines, in general the terms $\cos\theta$ and $\cos\theta''$ are interchanged, so that

$$\rho_1(\text{TE}) = \frac{\sqrt{\varepsilon_1}\cos\theta - \sqrt{\varepsilon_2}\cos\theta''}{\sqrt{\varepsilon_1}\cos\theta + \sqrt{\varepsilon_2}\cos\theta''} = -\rho_2(\text{TE}),$$

$$\tau_{12}(\text{TE}) = \frac{2\sqrt{\varepsilon_1}\cos\theta}{\sqrt{\varepsilon_1}\cos\theta + \sqrt{\varepsilon_2}\cos\theta''},$$

$$\tau_{21}(\text{TE}) = \frac{2\sqrt{\varepsilon_2}\cos\theta''}{\sqrt{\varepsilon_1}\cos\theta + \sqrt{\varepsilon_2}\cos\theta''}. \tag{7.124}$$

Using the same equation as (7.119) but with the A and B redefined as

$$A = \frac{1}{2}\sqrt{\frac{\varepsilon_1\cos\theta}{\varepsilon_2\cos\theta''}} \text{ and } B = \frac{1}{2}\sqrt{\frac{\varepsilon_2\cos\theta''}{\varepsilon_1\cos\theta}}$$

and for the resistive film surrounded by a medium with a relative dielectric constant ε_1, the equation for D to be used in Equation (7.120) is

$$D = \frac{\eta_0}{2R\sqrt{\varepsilon_1}\cos\theta}. \tag{7.125}$$

With these new equations for A, B and D the value of the resistance for broad-band transmission can be found to be

$$R_{\text{TE}} = \frac{\eta_0}{\sqrt{\varepsilon_2}\cos\theta'' - \sqrt{\varepsilon_1}\cos\theta}. \tag{7.126}$$

These two resistances, R_{TM} and R_{TE}, are plotted in Figure 7.19. It can be seen that for the TM incidence there is the Brewster angle at which the transmission is total. The value of the resistance at this point is infinite. So for the TM incidence there is a range of angles of incidence from 0 to the Brewster angle where broad-band transmission is possible. In Figure 7.20 the fraction of the incident power transmitted is shown against the angle of incidence. For TM polarisation, this fraction rises to unity at the Brewster angle. For TE polarisation, the resistance required for broad-band transmission is slowly decreasing as the angle of incidence changes. However, the fraction of the incident power transmitted decreases to zero at an angle of incidence of 90°.

This example has shown how the broad-band result derived in Example 7.11 can be extended to obliquely incident waves. However, it is not a straightforward matter as the sheet resistance required is now dependent on the polarisation and the angle of incidence. It could be used where both the polarisation and the angle of incidence are fixed. The main purpose of the example was to show how thin films can be analysed under oblique incidence, by summing all the multiple reflections within the film.

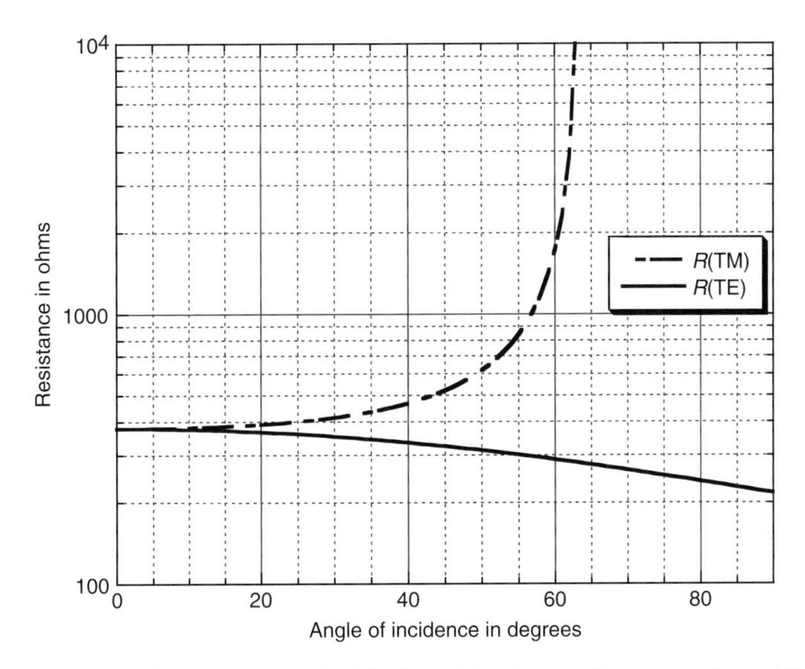

Figure 7.19 The resistance required for broad-band transmission at various oblique angles of incidence where the value of $\varepsilon_1 = 1$ and the value of $\varepsilon_2 = 4$. The Brewster angle for TM incidence occurs at 63.4°.

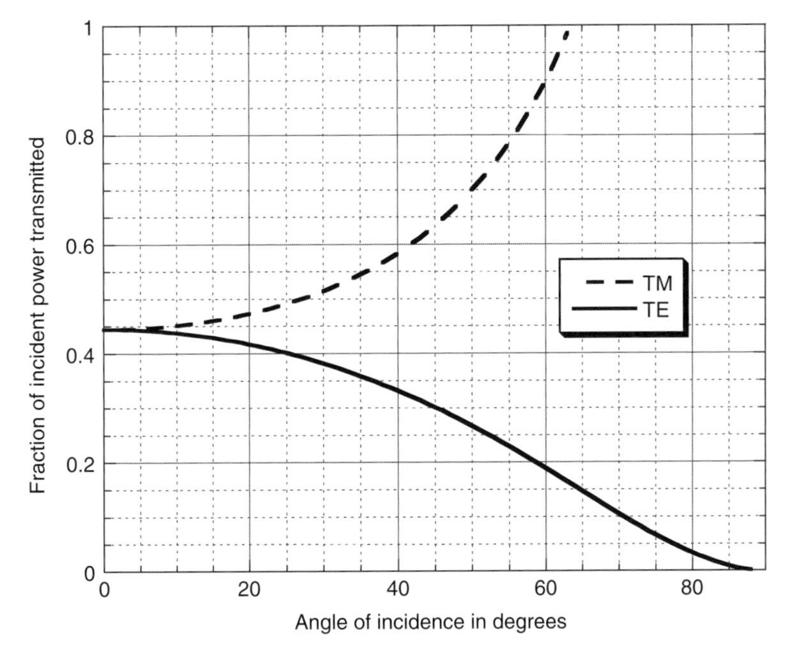

Figure 7.20 The fraction of the incident power which is transmitted through the resistive film and the dielectric sheet with a broad-band response. The fraction in the TM polarisation is enhanced by the Brewster angle, whereas the fraction in the TE polarisation is reduced with increasing angle.

7.7 Polarisation of electromagnetic waves

7.7.1 Linear polarisation

So far in this chapter, only linearly polarised waves have been considered. The direction of the linear polarisation is normally taken as the direction of the electric field. This may be because metallic antennae, which present a short circuit to the electric field, lie in the same direction. By adding together two linearly polarised waves, other polarisations can be produced. A simple example is the addition of two linearly polarised waves, one in the x direction and the other in the y. If the amplitudes of the waves are aE_0 and bE_0 in those directions respectively, then the resultant electric field, E_{LP}, will be of the form,

$$E_{\mathrm{LP}} = aE_0 \exp \mathrm{j}(\omega t - kz)\, \bar{x} + bE_0 \exp \mathrm{j}(\omega t - kz)\bar{y},$$

$$\text{so} \quad E_{\mathrm{LP}} = E_0 \exp \mathrm{j}(\omega t - kz)(a\bar{x} + b\bar{y}). \tag{7.127}$$

The resulting wave is a linearly polarised wave with a direction of polarisation at an angle $\theta = \arctan(b/a)$ to the x-axis.

7.7.2 Circular polarisation

The next type of polarisation occurs when two linearly polarised waves of equal amplitude, E_0, with one in the x direction and the other in the y, and also a phase difference between them of $\pi/2$ are combined. The resultant field, E_{CP}, is

$$E_{\mathrm{CP}} = E_0 \exp \mathrm{j}(\omega t - kz)\bar{x} \pm \mathrm{j}\, E_0 \exp \mathrm{j}(\omega t - kz)\, \bar{y},$$

$$\text{so} \quad E_{\mathrm{CP}} = E_0 \exp \mathrm{j}(\omega t - kz)(\bar{x} \pm \mathrm{j}\bar{y}). \tag{7.128}$$

Taking the real part of the first equation gives a wave:

$$R(E_{\mathrm{CP}}) = E_0 \cos(\omega t - kz)\bar{x} \mp E_0 \sin(\omega t - kz),$$

and in polar form this becomes

$$R(E_{\mathrm{CP}}) = E_0 \angle \mp (\omega t - kz). \tag{7.129}$$

The resultant wave is a vector of constant amplitude, E_0, which rotates around the z-axis at an angular frequency ω. The positive sign in Equation (7.128) relates to the negative sign in Equation (7.129) and is a wave which rotates in an anti-clockwise direction when viewed in the positive z direction. The other sign refers to the clockwise rotation. The confusion between these two is very common. It can be seen that the addition of these two circularly polarised waves results in a linearly polarised wave, E_{LP}, as follows:

$$E_{\mathrm{LP}} = E_{\mathrm{CCW}} + E_{\mathrm{CW}} = E_0 \exp \mathrm{j}(\omega t - kz)(\bar{x} + \mathrm{j}\bar{y}) + E_0 \exp \mathrm{j}(\omega t - kz)(\bar{x} - \mathrm{j}\bar{y}),$$

$$E_{\mathrm{LP}} = 2E_0 \exp \mathrm{j}(\omega t - kz)\bar{x}. \tag{7.130}$$

8.1 The velocity of photons and electrons

Most of the fundamental particles have a rest mass, m_0, i.e. the mass when their velocity is zero. However, when a particle is not at rest, the mass is given by the relativistic equation

$$m = \frac{m_0}{\sqrt{1 - \left(\frac{v}{c}\right)^2}}, \tag{8.1}$$

which means that as their velocity, v, is increased towards the velocity of light, their relativistic mass, m, tends to an infinite value and so the force necessary to accelerate them also tends towards infinity. In the special case of the photon, the rest mass is considered to be extremely small. Despite many attempts to measure this rest mass, so far it has been assumed to be less than 10^{-52} g and can therefore be reasonably neglected. If this were not so, Maxwell's equations and many others would need modification. So for a photon, since $v = c$ and assuming $m_0 = 0$, the right-hand side of Equation (8.1) is indeterminate. The left-hand side of Equation (8.1) does have a value, m_p, and this can be obtained from the energy of a photon, E, since:

$$E = hf = m_p c^2 \text{ so that } m_p = \frac{hf}{c^2}, \tag{8.2}$$

which for an optical frequency of, say, 10^{14} Hz is only 7.36×10^{-34} g and is about a factor of 10^6 smaller than an electron at rest. Besides a fixed velocity in free space of c, a photon also has a momentum, p, given by

$$p = m_P c = \frac{hf}{c}. \tag{8.3}$$

By comparison, electrons do not travel at the velocity of light as this would require an infinite force. Typical velocities for the conduction electrons, the so-called Fermi electrons, are about 10^6 $\mathrm{ms^{-1}}$on average. So here begins the first problem with transmission lines. The signals on transmission lines are capable of travelling at velocities up to and including the velocity of light. Yet in the first three chapters of this book, it was assumed that the currents in the wires were travelling at these velocities. Clearly, it is much more likely that the signals on a transmission line consist of photons making up an electromagnetic wave rather than the conduction electrons in the wires. The role of the electrons in conductors is therefore one of guiding the photons along the transmission lines by forcing the electric fields of the photons to be nearly normal to the surface of any conductors. Similarly, in dielectric guides, the fixed electrons in the two distinct parts of the dielectric structure move to form a reflecting barrier to the incident photons. This view of the signal propagation makes consistent the propagation along either metallic structures, or dielectric structures or in free space. It does, however, change the nature of electricity. The currents induced in the conductors as a result of the photons passing by the surface are mainly the movement of the electrons towards the surface or away from it, depending on the direction of the normal electric field, rather than the movement of the electrons in the direction of the signal itself. First of all, when an electromagnetic wave arrives at a conducting surface there will be a response time for the electrons to move to compensate for the normal electric field. Beginning with the Lorentz force on a charge,

$$\overline{F} = Q\left(\overline{E} + \overline{v} \times \overline{B}\right), \tag{8.4}$$

and taking just the amplitudes of $\overline{v} \times \overline{B}$ as $v_e \times \mu H$ or $v_e \times \sqrt{\mu\varepsilon}E$ or v_e/cE, then the ratio of the two terms in the brackets of Equation (8.4) is the ratio of the velocity of electrons, v_e, to the velocity of light, c. In other words, the magnetic force is only about 1% at most of the electric force and can reasonably be neglected. Using the classical Drude model for conduction, the force on an electron with a charge e and mass m is given by

$$m\frac{\mathrm{d}v}{\mathrm{d}t} = -eE - mf_c v, \tag{8.5}$$

where f_c is the mean frequency of collisions, and hence the term represents the loss of momentum. When there is no external electric field, these collisions both lose and gain energy so as to maintain the whole ensemble in thermal equilibrium. The presence of an external electric field disturbs this equilibrium as the electrons absorb energy from the field and pass it to the lattice of atoms via these collisions. As a result the temperature of the lattice is increased and a new equilibrium is achieved. This is the basic mechanism behind Ohm's Law and, despite its complexity, it is surprisingly linear in many conductors. Now if this field is changing sinusoidally with time at a frequency ω, the velocity will be changing in the same way and Equation (8.5) becomes

$$\mathrm{j}\omega m v = -eE - mf_c v \text{ and } v = \frac{-eE}{m(\mathrm{j}\omega + f_c)}. \tag{8.6}$$

Now if the density of free electrons is n_e, then the current density is given by

$$J = -n_e e v = \frac{n_e e^2 E}{m(\mathrm{j}\omega + f_c)}. \tag{8.7}$$

Since microwave frequencies are usually much less than f_c at room temperatures,

$$J = \frac{n_e\, e^2 E}{mf_c} = \sigma E \quad \text{and} \quad \sigma = \frac{n_e\, e^2}{mf_c}. \tag{8.8}$$

This gives the classical equation for conductivity attributed to Drude. The frequency, f_c, is typically about 2.5×10^{13} Hz for copper at room temperature, where the mean free path, l_e, is approximately 4×10^{-8} m and the velocity, v_e, is 10^6 ms^{-1}, as

$$f_c = \frac{v_e}{l_e}. \tag{8.9}$$

Below this frequency, the electrons can be considered as moving in a 'viscous medium'. It is useful to note how small the velocities of electrons are in applied electric fields. Using Equation (8.6) and the condition $\omega \ll f_c$, for a typical field of 1 Vm^{-1} the resultant velocity is only 1 cm s^{-1}. This velocity is called the drift velocity and is well below the value of v_e quoted above. Furthermore, the mean free path is strongly dependent on temperature, so that at low temperatures f_c could be well below microwave frequencies for a bulk conductor. In this case Equation (8.7) shows that the conductivity equation becomes imaginary:

$$J = \frac{n_e e^2 E}{j\omega m} \quad \text{or} \quad \sigma = -j\frac{n_e e^2}{\omega m}. \tag{8.10}$$

In this case the electrons behave like a plasma and move as if they were free, with all the oddity of a dielectric constant less than unity. Since

$$\varepsilon = \varepsilon' - j\varepsilon'', \ j\omega\varepsilon = j\omega\varepsilon' + \omega\varepsilon'' = j\omega\varepsilon' + \sigma, \tag{8.11}$$

$$\text{and so } j\omega\varepsilon = j\omega\varepsilon' - j\frac{n_e e^2}{\omega m} = j\omega\left(e' - \frac{n_e e^2}{\omega^2 m}\right) \text{ or } \varepsilon = \varepsilon' - \frac{n_e e^2}{\omega^2 m}. \tag{8.12}$$

The frequency at which the dielectric constant goes to zero is called the plasma frequency and is given by

$$\omega_p = \sqrt{\frac{n_e e^2}{m\varepsilon'}}, \tag{8.13}$$

and is approximately 1.5×10^{15} Hz in metals. So, in summary, a metal conductor has a conductivity which is reasonably constant from very low frequencies right through all the microwave frequencies and it is given by Equation (8.8). There may be small deviations from this equation depending on the atomic structure. However, at optical frequencies, it can no longer be assumed that a metal is still a conductor and that the electromagnetic fields will only penetrate a very small distance into the surface.

So when an electromagnetic wave begins to travel down a transmission line composed of metallic guides, there will be a brief transition period when the first part of the wave will be lost. This process continues for a very short time until the electrons have begun to move to compensate for the electric fields in the conductors. After this, positive and negative charges will appear on the surface of the conductors like a 'Mexican wave' as the photons pass by. So the electrons in the surface are oscillating in and out of the surface at only the frequency of the passing electromagnetic wave. Although the drift velocity is very small, the amount of movement required to cancel or compensate for the normal electric field is also very small and so the response time is adequate for this process. However, since they are all at slightly different phases along the transmission line, they may give the appearance of a wave travelling at the velocity of light. Two problems still remain with this approach. The first is to do with the current in a termination and the second is concerned with the current along the conductors themselves. These will be discussed in the next paragraphs.

In the book listed at the end of this chapter, Brillouin discusses this phenomenon in great detail. In a similar way that the electrons in a conductor respond eventually to the arrival of photons, the electrons in a dielectric also respond to the photons' arrival. The curious result is that the initial part of a 'step wave' of photons travels through everything at the velocity of light. There is no wave guidance by either conductors or dielectrics. These early photons arrive at the detector somewhat diminished in numbers and are called 'leaders' or 'forerunners'. Then, as the electrons reach a steady state with the sinusoidal electric fields, the main part of the signal arrives at the detector. In other words, initially the photons are free to pass through both conductors and dielectrics at the speed of light. Only when the electrons in these structures begin moving in response to the electric fields set up by the photons will there be any of the guidance or slowing down of the photons. In the case of a dielectric filled waveguide or transmission line, the forerunners or leaders will arrive during a time t, given by

$$t = \frac{D}{c}(\sqrt{\varepsilon_R} - 1), \tag{8.14}$$

where D is the length of the line, ε_R is the relative dielectric constant and c is the velocity of light. During this period, the detected waveform will be small and have a changing frequency, as shown by Brillouin. Similar phenomena will occur when the photons are switched off and the electrons return to their thermal equilibrium position.

When an electromagnetic wave is incident on a resistive termination, then the tangential electric fields will set up currents according to the usual Ohm's Law. So, in both cases, the currents respond mainly to the direction of the electric field. The final problem concerns the motion of the electrons along the conductors of a transmission line. The force that produces this motion is the one due to the magnetic field that was neglected earlier on. As a result the electrons in the surface region of the conductors move in an elliptical motion as the photons pass by. The major axis is aligned with the photons' electric field. So, in summary, it is not the currents in the conductors that convey the energy down a transmission line, but the photons in the space between the wires. In the books listed at the end of the chapter, it can be seen that this view of transmission lines has been held for well over a century. This means that the popular concept of electricity as a flow of electrons is a simplified picture which omits the key principle that only photons can carry the energy fast enough to explain the phenomena; electrons are a bit too slow. However, electrons in the conductors play a vital role in guiding photons along the transmission line. At the point where these lines end, which may be an antenna, some of these photons will be free to move on unguided into space.

8.2 The momentum of photons

When an electromagnetic wave is incident on a termination, there are two things taking place. One is that part or all of the energy is absorbed by the electrons in the termination, with the extra comment from the previous paragraph that this may

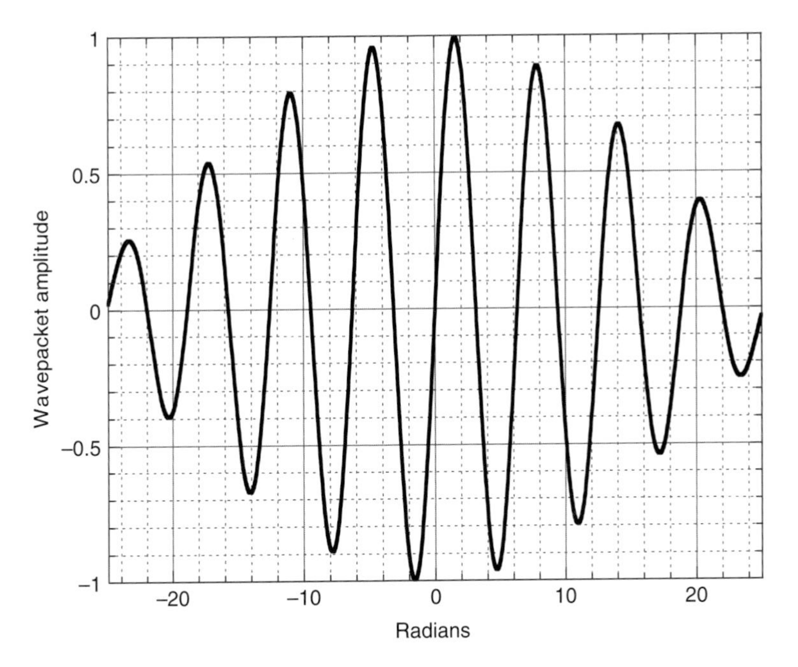

Figure 8.1 A typical wave packet of 21 sine waves. The frequency of each wave is equally spaced over a frequency range $\pm 10\,\%$ of the centre frequency.

$$\Delta E = \Delta n \times hf \text{ and so } \Delta n \times \Delta t \geq \frac{1}{2\pi f}. \tag{8.30}$$

Finally, the uncertainty in phase is $\Delta\varphi$ which is related to the uncertainty in time via

$$\Delta\varphi = \omega\Delta t \text{ so } \Delta n \times \Delta\varphi \approx 1. \tag{8.31}$$

A useful result from quantum mechanics is that the uncertainty in number for coherent photons, Δn, is equal to $\sqrt{n}$ if $n >$ about 400. Now in Chapter 7 a 1 mW continuous signal at 10^{15} Hz contains $1.51.10^{15}$ photons, giving an uncertainty of $3.88.10^{7}$ which is a factor of $3.89.10^{7}$ smaller than the total number. Two factors arise from these numbers. The first is that where the number of photons is large, the uncertainty in the amplitude of the resultant electromagnetic wave is negligible. The second uses Equation (8.31) and the uncertainty in phase must also be very small. In other words, if the photon numbers are high, then the phase and amplitude of electromagnetic waves can be measured with great precision.

Now when an electron emits a photon, there will be a small uncertainty in the frequency of the individual photon, Δf_{p}, and this will also be related via Heisenberg's uncertainty principle to Planck's constant. Start with the uncertainty principle in terms of the uncertainty in position, Δx_{p}, and the uncertainty in momentum, Δp_{p}, which is given by

$$\Delta x_{\mathrm{p}} \times \Delta p_{\mathrm{p}} \geq \frac{h}{2\pi}. \tag{8.32}$$

Then use Equation (8.3):

$$p_\mathrm{p} = \frac{hf}{c} \text{ and so } \Delta p_\mathrm{p} = \frac{h}{c}\Delta f \text{ and hence } \Delta x \times \Delta f \geq \frac{c}{2\pi}. \tag{8.33}$$

This has the following implications for photons. As $\Delta f \to 0$, $\Delta x \to \infty$, which means that a single frequency plane wave is infinite in extent in the direction of propagation. It may be limited in the transverse directions by the boundary conditions of the transmission line. However, since $\Delta f \neq 0$ for all sources, even the most precise atomic sources used for the definition of time itself, i.e. caesium atoms, an individual photon will have a range of frequencies within a small bandwidth. As a result it will not be infinite in extent. Although Fourier analysis uses perfect plane waves, these do not exist in practice and they are truncated into a wave packet similar to the one described above. Finally, using $\Delta x = c\Delta t$,

$$\Delta t \times \Delta f \geq \frac{1}{2\pi}. \tag{8.34}$$

So, taking the example of a microwave source, where Δf might be 10^5 Hz, then the value of $\Delta t \geq 1.6$ μs. This begins to show the limits for precise frequency measurements. In order to obtain an accuracy of 10^5 Hz in 10^{10} Hz then the measurement time must be greater than 1.6 μs to allow equilibrium to be achieved.

In most circuits today, the signals are pulses and these have a wide frequency spectrum. Figure 8.2 shows a typical isolated pulse made up from sine waves using a Fourier integral:

$$\int_0^a \frac{\cos(wx)\sin(w)}{w}\,\mathrm{d}w. \tag{8.35}$$

Figure 8.2 shows the Gibbs' phenomenon or overshoot near the discontinuity. This can be avoided by using other functions that are less steep at $x = \pm 1$.

Now the Fourier series usually takes plane waves of infinite extent. However, since the resultant pulse is clearly not of infinite extent, it is possible to do a similar analysis with wave packets rather than infinite waves, so long as these packets are much more extensive in time than the pulse itself. Effectively, the waveform is resolved into orthogonal wave packets, i.e. the centre frequencies of the wave packets are harmonic.

All that remains is to discuss a conceptual problem. When a photon is released by an atom by radiation from an accelerating electron, it must be initially very small. However, since it is non-locatable it must expand to fill the available space. Since a photon wave-function would be the probability density of detecting a photon, then in transmission lines this will be related to the Poynting's vector. For example, in a rectangular waveguide it is not unreasonable to assume the photon is trapped inside the guide, but free to move down the guide as a set of plane waves, as described in Chapter 7. When the photon reaches the end of the guide, there may be a detector which could be anywhere across the guide. For the TE_{10} mode, the Poynting's

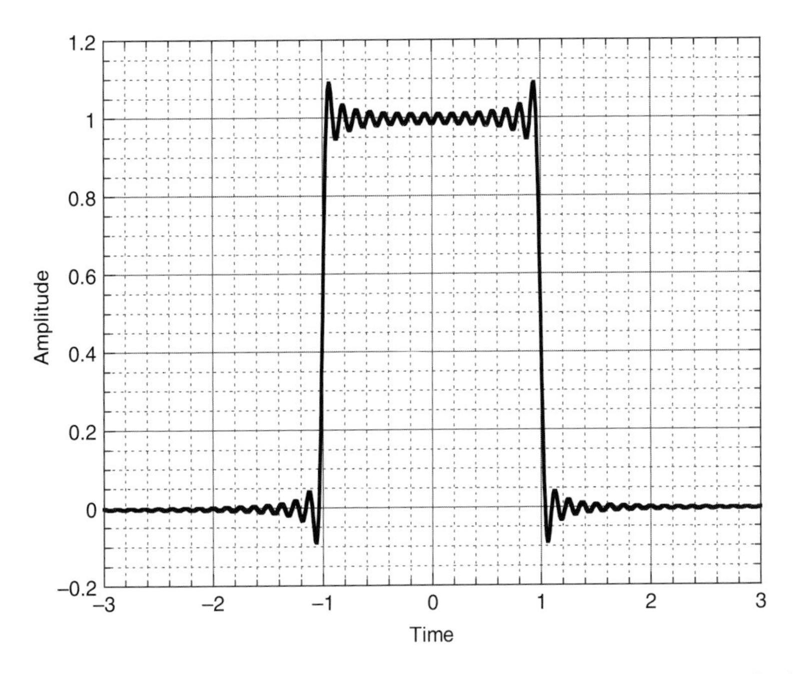

Figure 8.2 A square pulse made up of a large number of sine waves. $a = 50$ in the integral of Equation (8.34).

vector amplitude has been plotted in Figure 8.3. The amplitude has been normalised so that the total probability is one. So this figure shows the probability of detecting a photon across the guide. Obviously, in the y direction the Poynting's vector is constant, so the probability does not vary in that direction. Now when a photon is detected by an electron, it is totally absorbed. So that even though its probability density was widespread across the waveguide, at the place of detection it will suddenly have a probability of unity. This phenomenon is called the collapse of the wave-function (or probability density function) and it takes place instantaneously, irrespective of how widespread the wave-function was beforehand. In other words, it is saying that there is a probability that the photon might be anywhere across the guide given by the Poynting's vector, but as soon as it is detected, then the probability must be zero everywhere else, other than the point of detection.

8.5 Photon absorption and reflection from a capacitor

It is a useful illustration of photons to reconsider Example 1.9 in Chapter 1. The circuit diagram is reproduced in Figure 8.4.

Here a step wave of amplitude 5 V was incident on a capacitor after a delay of 10 ns. Just for the moment, assume that the zero frequency part of the electromagnetic wave which follows the step consists not of zero frequency photons with all the problems of zero energy and infinite numbers, but of photons forming short

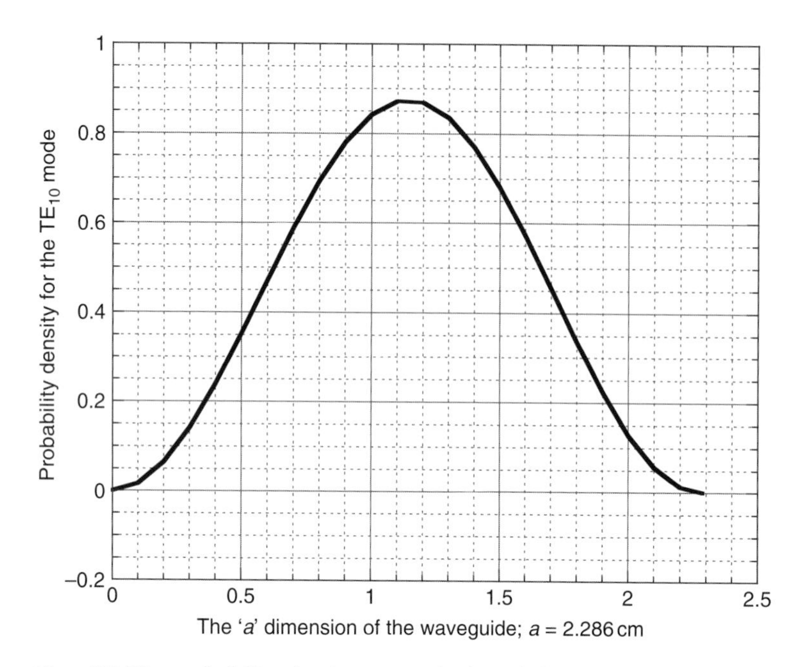

Figure 8.3 The probability density across the broad dimension of a rectangular waveguide supporting the TE_{10} mode. The density has been normalised to give an overall probability of unity.

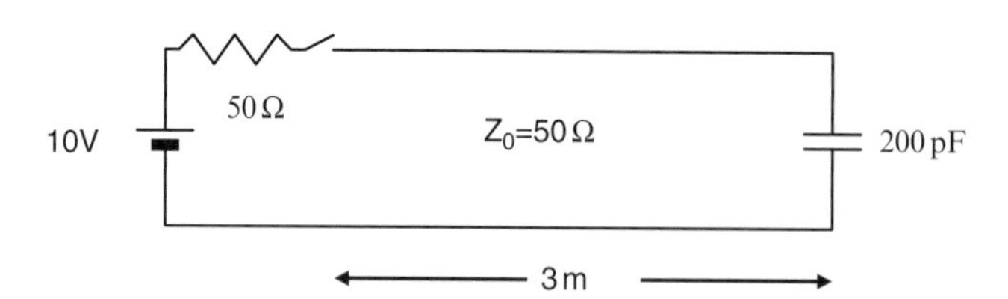

Figure 8.4 The circuit diagram for Example 1.9 in Chapter 1. The velocity on the line is $3.10^8\ \mathrm{ms}^{-1}$.

unidirectional pulses. These pulses combined in sufficient numbers randomly in time to form the 'DC' part of the wave that follows the step. The actual step is thus the beginning of these pulses. This concept is not too far removed from reality as a battery consists of electrons giving up their electrical energy in pulses. Now the reflection coefficient of the capacitor in Example 1.9 actually varies with time. Using the equations from Example 1.9:

$$V_- = 5 - 10\exp - \left(\frac{t-10}{10}\right) \text{ and } V_+ = 5, \text{ so}$$
$$\rho = \frac{V_-}{V_+} = 1 - 2\exp - \left(\frac{t-10}{10}\right). \tag{8.36}$$

not dissipate energy. In simple waveguides, the propagation constant is either real for propagating modes, or imaginary for evanescent fields, and complex propagation constants only arise when losses are present. These complex modes exist in more complex waveguides, where there is a mixture of dielectrics present, for instance, in a circular waveguide partially filled with a dielectric rod or a microstrip line or coplanar waveguide. The complex solutions in these cases have propagation constants in pairs of complex conjugates. The group velocity can be negative and so the structure supports backward waves. These complex modes usually occur in higher order modes and do not normally affect the mono-mode bandwidth discussed in this book. However, they are important when analysing various structures in the guides where energy might be stored in these higher order modes. In some cases, the Poynting vector may be positive or negative in separate parts of the structure so that the net power flow is zero.

8.8 Metamaterials

One of the exciting recent developments in electromagnetic theory has been the advent of metamaterials. These are materials not found in nature, but ones which have been engineered to have special properties. For most ordinary materials, the relative permeability and permittivity are positive. If losses are combined, they are usually in the form of a negative imaginary part, i.e.

$$\varepsilon = \varepsilon' - j\varepsilon'' \text{ and } \mu = \mu' - j\mu''.$$

However, as Equation (8.12) shows, even in a metal conductor at high enough frequency, there is the possibility of both zero and negative permittivity. The same can occur under special conditions in both dielectric and magnetic materials. Neglecting losses for the moment, this would give equations of the form

$$v_\mathrm{p} = \frac{1}{\sqrt{-\varepsilon_\mathrm{R}\,\varepsilon_0 \times -\mu_\mathrm{R}\,\mu_0}} = \pm\frac{c}{\sqrt{\varepsilon_\mathrm{R}\,\mu_\mathrm{R}}} \text{ and } \eta = \sqrt{\frac{-\mu_\mathrm{R}\,\mu_0}{-\varepsilon_\mathrm{R}\,\varepsilon_0}} = \eta_0\sqrt{\frac{\mu_\mathrm{R}}{\varepsilon_\mathrm{R}}}, \tag{8.46}$$

which leads to a dilemma in interpretation of the first equation for the velocity: it is possible to choose the solution with a negative sign. The second equation is normally interpreted, as expected, with a positive sign. The negative sign gives waves with a negative phase velocity and, rather curiously, at an interface between dielectrics can give negative angles of refraction. This has developed into many interesting areas of dispersion compensation in circuits and making very small antennae and many more. Despite the negative sign, the principle of causality is not broken and the group velocity usually has the opposite sign. In addition to these 'double negative' materials are the 'single negative' materials which have both an imaginary intrinsic impedance and an imaginary phase velocity. This is characteristic of a medium which supports only evanescent waves.

8.9 Photonic bandgap materials

It is possible to construct a material with regularly spaced defects, as shown in Figure 8.7.

As long as the separation of the defects, a, is much smaller than the wavelength of an electromagnetic wave, then the defects change the dielectric constant. So, for example, if the main material had a dielectric constant, ε_1, were 80% of the total, and the defects had a dielectric constant, ε_2, then the effective dielectric constant would be given by

$$\varepsilon_{\text{eff}} = 0.8\varepsilon_1 + 0.2\varepsilon_2. \tag{8.47}$$

This effect has been used for many years to manufacture dielectrics with prescribed dielectric constants.

Now if the frequency is increased to a point where the wavelength is about equal to a, then the defects begin to scatter the electromagnetic wave in a coherent manner. In particular, there are bands of frequencies where propagation is not allowed. This is the analogy of the electron bandgap in semiconductor theory. So these materials can be used to form perfect reflectors. This is particularly useful at optical frequencies, where metallic conductors are not perfect reflectors. By removing a line of defects, transmission lines can be made which have perfectly reflecting walls, i.e. they form quasi-metallic waveguides. If the frequency is further increased, the coherent scattering becomes incoherent and the reflective properties cease.

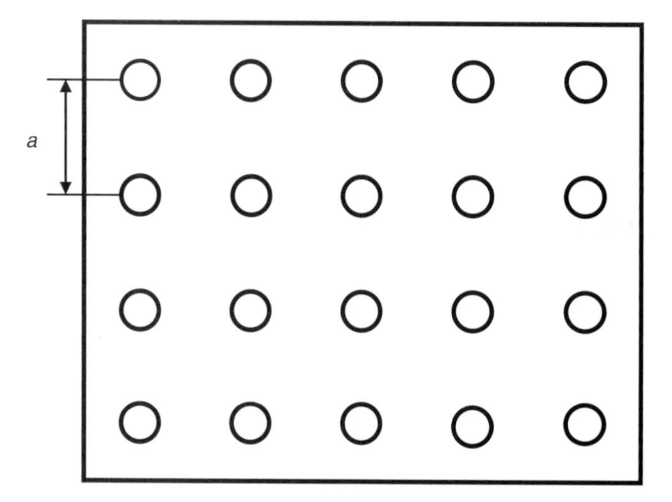

Figure 8.7 A sample of a photonic bandgap material with holes spaced a distance a apart. The defects could be small holes or a small quantity of another material; usually both materials are dielectrics.

8.10 Conclusion

This chapter has attempted to give a somewhat sketchy introduction to the photon and its role in transmission lines. There are many unanswered questions in this chapter and it will be interesting to see how this part of the subject develops. The author has not covered the aspect of spin, which is currently much discussed, as so far this does not appear, except as perhaps circularly polarised waves, in transmission lines. There will be many advances in this area, especially as single photon communication at optical frequencies advances.

8.11 Further reading

Photons

C. Roychoudhuri, A. F. Kracklauer and K. Creath *The Nature of Light: What is a Photon?*, Boca Raton, CRC Press, 2008.

J. D. Jackson *Classical Electrodynamics*, Second edition, New York, Wiley, 1975. Chapter 1, section 1.2 on the rest mass of a photon and chapter 7, section 7.11 gives some experimental evidence for Brillouin's theory.

L. Brillouin *Wave Propagation and Group Velocity*, New York, Academic Press, 1960.

C. A. Brau *Modern Problems in Classical Electrodynamics*, Oxford, Oxford University Press, 2004. Chapter 2, section 2.3 on the photon rest mass and chapter 7, section 7.2.3 on 'viscous electron motion'.

J. A. Stratton *Electromagnetic Theory*, New York, McGraw-Hill, 1941. Chapter 5, section 5.18, Phase and group velocities.

F. A. Benson *Millimetre and Submillimetre Waves*, London, Iliffe Books Ltd, 1969. See chapter 1 by A. L. Cullen on classical and quantum theories.

R. Loudon *The Quantum Theory of Light*, Oxford, Clarendon Press, 1973. Page 153 shows the amplitude and phase uncertainty for low photon numbers.

S. P. Thompson *Elementary Lessons in Electricity and Magnetism*, London, Macmillan, 1905. Energy not in the wires – see Preface and section 519, page 557, Radiation pressure.

K. F. Sander and G. A. L. Reed *Transmission and Propagation of Electromagnetic Waves*, Second edition, Cambridge, Cambridge University Press, 1986. Pages 46–48, Radiation pressure.

B. I. Bleaney and B. Bleaney *Electricity and Magnetism*, Oxford, Clarendon Press, 1957. Section 10.8, Radiation pressure and section 17.3, Noise and Planck's radiation law.

A. Sommerfeld *Electrodynamics, Lectures on Theoretical Physics*, Vol. 3, New York, Academic Press, 1964. Section 13, Radiation pressure.

W. K. H. Panofsky and M. Phillips *Classical Electricity and Magnetism*, Reading, MA, Addison-Wesley, 1964. Section 11–13, Radiation pressure.

R. Loudon *The Quantum Theory of Light*, Oxford, Clarendon Press, 1973. Page 152 gives an equation similar to (8.21).

Metamaterials

N. Engheta and R. W. Ziolkowski *Metamaterials: Physics and Engineering Explorations*, New York, Wiley InterScience, 2006.

Anomalous skin effect

A. H. Wilson *The Theory of Metals*, Second edition, Cambridge, Cambridge University Press, 1953. Section on anomalous skin effect.

R. E. Matrick *Transmission Lines for Digital and Communication Networks*, New York, McGraw-Hill, 1969. Chapter 4.

Complex modes

M. Mrozowski *Guided Electromagnetic Waves: Properties and Anaylsis*, Taunton, UK, Research Studies Press Ltd, 1997. Complex modes in chapter 12.

Photonic bandgap materials

M. Skorobogatiy and J. Yang *Fundamentals of Photonic Crystal Guiding*, Cambridge, Cambridge University Press, 2009.